BUILDING ELECTRIFICATION

AUTHOR

Dr. Arnab Ghosh

Assistant Professor

Department of Electrical Engineering (EE)

National Institute of Technology Rourkela, Odisha

REVIEWER

Dr. M. Vijay Karthik

Associate Professor

Dept. of Electrical & Electronics Engineering (EEE)

CMR Engineering College, Kandlakoya, Medchal Road, Hyderabad, Telagana-501401

KHANNA BOOK PUBLISHING CO. (P) Ltd.

4C/4344, Ansari Road, Darya Ganj, New Delhi, India

All India Council for Technical Education

Nelson Mandela Marg, Vasant Kunj, New Delhi, India

Dear Readers,

To prevent the piracy, this book is secured with HIGH SECURITY HOLOGRAM on the front title cover. In case you don't find the hologram on the front cover title, please write us to at contact@khannabooks.com or whatsapp us at +91-99109 09320 and avail special gift voucher for yourself.

Specimen of Hologram on front Cover title:

Moreover, there is a SPECIAL DISCOUNT COUPON for you with EVERY HOLOGRAM.
How to avail this SPECIAL DISCOUNT:
Step 1: Scratch the hologram
Step 2: Under the scratch area, your "coupon code" is available
Step 3: Logon to www.khannabooks.com
Step 4: Use your "coupon code" in the shopping cart and get your copy at a special discount
Step 5: Enjoy your reading!

ISBN: 978-93-5538-328-0
Book Code: DIP506EN

Building Electrification
by Dr. Arnab Ghosh

First Edition: 2026

Originally Published by:
All India Council for Technical Education, Ministry of Education, GOI

Exclusively Published & Distributed by:
Khanna Book Publishing Co. (P) Ltd.
4C/4344, Ansari Road, Daryaganj,
New Delhi (INDIA)

Visit us at: **www.khannabooks.com**
Write us at: **contact@khannabooks.com**

To view complete list of books,
Please scan the QR Code:

Printed in Bharat.

OUTCOME BASED EDUCATION

For the implementation of an outcome based education the first requirement is to develop an outcome based curriculum and incorporate an outcome based assessment in the education system. By going through outcome based assessments evaluators will be able to evaluate whether the students have achieved the outlined standard, specific and measurable outcomes. With the proper incorporation of outcome based education there will be a definite commitment to achieve a minimum standard for all learners without giving up at any level. At the end of the programme running with the aid of outcome based education, a student will be able to arrive at the following outcomes:

PO1. **Basic and Discipline specific knowledge:** Apply knowledge of basic mathematics, science and engineering fundamentals and engineering specialization to solve the engineering problems.

PO2. **Problem analysis:** Identify and analyses well-defined engineering problems using codified standard methods.

PO3. **Design/development of solutions:** Design solutions for well-defined technical problems and assist with the design of systems components or processes to meet specified needs.

PO4. **Engineering Tools, Experimentation and Testing:** Apply modern engineering tools and appropriate technique to conduct standard tests and measurements.

PO5. **Engineering practices for society, sustainability and environment:** Apply appropriate technology in context of society, sustainability, environment and ethical practices.

PO6. **Project Management:** Use engineering management principles individually, as a team member or a leader to manage projects and effectively communicate about well-defined engineering activities.

PO7. **Life-long learning:** Ability to analyse individual needs and engage in updating in the context of technological changes.

COURSE OUTCOMES

After completion of the course the students will be able to:

CO-1: Have a good understanding of selecting accessories, wires, cables and wiring systems for electrification.

CO-2: Have a good understanding of the designing of electrical wiring installation system for residential unit.

CO-3: Student will be able to design proper illumination scheme for residential unit.

CO-4: Student will be able to prepare wiring layouts on wiring board.

CO-5: Student will be able to locate and diagnose faults in electrical wiring installation.

CO-6: Student will be able to do proper earthing for building electrification.

Mapping of Course Outcomes with Programme Outcomes to be done according to the matrix given below:

Course Outcomes	Expected Mapping with Programme Outcomes *(1- Weak Correlation; 2- Medium correlation; 3- Strong Correlation)*						
	PO-1	**PO-2**	**PO-3**	**PO-4**	**PO-5**	**PO-6**	**PO-7**
CO-1	3	3	3	3	1	2	3
CO-2	3	3	3	3	1	2	3
CO-3	3	3	3	3	1	2	3
CO-4	3	3	2	3	1	2	3
CO-5	3	3	2	3	1	1	3
CO-6	3	3	2	3	1	1	3

ABBREVIATIONS AND SYMBOLS

List of Abbreviations

General Terms			
Abbreviations	**Full form**	**Abbreviations**	**Full form**
A	Ampere	EMT	Electrical metallic tube
AB Switch	Air Break Switch	EMI	Electromagnetic interference
AC	Alternating Current	FMC	Flexible metal conduit
ACCB	Air Case Circuit Breaker	FBT	Flyback Transformer
ACDB	Alternating Current Distribution Board	GFCI	Ground Fault Circuit Interrupter
		GFI	Ground Fault Interrupter
AFC	Automated Frequency Control	GIS	Gas Insulated Switchgear
AFC	Available Fault Current	GO	Gang Operated
AFCI	Arc Fault Circuit Interrupter	HT	High Tension
AH	Ampere Hour	HVDC	High-voltage direct current
AHU	Air Handling Unit	ICDB	Iron Clad Distribution Board
AIC	Ampere Interrupting Capacity	ICDP	Isolator with Circuit Disconnecting Position
APFC	Automatic Power Factor Control		
ATS	Automatic Transfer Switch	IMC	Intermediate Metal Conduit
BLDC	Brushless DC	IPMSM	Interior Permanent Magnet Synchronous Motor
BLDM	Brushless DC Motor		
CT	Current Transformer	ICTP	Isolator with Circuit Test Position
CFLs	Compact Fluorescent Lamps	IDCs	Insulation displacement connectors
DB	Distribution Board	LDPE	Low-density polyethylene
DC	Direct Current	LA	Lightning Arrester
DCDB	Direct Current Distribution Board	LANs	Local area networks
DO Fuse	Drop Out Fuse	LOPT	Line Output Transformer
DOL	Direct ON Line	LPS	Lightning Protection System
DP	Double Pole	LRA	Locked Rotor Amperes
DPIC	Double Pole Iron Clad	LT	Low Tension
EHT	Extra High Tension	LVDT	Linear Variable Differential Transformer
ELCB	Earth Leakage Circuit Breaker		
OLTC	On-load Tap Changer	LED	Light-Emitting Diode
PB	Push Button	MCB	Miniature Circuit Breaker
PPE	Personal protective equipment	MCC	Motor Control Center Panel
PSC	Predicted short-circuit current	MCCB	Moulded Case Circuit Breaker
PCC	Power Control Centre Panel	MCP	Motor Circuit Protection

General Terms			
Abbreviations	**Full form**	**Abbreviations**	**Full form**
PDP	Power Distribution Panel	NC	Normally Closed
PT	Potential Transformer	NCT	Neutral Current Transformer
PVC	Polyvinyl chloride	NO	Normally Open
RCBO	Residual Current Circuit Breaker with Overload Protection	ONCT	On-circuit Tap Changer
		SPDT	Single Pole Double Throw
RFID	Radio Frequency Identification		
RCCB	Residual Current Circuit Breaker	SPST	Single Pole Single Throw
RLA	Rated Load Amperes	SSR	Solid State Relay
RMC	Rigid Metal Conduit	ST	Super Tension
RMG	Ring Main Gear	SWG	Standard Wire Gauge
RMU	Ring Main Unit	TMCB	Thermal Magnetic Circuit Breaker
RNC	Rigid Non Metallic Conduit	TPIC	Triple Pole Iron Clad
RPR	Reverse Power Relay	TPS	Triple Pole Switch
RMC	Rigid metal conduit	TRS	Tough Rubber Sheathed
SB	Switch Board	TPE	Thermoplastic elastomer
SCIM	Squirrel Cage Induction Motor	UART	Universal Asynchronous Receiver and Transmitter
SFU	Switch Fuse Unit		
XLPE	Cross linked Polyethylene	UDS	Uniform Drawing System
VCB	Vacuum Circuit Breaker	UG	Under Ground
VIR	Vulcanised rubber	USART	Universal Synchronous and Asynchronous Receiver and Transmitter
VFD	Variable Frequency Drive		

List of Units

General Terms					
Abbreviations	**Full form**	**Unit**	**Abbreviations**	**Full form**	**Unit**
V	Voltage	Volt	I	Current	Ampere
R	Resistance	ohm	P	Power	watt

List of Symbols

Symbols	Description	Symbols	Description
+	Addition	X	Multiplication
L	Length	η	*Efficiency*
/	Division	V_d	Permissible voltage drop
A	Cross-sectional area of the conductor	–	Subtraction
*	Multiplication	E	Illuminance in lux
r	Distance from the light source	Φ	Luminous in lumens
π	Mathematical constant		

GUIDELINES FOR TEACHERS

To implement Outcome Based Education (OBE) knowledge level and skill set of the students should be enhanced. Teachers should take a major responsibility for the proper implementation of OBE. Some of the responsibilities (not limited to) for the teachers in OBE system may be as follows:

- Within reasonable constraint, they should manoeuvre time to the best advantage of all students.
- They should assess the students only upon certain defined criterion without considering any other potential ineligibility to discriminate them.
- They should try to grow the learning abilities of the students to a certain level before they leave the institute.
- They should try to ensure that all the students are equipped with the quality knowledge as well as competence after they finish their education.
- They should always encourage the students to develop their ultimate performance capabilities.
- They should facilitate and encourage group work and team work to consolidate newer approach.
- They should follow Bloom's taxonomy in every part of the assessment.

Bloom's Taxonomy

Level	Teacher should Check	Student should be able to	Possible Mode of Assessment
Create	Students ability to create	Design or Create	Mini project
Evaluate	Students ability to justify	Argue or Defend	Assignment
Analyse	Students ability to distinguish	Differentiate or Distinguish	Project/Lab Methodology
Apply	Students ability to use information	Operate or Demonstrate	Technical Presentation/ Demonstration
Understand	Students ability to explain the ideas	Explain or Classify	Presentation/Seminar
Remember	Students ability to recall (or remember)	Define or Recall	Quiz

GUIDELINES FOR STUDENTS

Students should take equal responsibility for implementing the OBE. Some of the responsibilities (not limited to) for the students in OBE system are as follows:

- Students should be well aware of each UO before the start of a unit in each and every course.
- Students should be well aware of each CO before the start of the course.
- Students should be well aware of each PO before the start of the programme.
- Students should think critically and reasonably with proper reflection and action.
- Learning of the students should be connected and integrated with practical and real life consequences.
- Students should be well aware of their competency at every level of OBE.

LIST OF FIGURES

CONTENTS

UNIT 1 WIRING TOOLS AND ACCESSORIES

UNIT SPECIFICS

This unit elaborately discusses the following topics:

- ✓ Introduction to wiring tools and accessories required for wiring
- ✓ Classification of electrical accessories and essential wiring tools
- ✓ Exploring the diverse spectrum of electrical switches
- ✓ Diverse variety of electrical holder types
- ✓ Various types of socket outlets and plugs
- ✓ Exploring various types of connectors, adaptors, and ceiling roses

This unit overviews the importance of wiring tools and accessories in electrical installations. It discusses the significance of using the right tools for the job and the impact of quality tools on project outcomes. The unit introduces safety practices and precautions when working with wiring tools, emphasizing the importance of proper training and equipment maintenance. This unit focuses on essential wiring tools used in electrical installations, including wire strippers, pliers, screwdrivers, and measuring devices. It provides detailed descriptions of each tool and practical demonstrations of how they are used in wiring projects. The unit offers tips for selecting high-quality tools and maintaining them for optimal performance and longevity. This unit explores specialized wiring tools for specific tasks, such as crimping tools, conduit benders, cable cutters, and fish tapes. Each tool is examined in detail, highlighting its unique features, applications, and best practices. Practical examples and demonstrations illustrate how specialized tools can improve efficiency and accuracy in wiring installations. This unit examines wire connectors and terminals for joining and terminating electrical wires. Common types of connectors, including wire nuts, terminal blocks, crimp connectors, and soldered connections, are described in detail. Practical guidance is provided on selecting the appropriate connector for different wiring applications and ensuring reliable electrical connections.

RATIONALE

This fundamental unit on electrical wiring tools and accessories stems from the importance of utilizing the right tools and equipment in electrical installations. In electrical engineering and construction, properly selecting, using, and maintaining wiring tools and accessories are critical factors that significantly impact electrical systems' safety, efficiency, and reliability. High-quality tools and accessories are essential for achieving reliable electrical connections and installations that meet or exceed industry standards. Through detailed explanations of various types of wiring tools and accessories, their features, applications, and best practices, this unit seeks to instill in learners the importance of quality and reliability in electrical work, fostering a culture of excellence and professionalism.

PRE-REQUISITES

Basics of Electrical Engineering

UNIT OUTCOMES

The list of outcomes of this unit is as follows:

U1-O1: Describe various tools required for wiring.

U1-O2: Explain the classification of electrical accessories.

U1-O3: Explain the diverse spectrum of electrical switches.

U1-O4: Describe holders, socket outlets, plugs, connectors, adaptors, etc.

Unit-1 Outcomes	Expected Mapping with Programme Outcomes (*1- Weak Correlation; 2- Medium correlation; 3- Strong Correlation*)					
	CO-1	**CO-2**	**CO-3**	**CO-4**	**CO-5**	**CO-6**
U1-O1	3	3	2	3	2	1
U1-O2	3	3	2	3	1	2
U1-O3	3	3	1	3	1	1
U1-O4	3	3	2	3	1	2

Introduction

Electricians must have adequate training and experience to effectively carry out their professional responsibilities. To acquire these materials, learners must enroll in a trade school or vocational program and serve as apprentices under a proficient electrician. By engaging in these actions, individuals will familiarise themselves with the techniques necessary to carry out the diverse tasks integral to their professions successfully. Nevertheless, electrical tools are an essential element that electricians must own for their profession. Electricians can significantly facilitate their work by possessing the appropriate equipment, mastering its usage, and acquiring needed expertise through training and experience. Furthermore, upon course completion, they can securely ensure their ability to execute the required activities. Moreover, by showcasing their proficiency in the job, they will earn clients' confidence and have the opportunity to pick up additional responsibilities.

1.1 Various Tools Required for Wiring

Moreover, electrical tools benefit individuals who are not necessarily electricians, as they assist in many activities. Furthermore, it provides additional measures to enhance their safety. Their productivity and job quality will improve with suitable equipment. Numerous emerging trends in the construction industry can significantly improve the quality of work carried out by electricians. These trends involve using advanced tools and equipment to enhance efficiency and productivity. Project management software is an excellent tool that may aid individuals and improve the convenience of accomplishing various tasks. Occasionally, the organization faces numerous tasks requiring prompt completion as a part of their responsibilities. Within two weeks, or conceivably even a solitary week, they could be assigned numerous jobs. They need plans to carry out their jobs with precision. Planning involves monitoring staff schedules, project timelines, and other transactions, such as purchases. All of these duties must be executed systematically to mitigate miscommunication and disputes that may result in reduced work hours or a decline in the quality of the deliverables. Project managers, or individuals in a position of principal responsibility, are accountable for monitoring their team and ensuring that all tasks are effectively managed and accomplished. Project management tools such as Pro Crew Schedule facilitate the efficient completion of team communication, multiple work scheduling, crew management, and other crucial project completion activities.

1.1.1 Screwdriver

Screwdrivers are vital tools for electricians. A variety of screwdrivers are utilized for tightening and loosening hardware. Electricians employ screwdrivers to install and remove outlets and cover plates, switches, and other electrical components. Furthermore, screwdrivers are equipped with insulating rubber jackets on their handles, enhancing safety while working with electricity. A screwdriver is a manual tool designed explicitly for spinning screws. Typically, it includes a handle and a shaft, finishing with a tip that securely fits into the head of a screw. The tip can exhibit diverse configurations and dimensions to correspond with various screw types. Screwdrivers are crucial for various applications, such as furniture assembly, electronics, and simple household repairs. Here are some common types of screwdrivers and their uses:

a) **Flathead Screwdriver:** Also known as a slotted screwdriver, this tool features a flat, single-bladed tip that perfectly matches the straight slot of a flathead (slotted) screw. Flathead screwdrivers (Fig.1.1(a)) are frequently employed for basic and universal tasks.

b) **Phillips Screwdriver:** Features a cross-shaped tip that fits into the corresponding cross-shaped slot of a Phillips screw. Phillips screwdrivers (Fig.1.1(b)) are extensively utilized in various applications and are renowned for their superior grip compared to flathead screwdrivers.

c) **Robertson (Square) Screwdriver:** This tool features a square-shaped tip that perfectly matches the square-shaped opening found in Robertson screws. Robertson screwdrivers (Fig.1.1(c)) are widely used in Canada and are renowned for providing a secure and reliable grip.

d) **Torx Screwdriver:** Specifically engineered for Torx screws, which feature a recess in the shape of a star. Torx screwdrivers (Fig.1.1(d)) are available in many sizes and are commonly employed in electronics, automotive applications, and various other industries.

e) **Hex (Allen) Screwdriver:** It is equipped with a hexagonal tip, designed explicitly for hex screws. Hex screwdrivers (Fig.1.1(e)) are frequently employed to assemble furniture, bicycles, and machinery.

f) **Precision Screwdriver:** This tool (Fig.1.1(f)) features a small and highly accurate point, making it ideal for complex tasks involving vulnerable electronics, watches, and compact devices.

g) **Multi-Bit Screwdriver:** This screwdriver (Fig.1.1(g)) is equipped with interchangeable tips or bits, enabling the user to switch between various types and sizes of screwdriver tips effortlessly. This feature offers flexibility and minimizes the requirement for different screwdrivers.

h) **Electric Screwdriver:** Electric screwdrivers (Fig.1.1(h)), which are powered by batteries or electricity, automate the process of screw turning, thereby simplifying duties and reducing physical effort.

i) **Impact Screwdriver:** This is intended to loosen or tighten screws that are difficult to move. Impact screwdrivers (Fig.1.1(i)) usually have a built-in impact mechanism, enabling the user to apply extra force without unnecessary effort.

j) **Offset Screwdriver:** Offset screwdriver (Fig.1.1(j)) includes a curved or inclined shaft, enabling the user to access screws in restricted or narrow areas where a linear screwdriver may not be suitable.

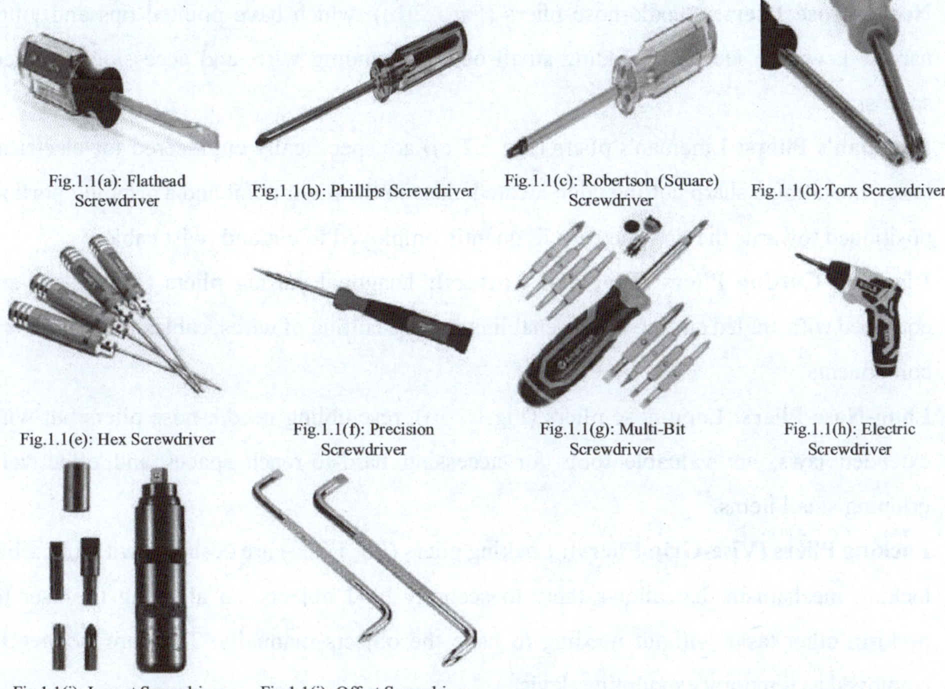

Fig.1.1(a): Flathead Screwdriver

Fig.1.1(b): Phillips Screwdriver

Fig.1.1(c): Robertson (Square) Screwdriver

Fig.1.1(d):Torx Screwdriver

Fig.1.1(e): Hex Screwdriver

Fig.1.1(f): Precision Screwdriver

Fig.1.1(g): Multi-Bit Screwdriver

Fig.1.1(h): Electric Screwdriver

Fig.1.1(i): Impact Screwdriver

Fig.1.1(j): Offset Screwdriver

Figure 1.1 Various screwdrivers

When choosing a screwdriver for a specific job, it is essential to select the appropriate type and size of the screwdriver tip that corresponds to the screw head. Incorrect selection of screw type or size might result in screw damage or restrict the ability to tighten or loosen it. Furthermore, consider the handle's quality and comfort for effortless usage throughout longer work sessions.

1.1.2 Pliers

Another crucial tool for electricians is a set of pliers. Various pliers include needle nose, linesman, diagonal cutting, and tongue-and-groove pliers. Wires can undergo twisting, bending, straightening, cutting, and gripping using pliers. Pliers are multifunctional hand tools engineered explicitly for gripping, bending, cutting, and manipulating diverse materials. They are available in myriad forms and dimensions, each intended for a specific purpose. Here are some common types of pliers and their uses:

a) **Slip-Joint Pliers:** These pliers feature a pivot point that can be adjusted, enabling the user to modify the range of the jaws. Slip-joint pliers (Fig.1.2(a)) are versatile and suitable for gripping and turning various objects.

b) **Needle-Nose Pliers:** Needle-nose pliers (Fig.1.2(b)), which have pointed tips and long, narrow jaws, are ideal for holding small objects, bending wire, and accessing confined spaces.

c) **Lineman's Pliers:** Lineman's pliers (Fig.1.2(c)) are specifically engineered for electrical tasks, including a sharp cutting edge located close to the pivot point and a gripping portion positioned towards the tip. They are frequently employed to cut and twist cables.

d) **Diagonal Cutting Pliers (Diagonal Cutters):** Diagonal cutting pliers (Fig.1.2(d)) are equipped with angled cutting edges, enabling precise cutting of wires, cables, and miniature components.

e) **Long-Nose Pliers:** Long-nose pliers (Fig.1.2(e)), resembling needle-nose pliers but with extended jaws, are valuable tools for accessing hard-to-reach spaces and effectively gripping small items.

f) **Locking Pliers (Vise-Grip Pliers):** Locking pliers (Fig.1.2(f)) are designed with a flexible locking mechanism that allows them to securely hold objects, so allowing the user to perform other tasks without needing to have the objects manually. They are frequently employed as temporary mounting devices.

g) **Channellock Pliers:** Channellock pliers (Fig.1.2(g)) are easily identifiable due to their unique tongue-and-groove shape, which allows them a tight grip on items of different sizes. The flexible jaw enables freedom in gripping.

h) **Crimping Pliers:** Crimping pliers (Fig.1.2(h)) are exclusively engineered for attaching connections to cables using crimping. They are frequently employed in electrical tasks and in establishing reliable connections for wires.

i) **Wire Stripping Pliers:** Wire stripping pliers (Fig.1.2(i)), equipped with precise cutting and stripping blades, are specifically designed to gently remove insulation from electrical wires while ensuring the conductor remains undamaged.

j) **Flat-Nose Pliers:** Flat-nose pliers (Fig.1.2(j)) are characterized by their broad and flat surface jaws and are well-suited for gripping and bending square or flat objects. They provide a secure grip and are commonly used in jewelry making and metalworking.

k) **Round-Nose Pliers:** Round-nose pliers (Fig.1.2(k)), with their curved jaws, are well-suited for precise tasks like jewelry creation, as they are perfect for forming loops, bends, and shaping wire.

l) **Snap Ring Pliers (Circlip Pliers):** Snap ring pliers (Fig.1.2(l)) are designed for the installation or removal of snap rings or circlips. There are two types of snap ring pliers, external and internal, which are used depending on the type of snap ring being handled.

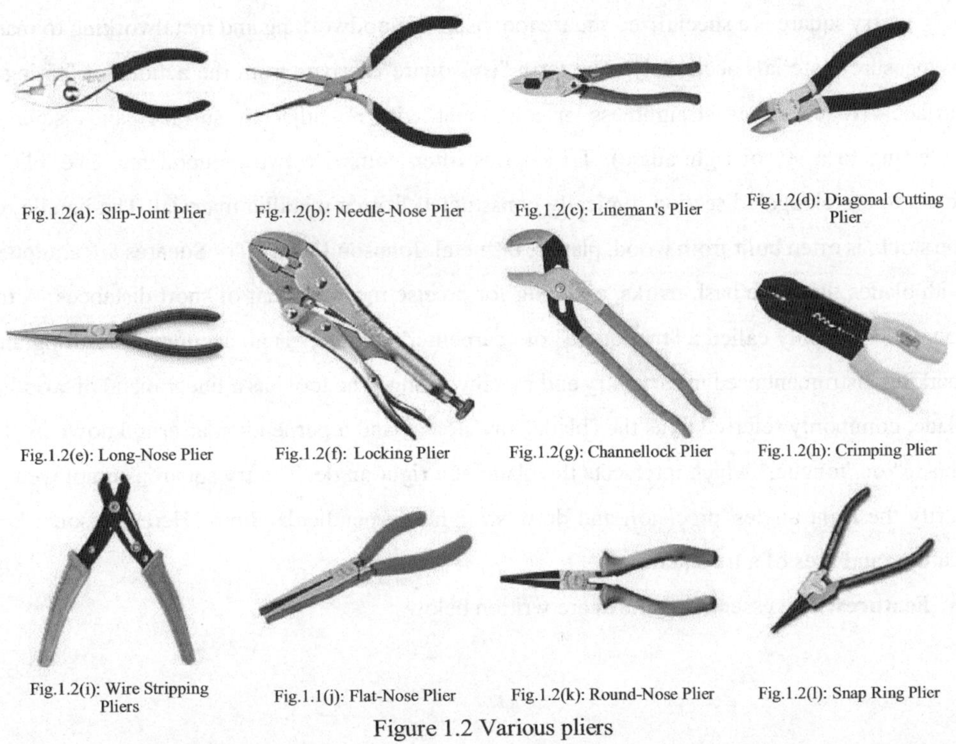

Fig.1.2(a): Slip-Joint Plier Fig.1.2(b): Needle-Nose Plier Fig.1.2(c): Lineman's Plier Fig.1.2(d): Diagonal Cutting Plier

Fig.1.2(e): Long-Nose Plier Fig.1.2(f): Locking Plier Fig.1.2(g): Channellock Plier Fig.1.2(h): Crimping Plier

Fig.1.2(i): Wire Stripping Pliers Fig.1.1(j): Flat-Nose Plier Fig.1.2(k): Round-Nose Plier Fig.1.2(l): Snap Ring Plier

Figure 1.2 Various pliers

It is crucial to select the appropriate type of pliers for the specific operation and employ them effectively to prevent harm to the tool or oneself. High-quality pliers featuring comfortable grips can significantly enhance user experience, particularly over extended periods of operation.

1.1.2.1 Tongue-and-groove pliers:

Tongue-and-groove pliers regularly referred to as groove joints, are usually used in plumbing tasks but can also be advantageous in electrical work. It can be utilized to fine-tune the expansion-type ceiling fan box pliers, secure cable clamps, detach knockouts from metal electrical boxes, ream pliers to tighten locknuts, and do other tasks.

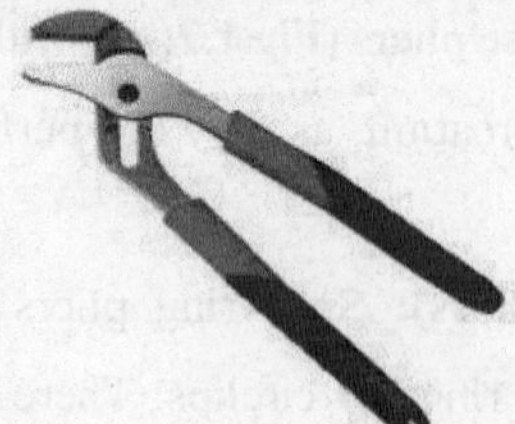

Figure 1.3 Tongue-and-groove pliers

1.1.3 Try Square

A try square is a specialized square tool used in woodworking and metalworking to mark or measure materials accurately. The term "try square" derives from the actions of "trying a surface" (to check its straightness or alignment with an adjacent surface) and "square" (referring to a 90° or right angle). Tri squares often comprise two components. The 'blade' refers to the elongated section, typically constructed from a metallic material. The handle, or the stock, is often built from wood, plastic, or metal. Johnson Level's Try Squares are equipped with blades that have hash marks, allowing for precise measurement of short distances. A try square, commonly called a "try square" or "carpenter's square," is an accurate measuring and marking instrument used in carpentry and metalworking. The tool has a linear metal or wooden blade, commonly referred to as the "blade" or "stock," and a perpendicular arm, known as the "blade" or "tongue," which intersects the blade at a right angle. The try square is employed to verify the right angles' precision and draw straight, perpendicular lines. Here are some key features and uses of a try square:

- **Features:** The essential features are written below

a) **Blade and Tongue:** The two arms of the try square consist of the tongue and blade. They form a right angle by assembling at a 90-degree angle. Although it may vary, the blade's typical length is between 6 and 12 inches.

b) **Material:** Wood, metal, or a composite of the two may be utilized to construct try squares. Steel is frequently used to build the blade for its durability and precision.

c) **Graduations:** Certain try squares feature graduated markings on the blade, enabling the user to measure, mark, and verify correct angles.

- **Uses:** The significant uses are given below

a) **Checking Right Angles:** A try square is the principal tool for verifying the precision of right angles in woodworking and metalworking activities. Woodworkers and carpenters utilize it to verify the perpendicular joints and corners.

b) **Marking Square Lines:** The try square marks straight, perpendicular lines on workpieces. This is particularly useful for tasks such as laying out cuts or marking the edges of a board.

c) **Measuring:** The try square outlines accurate, orthogonal lines on workpieces. This is especially advantageous for tasks such as setting up cuts or defining the boundaries of a board.

d) **Setting Up Tools:** Carpenters and woodworkers use try squares to establish the proper alignment of other instruments, such as saws and chisels, verifying that they are precisely perpendicular.

e) **Quality Control:** Try squares, frequently employed in manufacturing and construction as a quality control instrument, to verify the perpendicular of completed components.

- **How to Use:**

To utilize a try square, position the blade in contact with one side of the workpiece and the tongue against the neighbouring side. Verify that the intersection of the two arms creates a precise right angle of 90 degrees. If not, modifications may be necessary to guarantee precision.

The try square (Fig.1.4) is a fundamental piece of equipment for attaining exactness and precision in a wide range of wood and metalworking tasks. The tool's simplistic form facilitates user-friendliness, rendering it an excellent asset for carpenters, woodworkers, and DIY enthusiasts.

Figure 1.4 Try square

1.1.4 Saws

Multiple categories of saws exist that are specifically engineered for applications and materials. Below is a summary of some popular saw types, along with their principal applications:

a) ***Circular Saw:***

✓ **Use:** Optimal for making linear cuts in wood, plywood, medium-density fibreboard, and additional thin layers of material. It is frequently employed in the fields of construction and woodworking.

✓ **Features:** Circular saws (Fig.1.5(a)) are equipped with a circular blade with rotating teeth used for cutting purposes. They can be held in the hand or affixed to a table.

b) **Miter Saw:**

✓ **Use:** Specifically engineered for precise crosscuts and miter cutting in wood, commonly employed in framing, molding, and trim applications.

✓ **Features:** Miter saws (Fig.1.5(b)) feature a circular blade affixed to a rotating arm, enabling users to establish accurate cutting angles.

c) **Jigsaw:**

✓ **Use:** Utilised to precisely cut complex curves and non-uniform forms in wood, metal, plastic, and various other substances.

✓ **Features:** Jig saws (Fig.1.5(c)) have a slender, oscillating blade that moves vertically, rendering them well-suited for precise cuts.

d) **Reciprocating Saw:**

✓ **Use:** Suitable for performing demolition tasks and effortlessly cutting through various materials, such as wood, metal, plastic, and drywall.

- ✓ **Features:** Reciprocating saws (Fig.1.5(d)) are equipped with a blade that oscillates linearly, enabling forceful cutting.

e) **Band Saw:**

- ✓ **Use:** Highly proficient in the precision cutting of complex curves and irregular shapes in various materials such as wood, metal, and plastic. Frequently employed in woodworking and metalworking workshops.
- ✓ **Features:** Band saws (Fig.1.5(e)) consist of a continuous loop of toothed blades stretched over two wheels, enabling the execution of accurate and sensitive cuts.

f) **Table Saw:**

- ✓ **Use:** The primary device for creating precise, linear cuts in large wooden objects. Commonly employed in the fields of woodworking and building.
- ✓ **Features:** Table saws (Fig.1.5(f)) consist of a level platform with a circular blade extending outward from the surface. They are adaptable to various types of cutting.

g) **Hand Saw (Crosscut Saw, Rip Saw, Panel Saw):**

- ✓ **Use:** Intended for the manual process of cutting wood. Various categories comprise crosscut saws, used for cutting perpendicular to the wood grain; rip saws, used for cutting parallel to the grain; and panel saws, which are versatile tools for broad cutting purposes.
- ✓ **Features:** Hand saws (Fig.1.5(g)) have a long, straight blade with teeth for cutting wood.

h) **Chainsaw:**

- ✓ **Use:** Mainly employed to fell trees, prune branches, and prepare firewood.
- ✓ **Features:** Chainsaws (Fig.1.5(h)) consist of a guide bar equipped with a revolving chain fitted with teeth. These tools are frequently employed in the fields of forestry and arboriculture.

i) **Hacksaw:**

- ✓ **Use:** Designed to precisely cut metal, plastic, and other robust elements. Frequently employed in the fields of metalworking and plumbing.
- ✓ **Features:** Hacksaws (Fig.1.5(i)) consists of a blade with closely spaced teeth that is tensioned between two handles. These instruments are manually operated and designed for precise cutting.

j) **Japanese Pull Saw:**

- ✓ **Use:** Renowned for their exceptional precision and expertise in producing perfect cuts in timber. Utilized in the field of woodworking for precise carpentry tasks.

✓ **Features:** Japanese pull saws (Fig.1.5(j)) are equipped with a slim and flexible blade that performs cutting action during the pull stroke instead of the push stroke.

k) **Coping Saw:**

✓ **Use:** Ideal for making intricate curved cuts and detailed work in wood.

✓ **Features:** Coping saws (Fig.1.5(k)) include a slender, narrow blade tightly stretched within a U-shaped metal structure, enabling easy movement in confined areas.

l) **Panel Saw:**

✓ **Use:** Designed for cutting large panels of wood or sheet materials.

✓ **Features:** Panel saws (Fig.1.5(l)) feature an elongated, uncurved blade with massive teeth, making them well-suited for creating straight cuts throughout a panel.

The choice of an appropriate saw for a particular operation is contingent upon the material's nature, the desired type of cut, and the desired level of accuracy. It is imperative to adhere to safety protocols consistently and don suitable protection equipment while operating saws.

Fig.1.5(a): Circular saw Fig.1.5(b): Miter saw Fig.1.5(c): Jig saw Fig.1.5(d): Reciprocating saw

Fig.1.5(e): Band saw Fig.1.5(f): Table saw Fig.1.5(g): Hand saw Fig.1.5(h): Chainsaw

Fig.1.5(i): Hacksaw Fig.1.5(j): Japanese pull saw Fig.1.5(k): Coping saw Fig.1.5(l): Panel saw

Figure 1.5 Various saws

1.1.5 Hacksaw

Hacksaws are suitable for cutting metals of small and medium sizes as well as insulated cable and metal pipes.

A hacksaw is a multifunctional physical device used for a wide range of materials, primarily focusing on metal and plastic. The tool is made up of a slim blade with closely spaced teeth that is firmly positioned between two handles. The interchangeable blade usually includes finely crafted teeth specifically engineered for precise metal cutting. The following are a few features and applications of a hacksaw:

- **Features:** The essential features are written below

a) **Frame:** A hacksaw frame's purpose is to securely hold and apply tension on the blade. The typical shape of the object is C-shaped, which facilitates convenient blade replacement.

b) **Blade:** Hacksaw blades possess finely sharpened teeth engineered explicitly for cutting down metal and other rigid materials. The blades are available in varying lengths and tooth counts, providing adaptability for cutting diverse materials.

c) **Tensioning Mechanism:** Hacksaws commonly feature a tensioning mechanism that enables the user to modify the tension applied to the blade precisely. Optimal tension is crucial for effective and precise cutting.

d) **Adjustable Angle:** Certain hacksaws feature a modifiable frame that enables the operator to change the blade to various angles. This functionality is advantageous for precisely severing substances at diverse inclinations and confined areas.

e) **Comfortable Handle:** The handles of a hacksaw are scientifically made to provide a comfortable grip, enabling the user to apply sufficient force throughout the cutting process.

- **Uses:** The significant uses are given below

a) **Metal Cutting:** A hacksaw is mainly utilized for metal cutting. This applies to many metals, such as steel, aluminum, and copper. The blade's tiny teeth ensure a precise and perfect cut.

b) **Plastic Cutting:** Hacksaws are versatile tools that can be employed to cut plastic pipes, rods, and sheets. Nevertheless, it is crucial to utilize a suitable blade to cut plastics.

c) **DIY Projects:** Hacksaws are frequently employed in DIY projects to precisely cut metal pipes, bars, or threaded rods to a desired length.

d) **Metalworking and Fabrication:** Hacksaws are employed in metalworking workshops to achieve precise cuts on metal components and materials, mainly when alternative cutting instruments are unsuitable.

e) **Construction and Carpentry:** Hacksaws are commonly used to cut metal brackets, bolts, and other small metal components in construction and carpentry.

f) **Plumbing:** Plumbers employ hacksaws to precisely cut metal or plastic pipes to the required length during installation or repair.

- **How to Use:**

a) **Secure the Material:** Utilise clamps or alternative fastening methods to stabilize the cut material.

b) **Select the Right Blade:** Select a hacksaw blade with a suitable tooth count for your cutting material. Precision blades are appropriate for cutting metal, whereas rough blades can be utilized for less dense materials.

c) **Tension the Blade:** Adjust the tensioning mechanism on the hacksaw frame to tension the blade correctly.

d) **Start Cutting:** Ensure the hacksaw is held at an optimal angle and exerts moderate pressure during the cutting process. Allow the blade's teeth to perform the task while employing extended and fluid movements.

e) **Finish the Cut:** Keep cutting until the material is completely detached. Exercise caution when cutting along the line to guarantee precision.

Hacksaws are easy devices; however, they are vital for accurately cutting metal and other strong materials. Attaining precise and flawless cuts necessitates utilizing correct blade selection and attention to proper technique.

1.1.6 Chisel

Electricians utilize chisels to remove material from walls while installing pattress boxes. Typically, the angle of an electrician's chisel is less acute than that of a flooring chisel. Occasionally, they may be referred to as electrician's bolsters. A chisel is a tool used for cutting, consisting of a sharp metal blade at one end and a handle at the other. Chisels are utilized to sculpt, mold, or sever resilient substances such as wood, metal, stone, or concrete. They are available in many forms and sizes, each specifically tailored for particular uses. Below are several prevalent varieties of chisels, along with their respective applications:

a) **Wood Chisel:**

✓ **Use:** Utilized to cut, carve, and shape timber. Wood chisels are available in several widths and varieties, such as sharpened-edge, stronger, and hollow chisels.

✓ **Features:** A wood chisel's (Fig.1.6(a)) blade is commonly crafted from high-carbon steel or tool steel. Wood chisels often feature beveled edges to facilitate accurate and meticulous cutting.

b) Masonry Chisel:

✓ **Use:** Intended for cutting or shaping brick, stone, or concrete. Masonry chisels are frequently employed in the field of construction and masonry labor.

✓ **Features:** Masonry chisels (Fig.1.6(b)) possess a broader blade featuring a flat cutting edge, and they are frequently employed in conjunction with a hammer to enhance the cutting force.

c) Cold Chisel:

✓ **Use:** Utilised to cut or mold frigid metallic substances, such as iron or steel. Cold chisels are frequently used in the fields of metalworking and blacksmithing.

✓ **Features:** Cold chisels (Fig.1.6(c)) possess a broad, flat blade with a slanted edge. Metal is chopped or shaped by striking it with a hammer.

d) Bolster Chisel:

✓ **Use:** Bolster chisels are specialized tools for cutting or shaping brick, stone, or masonry, much like masonry chisels.

✓ **Features:** Bolster chisels (Fig.1.6(d)) possess a broader blade and are frequently employed for cutting or separating bricks.

e) Wood carving Chisel:

✓ **Use:** Explicitly designed for precise wood carving. Wood carving chisels are available in several forms, such as gouges, V-tools, and skew chisels.

✓ **Features:** Wood carving chisels (Fig.1.6(e)) have thin, specialized blades with various profiles to accomplish specific cutting outcomes.

f) Japanese Chisel (Nomi):

✓ **Use:** Japanese chisels commonly employed in woodworking and carpentry. Chisels are available in various forms and dimensions, such as bench and paring chisels.

✓ **Features:** Japanese chisels (Fig.1.6(f)) commonly feature a blade made of laminated steel and a handle crafted from wood. They are renowned for their sharpness and precision.

g) Sculptor's Chisel:

✓ **Use:** Intended for shaping and molding different substances, such as wood, stone, or metal. Sculptor's chisels are available in various shapes and sizes to accommodate different sculpting techniques.

✓ **Features:** Sculptor's chisels (Fig.1.6(g)) are equipped with more comprehensive, curved, or pointed blades designed to sculpt three-dimensional objects.

h) Gouge:

✓ **Use:** Gouges, like woodcarving chisels, possess a curved cutting edge and are employed to excavate or hollow out wood.

✓ **Features:** Gouges (Fig.1.6(h)) are available in a range of sweeps and widths, which can be used to generate diverse carving effects.

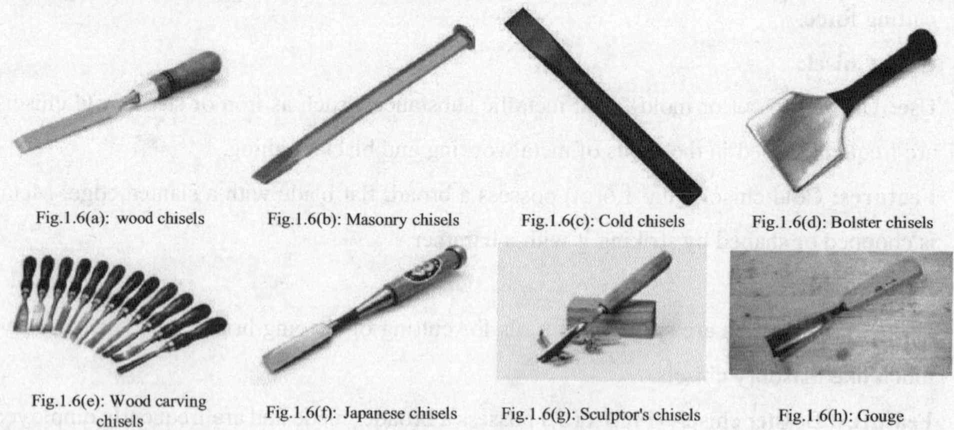

Fig.1.6(a): wood chisels Fig.1.6(b): Masonry chisels Fig.1.6(c): Cold chisels Fig.1.6(d): Bolster chisels

Fig.1.6(e): Wood carving chisels Fig.1.6(f): Japanese chisels Fig.1.6(g): Sculptor's chisels Fig.1.6(h): Gouge

Figure 1.6 Various chisels

How to Use:

✓ Secure the material you are working on.

✓ Hold the chisel with the bevel side down.

✓ Use a mallet or hammer to strike the chisel for cutting or shaping.

✓ Take care to control the depth and angle of the chisel to achieve the desired result.

✓ Regularly sharpen the chisel to maintain its effectiveness.

Chisels are multifunctional implements employed in several industries, such as woodworking, metalworking, masonry, and sculpture. The choice of chisel is contingent upon the particular use case and the material being manipulated. When engaging in chisel work, it is imperative to consistently employ suitable safety precautions, such as wearing eye protection and utilizing correct handling methods.

1.1.7 Hammer

Hammers are tools utilized to insert nails into a desired position forcefully. Two types are available: one with a soft face and one with a stern face. A hard-faced hammer has a cylindrical head for forcefully striking solid objects. The rewinding process utilizes a soft-faced hammer. Many soft-faced hammers are made from plastic or rubber materials. Multiple hammers exist, including mallets, ball peen hammers, and claw hammers. A hammer is a manual instrument with a weighted head affixed to a handle. Its primary function is to insert or extract nails from different materials forcefully, and it can also be employed for numerous other activities that involve hitting. There are multiple sorts of hammers, each specifically built for uses. Below are several prevalent varieties of hammers and their respective applications:

a) **Claw Hammer:**

✓ **Use:** Primarily utilized to insert nails into wood and perform various tasks related to woodworking. The posterior claw is utilized for the extraction of nails.

✓ **Features:** Claw hammers (Fig.1.7(a)) generally feature a slightly bent claw that offers enhanced leverage for extracting nails.

b) **Ball-Peen Hammer:**

✓ **Use:** Widely employed in metalworking to shape and mold metal and impact chisels and punches.

✓ **Features:** Ball-peen hammers (Fig.1.7(b)) are equipped with a curved peen on the opposite side of the striking face, specifically designed for operations like peening or shaping the edges of metal.

c) **Sledgehammer:**

✓ **Use:** Intended for demanding activities such as demolishing, driving stakes, or penetrating walls. It possesses the ability to deliver forceful strikes.

✓ **Features:** Sledgehammers (Fig.1.7(c)) have a substantial, flat metal head and a lengthy grip to enhance leverage.

d) **Dead Blow Hammer:**

✓ **Use:** Employed in jobs necessitating minimum recoil, such as automotive chores or when impacting delicate materials.

✓ **Features:** Dead blow hammers (Fig.1.7(d)) feature a hollow head filled with shot or sand, resulting in reduced bounce and a controlled impact.

e) **Rubber Hammer:**

✓ **Use:** Perfect for tasks requiring a gentler impact, such as furniture assembly or chisel striking, when using metal hammers could potentially result in damage.

✓ **Features:** Rubber hammers (Fig.1.7(e)) are equipped with a rubber or plastic head, which reduces the likelihood of causing scratches or dents on surfaces.

f) **Cross Peen Hammer:**

✓ **Use:** Utilised in metalworking to manipulate and mold metal, particularly in blacksmithing. The cross peen is positioned at a right angle to the striking face.

✓ **Features:** Cross peen hammers (Fig.1.7(f)) are available in several sizes, and the peen enhances their ability to shape objects.

g) **Bricklayer's Hammer (Masonry Hammer):**

✓ **Use:** Intended for cutting and placing bricks or masonry units. The chisel end is employed for cutting, whereas the flat end is utilized for hitting.

✓ **Features:** Bricklayer's hammers (Fig.1.7(g)) often include a chisel edge on one side and a flat face on the other.

h) **Tack Hammer:**

✓ **Use:** Commonly employed for intricate tasks such as upholstery work or fastening tiny tacks and brads.

✓ **Features:** Tack hammers (Fig.1.7(h)) often feature a compact and lightweight head and a magnetic surface that securely holds tacks in position.

i) **Club Hammer:**

✓ **Use:** Club hammers, which are smaller than hammers, are utilized for less intense demolition, chiseling, or the act of driving masonry nails.

✓ **Features:** Club hammers (Fig.1.7(i)) have a short handle and a double-faced head.

j) **Soft-Faced Hammer:**

✓ **Use:** Utilised when a gentler impact is necessary, such as in woodworking or while joining fragile materials.

✓ **Features:** Soft-faced hammers (Fig.1.7(j)) feature replaceable faces of rubber, plastic, or rawhide materials.

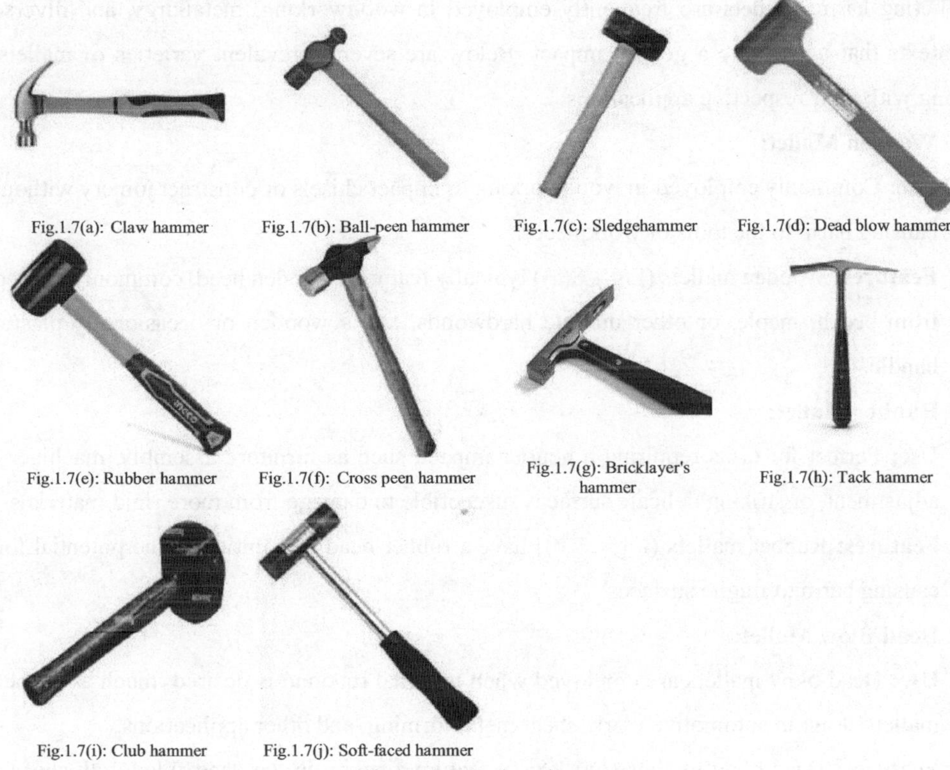

Fig.1.7(a): Claw hammer Fig.1.7(b): Ball-peen hammer Fig.1.7(c): Sledgehammer Fig.1.7(d): Dead blow hammer

Fig.1.7(e): Rubber hammer Fig.1.7(f): Cross peen hammer Fig.1.7(g): Bricklayer's hammer Fig.1.7(h): Tack hammer

Fig.1.7(i): Club hammer Fig.1.7(j): Soft-faced hammer

Figure 1.7 Various hammers

- **How to Use:**

- ✓ Choose the appropriate hammer for the task at hand.
- ✓ Ensure the head is securely attached to the handle.
- ✓ Hold the hammer near the end of the handle for better control.
- ✓ Strike with controlled and accurate blows.
- ✓ When driving nails, hold the hammer slightly to avoid bending the nails.
- ✓ Use proper safety gear, including eye protection, when using a hammer.

Hammers are essential in diverse vocations, and the meticulous choice of the appropriate kind for a particular undertaking is paramount. Performing routine maintenance, such as inspecting for loose components and verifying the handle's condition, is crucial to guarantee the equipment's safety and effectiveness.

1.1.8 Mallet

A mallet is a manual implement with a substantial head, usually wood, and a handle. It is utilized for striking or beating and is precisely engineered to administer blows without inflicting harm. Mallets are frequently employed in woodworking, metallurgy, and diverse contexts that necessitate a gentler impact. Below are several prevalent varieties of mallets, along with their respective applications:

a) **Wooden Mallet:**

✓ **Use:** Commonly employed in woodworking to impact chisels or construct joinery without causing harm to the tools or workpieces.

✓ **Features:** Wooden mallets (Fig.1.8(a)) typically feature a wooden head, commonly crafted from beech, maple, or other durable hardwoods, and a wooden or occasionally plastic handle.

b) **Rubber Mallet:**

✓ **Use:** Perfect for tasks requiring a gentler impact, such as furniture assembly, machinery adjustment, or striking delicate surfaces susceptible to damage from more rigid materials.

✓ **Features:** Rubber mallets (Fig.1.8(b)) have a rubber head that mitigates the potential for causing harm to fragile surfaces.

c) **Dead Blow Mallet:**

✓ **Use:** Dead blow mallets are employed when minimal rebound is desired, much as rubber mallets. Joint in automotive work, sheet metal forming, and other applications.

✓ **Features:** Dead blow mallets (Fig.1.8(c)) feature a cavity in the head filled with shot or sand, enabling precise impact control and reducing bounce-back.

d) **Rawhide Mallet:**

✓ **Use:** Commonly employed in metalworking, namely to shape or mold metal without causing any damage to its surface. Additionally, I was employed in jewelry fabrication and leather craftsmanship.

✓ **Features:** Rawhide mallets (Fig.1.8(d)) are equipped with an unprocessed animal hide head, which reduces the probability of leaving markings on the surface being struck.

e) **Plastic Mallet:**

✓ **Use:** Appropriate for situations that require a gentler impact to prevent damage. Frequently employed in building, woodworking, and automotive endeavors.

✓ **Features:** Plastic mallets (Fig.1.8(e)) feature a head composed of robust plastic material that exhibits resistance to harm.

f) **Nylon Mallet:**

- ✓ **Use:** Widely used in jewellery making, metalworking, and other precision work necessitating a non-damaging and accurate impact.
- ✓ **Features:** Nylon mallets (Fig.1.8(f)) have a head composed of nylon, which is softer than metal and won't damage or deform the striking surface.

g) **Brass Mallet:**

- ✓ **Use:** Commonly employed in metalworking, precisely to strike softer metals to avoid leaving any visible marks. Additionally, I am employed in the craft of jewellery fabrication.
- ✓ **Features:** Brass mallets (Fig.1.8(g)) feature a brass head with a hardness greater than plastic or rubber but less than steel.

h) **Leather Mallet:**

- ✓ **Use:** Utilised in leatherworking and jewellery making crafts, this tool shapes and molds materials without causing any harm or damage.
- ✓ **Features:** Leather mallets (Fig.1.8(h)) feature a leather head, typically packed with lead shot to increase weight.

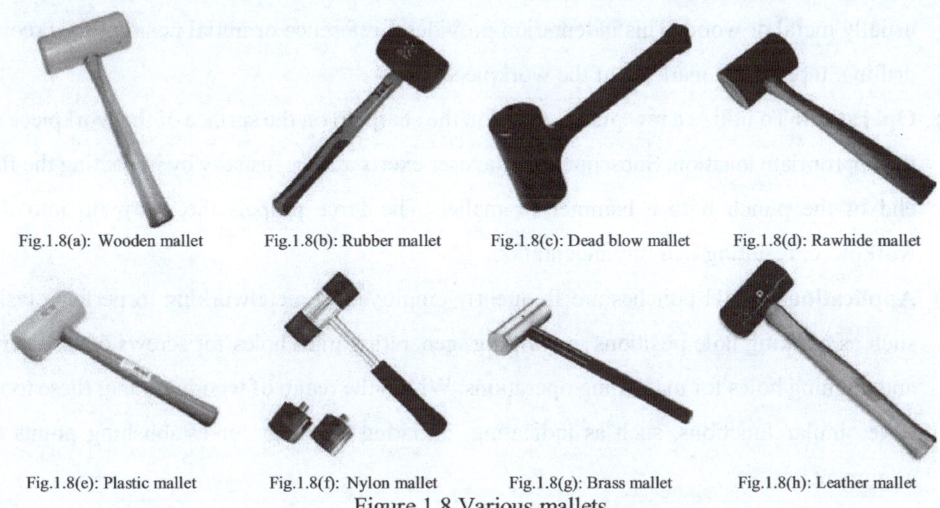

Fig.1.8(a): Wooden mallet Fig.1.8(b): Rubber mallet Fig.1.8(c): Dead blow mallet Fig.1.8(d): Rawhide mallet

Fig.1.8(e): Plastic mallet Fig.1.8(f): Nylon mallet Fig.1.8(g): Brass mallet Fig.1.8(h): Leather mallet

Figure 1.8 Various mallets

How to Use:

- ✓ Hold the mallet near the end of the handle for better control.
- ✓ Strike with controlled and accurate blows.
- ✓ Use the appropriate mallet type for the specific task to avoid damage to tools or workpieces.
- ✓ Wear appropriate safety gear, including eye protection, when using a mallet.

Mallets are multifunctional implements that deliver a gentler force than hammers, rendering them appropriate for jobs that require a delicate touch. The selection of a particular mallet type depends on the material being manipulated and the intended impact properties.

1.1.9 Rawl Punch

A rawl punch (Fig.1.9), referred to as a center punch, is a manual instrument employed in metalworking and woodworking to generate a slight depression or initial point in a workpiece. Below is a concise summary of a raw punch.

Figure 1.9 Rawl punch

a) **Design:** A rawl punch usually comprises a cylindrical metallic shaft with a sharp tip at one extremity and a flat or knurled surface at the other extremity, which facilitates grasping. The sharp end is toughened to endure frequent impacts without distorting.
b) **Function:** A rawl punch is primarily used to make a slight slope or dip in a work item, usually metal or wood. This indentation provides a reference or initial position for precise drilling, tapping, or marking of the workpiece.
c) **Operation:** To utilize a raw punch, position the sharp tip on the surface of the workpiece at the appropriate location. Subsequently, the user exerts a force, usually by impacting the flat end of the punch with a hammer or mallet. The force propels the sharp tip into the workpiece, resulting in a tiny indentation.
d) **Applications:** Rawl punches are frequently employed in metalworking to perform tasks such as marking hole positions for drilling, generating pilot holes for screws or fasteners, and aligning holes for machining operations. Within the realm of woodworking, these tools have similar functions, such as indicating, initiating openings, or establishing points of alignment.
e) **Types:** Rawl punches are offered in various sizes and designs to accommodate diverse workpiece applications and materials. The tip forms of these tools can vary, with options including conical or cylindrical, chosen based on the unique work needs.

1.1.10 Hand Drill Machine

This tool creates openings in wooden cross arms and studs to accommodate tightly fitting high-tension and low-tension wires originating from DP constructions or LT transformer bushings. A hand drill machine (Fig.1.10), a manual hand drill or hand-powered drill, is an essential equipment used to create holes in different materials. Hand drills, in contrast to electric or cordless drills, necessitate manual operation, wherein the user must crank a handle to turn the drill bit. These tools are adaptable and well-suited for small-scale drilling, carpentry, and other applications requiring accuracy and control.

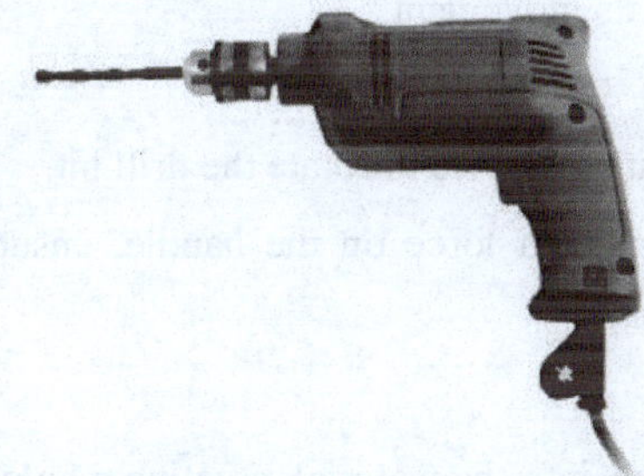

Figure 1.10 Hand drill machine

- **Key Components of a Hand Drill Machine:**

✓ **Handle:** The handle functions as a crank or knob that the user controls to revolve the drill bit. It supplies the necessary hand force for drilling.

✓ **Chuck:** The chuck is responsible for securing the drill bit. The chuck in a hand drill is commonly a three-jaw design that securely clamps onto the shank of the drill bit.

✓ **Frame or Body:** The hand drill's frame or body supports the other components and ensures drilling stability. It frequently has a gear system that converts the rotational action from the handle to the chuck.

✓ **Gear Mechanism:** Certain hand drills have a geared mechanism for higher rotational speed and torque. This is especially advantageous when penetrating more resistant materials.

✓ **Drill Bit:** The drill bit functions as the implement responsible for forming the aperture in the material. Hand drills may take a wide range of drill bits, varying in type and size depending on the specific task.

- **How to Use a Hand Drill Machine:**

a) **Insert the Drill Bit:**

✓ Open the chuck by turning the handle counterclockwise.

✓ Insert the shank of the drill bit into the chuck.

✓ Tighten the chuck by turning the handle clockwise until the drill bit is securely held.

b) **Position the Hand Drill:**

Position the apex of the drill bit precisely on the designated spot of the material where you intend to bore.

c) **Secure the Material:**

Whenever feasible, ensure the stability of the material being drilled by employing clamps or alternative methods to impede its movement.

d) **Turn the Handle:**

✓ Grasp the handle and turn it clockwise to rotate the drill bit.

✓ Exert a consistent and regulated force on the handle, ensuring it remains straight and perpendicular to the material.

e) **Drill the Hole:**

✓ As the drill bit rotates, it cuts into the material, creating a hole.

✓ Continue turning the handle until the hole reaches the desired depth.

f) **Release the Handle:**

✓ After finishing the drilling process, let go of the handle and rotate it in the opposite direction to the right to unlock the chuck.

✓ Remove the drill bit.

Advantages of Hand Drill Machines:

✓ Portable and lightweight.

✓ No dependence on electricity or batteries.

✓ Suitable for small-scale drilling tasks and precision work.

✓ Cost-effective and easy to maintain.

Limitations:

✓ Slower drilling speed compared to powered drills.

✓ Limited torque, making it less suitable for heavy-duty applications.

Hand drill machines are indispensable instruments for DIY enthusiasts, hobbyists, and woodworkers who need a manual and accurate method for drilling. They are also helpful in situations with limited access to power sources.

1.1.11 Portable Drilling Machine

A portable drilling machine (Fig.1.11), commonly called a hand-held or cordless drill, is a versatile and convenient power instrument used to drill holes in various materials. Unlike stationary drilling machines, portable drills are intended explicitly for mobility, enabling users to navigate their workspace freely. They operate using rechargeable batteries, obviating the necessity for a continuous electrical power supply. The following are the primary characteristics and applications of a portable drilling machine:

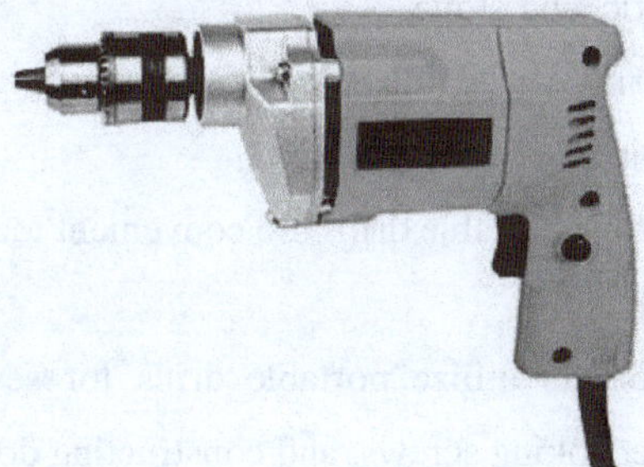

Figure 1.11 Portable drilling machine

Key Features:

✓ **Battery-Powered:** Portable drilling machines are commonly operated by rechargeable batteries, typically lithium-ion batteries. This feature offers portability and enables users to operate where power sockets are unavailable.

✓ **Variable Speed Control:** Numerous portable drills are equipped with adjustable speed settings, enabling users to modify the drilling speed according to the material being drilled and the size of the drill bit.

✓ **Chuck:** The chuck is the drill component that securely grips the drill bit. Portable drills typically feature a keyless chuck, enabling effortlessly changing drill bits without using a key or other supplementary gear.

✓ **Clutch:** Certain portable drills are equipped with a modifiable clutch mechanism. The clutch regulates the torque exerted on the drill bit, limiting excessive tightening of screws and enabling more precision in control.

✓ **Forward and Reverse Function:** Portable drills commonly feature a dual-direction capability, enabling users to drive or extract screws effortlessly.

✓ **LED Work Light:** Numerous contemporary portable drills are furnished with integrated LED work lights, which offer lighting in dimly lit or restricted areas.

✓ **Compact and Lightweight:** Portability is a crucial characteristic of portable drills, as they are specifically intended to be small and lightweight, facilitating effortless transportation and maneuverability.

✓ **Ergonomic Design:** The handle's ergonomic shape guarantees a pleasant grip and minimizes fatigue during extended use.

Common Uses:

- ✓ **Drilling Holes:** Portable drills are utilized to perforate holes in a wide range of materials, such as wood, metal, plastic, and stone.
- ✓ **Screwdriving:** Portable drills can be utilized to drive screws into various materials by using the suitable screwdriver bit.
- ✓ **Assembly and Disassembly:** Portable drills are convenient tools for assembling furniture, appliances, and other objects.
- ✓ **Woodworking:** Woodworkers utilize portable drills for several operations, including drilling pilot holes, countersinking screws, and constructing dowel joints.
- ✓ **Home Improvement Projects:** Portable drills are indispensable equipment for various home renovation chores, like installing shelves, cabinets, or curtain rods.
- ✓ **Light Construction Work:** Portable drills are utilized for various tasks in light construction work, such as the installation of plasterboard, framing, and other similar applications.
- ✓ **Automotive Repairs:** Portable drills are frequently used in vehicle repairs, such as drilling holes for brackets, removing or installing fasteners, and other related activities.

Portable drilling machines are multifunctional tools appropriate for skilled tradespeople and do-it-yourself enthusiasts. These tools provide adaptability and ease for various drilling and fastening tasks, making them a valuable asset to any collection of tools.

1.1.12 Files

In the realm of tools, the term "files" often pertains to manual tools employed to shape, refine, and polish the surfaces of different materials. These tools are frequently used in metallurgy, woodworking, and other crafts. Files possess a coarse texture featuring teeth or ridges that effectively eliminate material through friction with a workpiece. Below are several prevalent file formats and their respective purposes:

a) **Flat File:**

- ✓ **Shape:** Flat on both sides with parallel edges.
- ✓ **Use:** Flat files (Fig.1.12(a)) are used for general-purpose filing, smoothing, and leveling on flat surfaces.

b) **Half-Round File:**

- ✓ **Shape:** Flat on one side and curved (semi-circular) on the other.

✓ **Use:** Half-round files (Fig.1.12(b)) are used for filing concave surfaces, rounding edges, and shaping contours.

c) **Round File:**

✓ **Shape:** Circular or cylindrical in shape.

✓ **Use:** Round files (Fig.1.12(c)) expand apertures, form concave surfaces, and generate curved outlines.

d) **Square File:**

✓ **Shape:** Square or rectangular in cross-section.

✓ **Use:** Square files (Fig.1.12(d)) are used for filing slots, corners, and other square-shaped features.

e) **Triangle File:**

✓ **Shape:** Triangular in cross-section.

✓ **Use:** Triangle files (Fig.1.12(e)) are used for filing sharp angles, narrow corners, and other triangle characteristics.

f) **Rasp File:**

✓ **Shape:** Coarse is typically used for aggressive material removal.

✓ **Use:** Rasp files (Fig.1.12(f)) are used for shaping wood or soft materials, often used in woodworking.

g) **Needle File:**

✓ **Shape:** Slim and pointed, resembling a needle.

✓ **Use:** Needle files (Fig.1.12(g)) are used for meticulous tasks in confined or intricate areas, commonly employed in creating jewelry or the art of refined woodworking.

h) **Machinist's File:**

✓ **Shape:** Usually parallel in width and thickness.

✓ **Use:** Machinist's files (Fig.1.12(h)) are used for metalworking and machining for accurate and meticulous filing.

i) **Diamond File:**

✓ **Surface:** Embedded with diamond particles.

✓ **Use:** Needle files (Fig.1.12(i)) are used for filing hard materials such as carbide, ceramics, or glass.

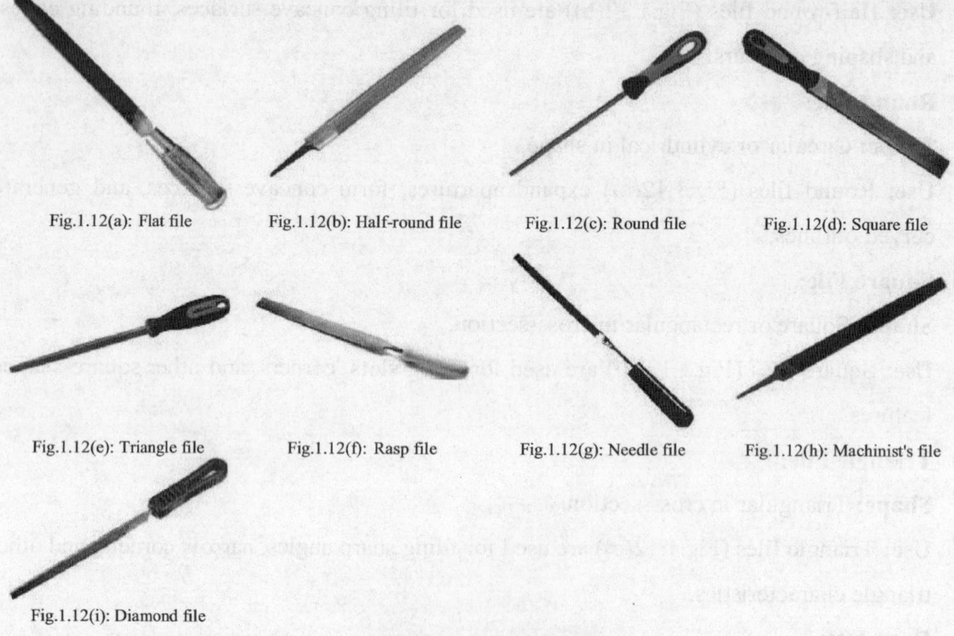

Figure 1.12 Various files

- **How to Use Files:**

✓ **Secure the Workpiece:** Ensure the workpiece is firmly secured using clamps or a vice.

✓ **Select the Right File:** Select the suitable format and dimensions of the file according to the content and the objective.

✓ **File in One Direction:** To achieve efficient material removal, file unidirectionally, away from your position. Refrain from employing a reciprocating sawing motion.

✓ **Apply Even Pressure:** Exert uniform force on the file while keeping a steady angle. Allow the file to perform the tasks.

✓ **Check Progress:** Regularly inspect the workpiece to assess your advancement and ensure you attain the desired shape and uniformity.

✓ **Use Multiple Files if Needed:** Depending on the activity's complexity, it may be necessary to utilize many files to attain the intended outcome.

Files are highly adaptable tools available in diverse forms and dimensions, enabling extensive uses. They are crucial in attaining accurate forms and smooth surfaces in various crafting and fabrication tasks.

1.1.13 Plumb Bob

A plumb bob (Fig.1.13) is a weighted object attached to a string. It is used to determine a vertical reference line called a plumb. A plumb is an instrument utilized to ascertain vertical alignment, akin to a spirit level employed for horizontal alignment. The origin of plumb bobs can be traced back to Ancient Egypt, where they have been utilized for millennia. Furthermore, this equipment is employed in surveying to ascertain the nadir, which denotes the specific vertical direction aligned with gravity. It is also used in instruments such as steel tape measures and theodolites. A plumb bob is a simple yet effective instrument used in construction, carpentry, and surveying to provide a vertical reference line. The device consists of a weight hung by a thread, usually a pointed metal or brass object. The plumb bob is hanging unconstrained, aligning effortlessly with the Earth's gravitational force, forming a vertical line.

Figure 1.13 Plumb bob

Key Components of a Plumb Bob:

- ✓ **Weight:** A plumb bob often consists of a metallic material, such as steel or brass, and is pointed at the bottom to ensure accurate suspension. The shape of the plumb bob enables its precise alignment with the force of gravity.
- ✓ **String or Line:** The weight is attached to a string or line, usually made of nylon or other flexible materials. The thread length determines the distance between the plumb bob and the surface below.

How to Use a Plumb Bob:

- ✓ **Secure the Plumb Bob:** Hang the plumb bob at a desired location to establish a vertical reference line. Ensure that the plumb bob remains undisturbed and can hang freely without any obstruction.

✓ **Allow it to Come to Rest:** After being suspended, the plumb bob will start swaying. Enable the object to achieve the complete cessation of movement, characterized by the downward settling of the pointing weight.

✓ **Mark the Reference Line:** Specify the precise position on the surface directly beneath it after the plumb bob has completely stopped. This designated line will function as a vertical benchmark.

✓ **Use for Alignment:** The plumb bob is commonly used to achieve vertical alignment. For example, it can be utilized in the construction domain to ensure the accurate vertical alignment of walls, posts, or columns.

Applications of Plumb Bob:

✓ **Construction:** Used to ensure accurate alignment in constructing walls, columns, and other structural elements.

✓ **Carpentry:** Facilitates the alignment and leveling of components in carpentry jobs, such as door hanging or cabinet installation.

✓ **Surveying:** Used in surveying for establishing vertical reference points.

✓ **Masonry:** Ensures precise alignment in the vertical direction while constructing walls or structures using bricks or blocks in masonry construction.

✓ **Metalworking:** Used to ensure the verticality of metal structures and components.

✓ **Alignment in Mechanical Work:** Useful in mechanical work for aligning machinery or components.

✓ **Geological Surveys:** These are applied in geological surveys to measure vertical angles.

Types of Plumb Bobs:

✓ **Traditional Plumb Bob:** Consists of a pointed metal weight and a string.

✓ **Dumpy Level Plumb Bob:** Utilised surveying devices such as dumpy levels to establish vertical benchmarks.

✓ **Optical Plumb Bob:** Includes a bubble level or optical device to enhance accuracy.

✓ **Self-Levelling Plumb Bob:** Equipped with a self-levelling mechanism for increased precision.

Plumb bobs, used for thousands of years, are crucial for ensuring accurate vertical alignment in various sectors and professions. They provide a straightforward and reliable vertical line creation method, especially when precision is essential.

1.1.14 Line Thread

In electrical wiring, the term "line thread" typically denotes a nylon cord or string employed to facilitate the drawing of electrical cables through conduits or raceways during the installation process. This procedure is widely recognized as "fishing" or "pulling" wires. The line thread (Fig.1.14) is commonly employed in electrical wiring applications.

a) **Preparation:** Before wire installation, it is essential to ensure that the conduit or raceway is correctly positioned and devoid of any obstructions or dirt.

b) **Attachment:** A secure connection is achieved by fastening the line thread to the end of the electrical cable or wire with an appropriate knot or tie.

c) **Fishing:** The line thread and the associated wire are subsequently inserted into one extremity of the conduit or raceway. The installer may utilize fish tape or a comparable item to direct the line thread through the pathway.

d) **Pulling:** After the line thread has reached the opposite end of the pipe or raceway, it is drawn through using consistent and regulated force. This operation involves pulling the electrical wire or cable along the designated pathway and finalizing the installation process.

e) **Securing:** Once the wire has been carried through, the line thread is separated and eliminated, and the wire is firmly fixed in position using suitable clamps, straps, or fasteners.

Employing line thread guarantees a seamless and effective electrical wire installation, especially in intricate or lengthy conduit lines. It aids in reducing friction and safeguarding the wires from harm throughout the pulling process, guaranteeing their correct installation and readiness for connection to electrical equipment or components.

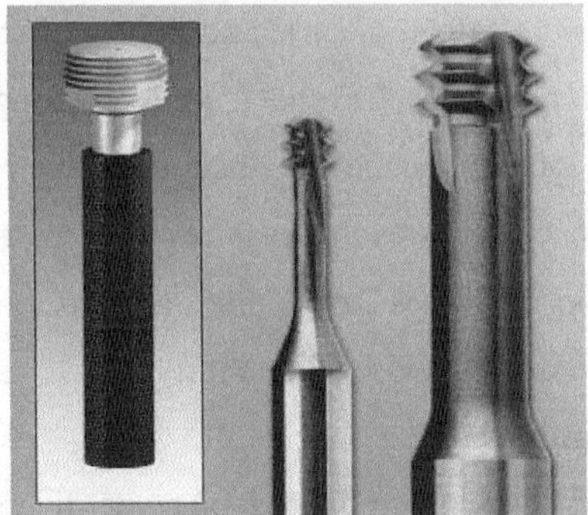

Figure 1.14 Line thread

1.1.15 Electricians knife

A utility knife or box cutter is a tool commonly employed by electricians to sever electrical tape, remove sheathing from non-metallic cables, and perform similar tasks. Linemen utilize these blades to remove the insulation from massive cables in high- and low-voltage transmission lines. An electrician's knife (Fig.1.15), referred to as a lineman's knife or an electrician's pocket knife, is a specialized implement specifically crafted for electricians and other professionals engaged in tasks related to electrical systems. It commonly includes a blend of cutting and stripping functionalities to ease many activities in the electrical domain. The electrician's knife possesses essential characteristics and capabilities, which include:

- **Key Features:**

- ✓ **Blade:** The knife typically features a linear, keen-edged blade suitable for various cutting tasks. The blade can possess partial grooves to enhance its adaptability when cutting various materials.
- ✓ **Sheepsfoot Blade:** Certain electrician's knives have a sheepsfoot blade pattern. The sheepsfoot blade features a linear cutting edge and a dull, curved tip, thereby minimizing the likelihood of inadvertently puncturing insulation while removing wire coverings.
- ✓ **Wire Stripping Notch:** The tool's blade is equipped with one or more notches or grooves specifically designed to remove insulation from electrical lines safely. The dimensions of these slots are often tailored to accommodate various wire gauges.
- ✓ **Electrician's Screwdriver:** Certain electrician's knives are equipped with an integrated screwdriver at the handle's end. This function is advantageous for the adjustment of screws on electrical equipment or junction boxes.
- ✓ **Insulated Handle:** To ensure safety, electrician's knives often feature an insulated handle that safeguards the operator from electric shock while working on active circuits.
- ✓ **Locking Mechanism:** A locking mechanism guarantees the blade's stable position during usage, eliminating unintended closures that may result in injury.
- ✓ **Pocket Clip:** A pocket clip facilitates convenient transportation of the knife in a pocket or attachment to a tool belt, ensuring swift accessibility.

- **How to Use an Electrician's Knife:**

- ✓ **Cutting:** The keen-edged blade is employed to sever diverse materials, encompassing insulation, cables, and packaging.

✓ **Stripping Wires:** Wire stripping notches are employed to safely remove insulation from electrical cables. The operator inserts the wire into the designated groove and drags the knife along the wire to remove the insulation.

✓ **Screwdriving:** The integrated screwdriver is suitable for rotating screws on electrical devices or components.

✓ **Safety Considerations:**

- It is imperative to consistently verify that the knife is in optimal condition, with a keen edge and a reliable locking mechanism.
- To prevent any harm to the conductor, ensure you utilize the correct wire stripping notch corresponding to the exact wire gauge.

Electrician's knives are indispensable implements for experts in the electrical industry, offering a small and adaptable option for cutting, stripping, and various other jobs. To avoid mishaps or harm, exercising caution and adhering to safety protocols are crucial when using these instruments.

Figure 1.15 Electrician's knife

1.1.16 Test Lamp

A test lamp, sometimes called a voltage tester or voltage detector, is a portable device to ascertain the existence or absence of electrical voltage in a wire, outlet, or circuit. Electricians, technicians, and homeowners frequently employ a safety instrument to detect active electrical circuits and prevent inadvertent contact with energized elements. The following are the primary characteristics and capabilities of a test lamp:

Key Features:

✓ **Probe or Tip:** The test lamp commonly features a sharp probe or tip to establish contact with the conductor or circuit under examination.

✓ **Indicator:** It incorporates an indicator, typically in the shape of a light or LED, which illuminates or provides a visual indication upon detecting the presence of electricity.

- ✓ **Handle:** The test lamp's handle is ergonomically built for comfortable grip and may have the required electronics for voltage sensing.
- ✓ **Insulated Shaft:** The probe or tip is usually attached to an insulated shaft or handle to safeguard the user from the risk of electric shock.

How to Use a Test Lamp:

- ✓ **Check the Test Lamp:** Before usage, verify that the test bulb is in optimal operational state, devoid of any apparent harm to the probe, insulation, or indication.
- ✓ **Insert the Probe:** Gently insert the sharp probe into the conductor or contact the examined circuit.
- ✓ **Observe the Indicator:** The presence of electrical voltage will be shown by the illumination or visual signal of the indicator if voltage is present.
- ✓ **Check for Absence of Voltage:** When conducting tests to verify that a circuit is not energized, carefully notice that the indicator remains off, indicating the absence of any electrical voltage.
- ✓ **Repeat as Needed:** The test lamp can be utilized to examine multiple locations within a circuit or to confirm the voltage condition of various conductors, depending on the specific application.

Safety Considerations:

- ✓ It is imperative to consistently adhere to appropriate safety protocols when handling electricity and utilize personal protective equipment (PPE) whenever required.
- ✓ Confirm that the test lamp suits the voltage range under examination.
- ✓ Before utilizing a test lamp, verify that it has been calibrated and that its indication is operating accurately.
- ✓ Utilise insulated tools and use caution when handling the test bulb to avoid the risk of electric shock.

Test lamps are essential for anyone engaged in electrical systems, offering a prompt and convenient means of detecting active circuits and guaranteeing the user's safety. They are frequently employed in electrical installations, maintenance, and troubleshooting endeavors.

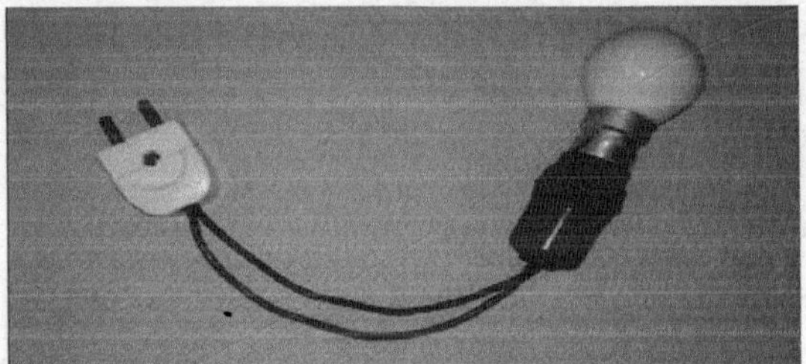

Figure 1.16 Test lamp

1.1.17 Tester

Qualified electricians utilize a variety of electrical testers to detect problems such as defective wiring, shorted or open circuits, continuity, voltage fluctuations, and amperage levels. This equipment may also benefit homeowners who prefer to conduct repairs and maintenance. To enhance your expertise in wiring, it is crucial to identify various types of electrical testers and understand their functioning. The term "tester" is a broad phrase encompassing multiple equipment or devices utilized for testing and measuring in multiple circumstances. Below are several prevalent categories of testers:

a) **Voltage Tester:** A voltage tester (Fig.1.17(a)) is utilized to ascertain the existence or nonexistence of electrical voltage in a wire or circuit. Typically, it consists of a probe that establishes contact with the circuit and an indicator, such as a light or LED, that indicates the existence of voltage.

b) **Multimeter:** Multimeter (Fig.1.17(b)) is an adaptable gadget that integrates multiple measurement functions into a single tool. Multimeters can quantify voltage, current, resistance, and occasionally additional electrical properties. They have extensive applications in electronics, electrical engineering, and problem-solving.

c) **Continuity Tester:** Continuity tester (Fig.1.17(c)) is exclusively engineered to assess the uninterrupted flow of electric current in a circuit. It often emits an auditory or visual indication when there is a channel with low resistance in the circuit, signifying continuity.

d) **Network Cable Tester:** A network cable tester (Fig.1.17(d)) is employed to verify the soundness and reliability of network cables, namely Ethernet cables. It verifies the correct wiring and detect issues such as open circuits, short circuits, or faulty connections.

e) **Battery Tester:** The battery tester (Fig.1.17(e)) assesses the battery's level of charge by monitoring its voltage or other relevant attributes. Battery testers are frequently employed to determine the condition of car batteries and other rechargeable batteries.

f) **Insulation Tester (Megohmmeter):** Insulation tester (Fig.1.17(f)) measures the impedance of insulation in electrical systems. This tester is essential for detecting any defects or vulnerabilities in insulation.

g) **Leak Tester:** Leak tester (Fig.1.17(g)) is employed across many sectors to identify leaks inside a system, such as gas leaks or fluid leaks. Various categories of leak testers are available, such as pressure decay testers, helium leak detectors, and bubble testers.

h) **Electronic Component Tester:** The electronic component tester (Fig.1.17(h)) is exclusively engineered for examining electronic components, such as resistors, capacitors, and diodes. These testers can quantify the attributes and parameters of electronic components.

i) **Blood Glucose Tester:** A blood glucose tester (Fig.1.17(i)) is utilized in medical facilities to monitor the levels of glucose in the blood of persons diagnosed with diabetes. These testers generally utilize a minute blood sample to furnish a glucose reading.

j) **Environmental Tester:** Environmental testers (Fig.1.17(j)) are devices utilized to quantify several environmental factors, including temperature, humidity, air quality, and radiation levels.

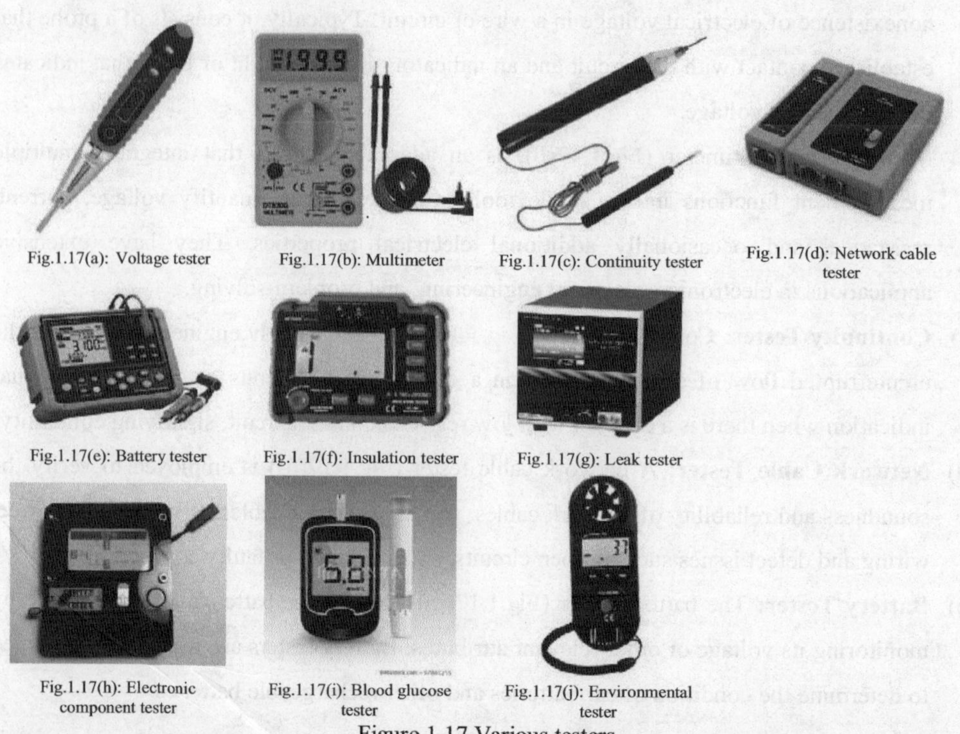

Fig.1.17(a): Voltage tester
Fig.1.17(b): Multimeter
Fig.1.17(c): Continuity tester
Fig.1.17(d): Network cable tester
Fig.1.17(e): Battery tester
Fig.1.17(f): Insulation tester
Fig.1.17(g): Leak tester
Fig.1.17(h): Electronic component tester
Fig.1.17(i): Blood glucose tester
Fig.1.17(j): Environmental tester

Figure 1.17 Various testers

The choice of tester utilized is contingent upon the nature of the application and the specific characteristics being assessed. Testers are vital instruments in diverse domains, facilitating experts and technicians to verify systems and gadgets' correct operation and security.

1.2 Classification of Electrical Accessories

Electrical accessories can be categorized into various groups according to their respective roles. Based on the categories you mentioned, here is a classification:

1.2.1 Controlling Accessories

- ✓ **Switches:** These components regulate the movement of electric current within a circuit. Some examples of switches are toggle switches, rocker switches, and dimmer switches.
- ✓ **Dimmer Switches**: These are used to adjust the intensity of lighting fixtures.
- ✓ **Timers:** Enable automated control of electrical devices at specific times.
- ✓ **Thermostats:** Control heating or cooling systems.

1.2.2 Holding Accessories

- ✓ **Clamps:** These hold and secure cables and wires in place.
- ✓ **Cable Ties:** Plastic or metal ties are bundled and secured cables.
- ✓ **Cable Clips:** Hold cables along walls or surfaces.
- ✓ **Conduit and Conduit Fittings:** Protective tubes that hold and protect wires.

1.2.3 Safety Accessories

- ✓ **Circuit Breakers:** Protect circuits from overloads and short circuits.
- ✓ **Fuses:** Safeguard against excessive current by breaking the circuit.
- ✓ **Surge Protectors:** Protect devices from voltage spikes.
- ✓ **Ground Fault Circuit Interrupters (GFCI):** Detect ground faults and prevent electrical shocks.

1.2.4 Outlet Accessories

- ✓ **Electrical Outlets:** Provide access to electrical power for devices.
- ✓ **Power Strips:** Multiple outlets in one unit for convenience.
- ✓ **Outlet Covers:** Enhance safety by covering unused outlets.
- ✓ **Plug Adapters:** Enable compatibility between different plug types.

1.3 Switches

Switches can be classified according to their construction and operational configuration. Below is an analysis of switch types based on their construction and operational configuration:

1.3.1 Types of Switches According to Construction

a) **Surface Switch:** Surface switches (Fig.1.18(a)) are mounted on the exterior of walls or panels. While their installation is straightforward, it is worth noting that they may possess a different level of visual appeal than flush-mounted switches.

b) **Flush Switch:** Flush switches (Fig.1.18(b)) are positioned in a manner that ensures they are aligned with the wall or panel, resulting in a neat and smooth visual presentation. They are frequently employed in contemporary interior design.

c) **Pull Switch:** Pull switches (Fig.1.18(c)) are engaged through the action of pulling a string or chain. They are frequently employed to regulate the operation of ceiling fans or bathroom light fixtures.

d) **Rotary Switch:** Rotary switches (Fig.1.18(d)) employ a revolving knob or dial to regulate different settings or operations. They are commonly present on appliances and technological devices.

e) **Knife Switch:** Knife switches (Fig.1.18(e)) include a lever-shaped blade that establishes or interrupts the electrical connection upon being manipulated. They are frequently employed in antiquated electrical systems or educational presentations.

f) **Pendant Switch:** Pendant switches (Fig.1.18(f)) are commonly employed in industrial or workshop environments. The objects are suspended from the ceiling and can be handled to operate machinery or lighting.

Fig.1.18(a): Surface switch

Fig.1.18(b): Flush switch

Fig.1.18(c): Pull switch

Fig.1.18(d): Rotary switch

Fig.1.18(e): Knife switch

Fig.1.18(f): Pendant switch

Fig.1.18(g): Leak tester

Figure 1.18 Various switches according to construction

g) **Main Switch (ICDP, ICTP):** Main switches (Fig.1.18(g)) serve as the principal means of disconnecting electrical infrastructure. ICDP (Isolator with Circuit Disconnecting Position) and ICTP (Isolator with Circuit Test Position) are primary switches employed for circuit isolation and testing.

1.3.2 Types of Switches According to Working Configuration

a) **Single Pole Switch:** A single-pole switch (Fig.1.19(a)) controls a single circuit and has only one set of contacts.

b) **Double Pole Switch:** Double-pole switches (Fig.1.19(b)) possess two sets of contacts and can simultaneously regulate two distinct circuits.

c) **Two-Way Switch:** Two-way switches (Fig.1.19(c)) are employed in lighting circuits and possess two switch positions, enabling control from two distinct locations.

d) **Two-Way Centre Off Switch:** Two-way center off switches (Fig.1.19(d)) feature a central position for turning off and two positions for turning on, typically employed in scenarios when one desires to operate a device from two different locations, with the option to turn it off in the middle.

e) **Intermediate Switch:** Intermediate switches (Fig.1.19(e)) are employed in three-way lighting circuits, enabling control of a light from three distinct locations and possessing more than two switch positions.

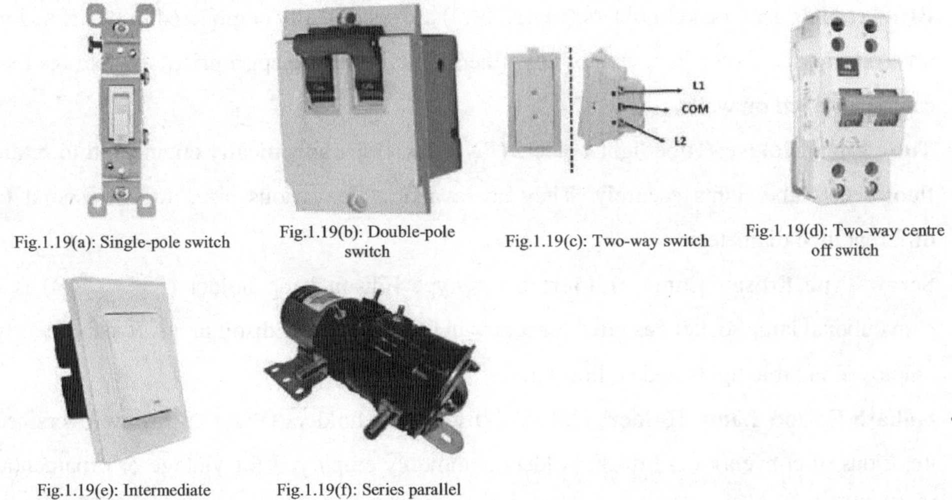

Fig.1.19(a): Single-pole switch

Fig.1.19(b): Double-pole switch

Fig.1.19(c): Two-way switch

Fig.1.19(d): Two-way centre off switch

Fig.1.19(e): Intermediate switch

Fig.1.19(f): Series parallel switch

Figure 1.19 Various switches according to working configuration.

f) **Series Parallel Switch:** Series parallel switches (Fig.1.19(f)) are employed to modify the connection configuration of electrical components between series and parallel, hence changing the behaviour of the circuit.

The following are some prevalent types of switches categorized by their construction and operational configuration. The selection of switch type is contingent upon the application and specifications of the electrical or electronic circuit being designed or worked on.

1.4 Holders

a) **Bayonet Cap Lamp Holder:** The bayonet cap lamp holder (Fig.1.20(a)) is specifically designed to accommodate bulbs with a bayonet cap base, which is commonly utilized in various household lamps.

b) **Pendant Holder:** Pendant lamp holders (Fig.1.20(b)) are commonly employed to suspend lamps from the ceiling. They are available in a variety of designs and can be highly attractive.

c) **Batten Lamp Holder:** Fluorescent tube lights are frequently equipped with batten holders. Batten lamp holders (Fig.1.20(c)) are specifically engineered to retain and connect tube lights securely.

d) **Angle Holder:** Angle holders (Fig.1.20(d)) are versatile and enable you to orient the light at various angles precisely. Spotlights and task illumination are their typical applications.

e) **Bracket Holder:** Bracket holders (Fig.1.20(e)) are specifically engineered to be affixed to a vertical surface, such as a wall or any other plane. They are appropriate for fixtures that can be installed on walls.

f) **Tube Light Holder:** Tube light holders (Fig.1.20(f)) are specifically engineered to retain fluorescent tube lights securely. They are available in various sizes to correspond to different tube diameters.

g) **Screw-Type Edison Lamp Holder:** Screw-type Edison lamp holder (Fig.1.20(g)) is a conventional lamp socket designed for screw-in bulbs, such as Edison lights. It is frequently employed in table lights and ceiling fixtures.

h) **Goliath Edison Lamp Holder:** Goliath Edison lamp holders (Fig.1.20(h)) are oversized iterations of conventional Edison holders commonly employed for vintage or ornamental bulbs.

i) **Swivel Lamp Holder:** Swivel holders (Fig.1.20(i)) provide the capability to be modified and turned to guide the illumination in different orientations precisely. They are beneficial for track lighting and fixtures that can be adjusted.

Every kind of holder has a distinct function and is suitable for various bulbs and fixtures. Selecting the appropriate lamp holder guarantees your lighting configuration's safety and optimal operation.

Fig.1.20(a): Bayonet cap lamp holder

Fig.1.20(b): Pendant lamp holder

Fig.1.20(c): Batten lamp holders

Fig.1.20(d): Angle holder

Fig.1.20(e): Bracket holder

Fig.1.20(f): Tube light holder

Fig.1.20(g): Screw-type Edison lamp holder

Fig.1.20(h): Goliath Edison lamp holder

Fig.1.20(i): Swivel holder

Figure 1.20 Various holders

1.5 Socket Outlets and Plugs

Socket outlets and plugs are available in several configurations, such as two-pin, three-pin, and multi-pin alternatives. These configurations dictate the quantity and functionalities of the electrical contacts. Here is a concise summary:

a) **Two-Pin Socket Outlet and Plug:** These are frequently employed for fundamental electrical equipment such as lamps and chargers. Typically, two pins are present, one designated for the live (hot) wire and the other for the neutral wire. They do not offer a ground connection for safety purposes. A two-pin socket is shown in Fig.1.21(a).

b) **Three-Pin Socket Outlet and Plug:** These models are more prevalent and offer more safety than two-pin alternatives. The three pins present are live (hot), neutral, and a ground pin, which serves for safety reasons. The purpose of the ground pin is to mitigate electrical failures by establishing a conduit for the safe dissipation of surplus current. Three-pin socket is shown in Fig.1.21(b).

c) **Multiple-Pin Socket Outlet:** Multi-pin socket outlets can possess diverse configurations and are frequently designed for specific uses. Additional pins may be included to cater to specific functions such as data transfer, audio/video interfaces, or power supply for appliances with distinct requirements. A multiple-pin socket is shown in Fig.1.21(c).

Fig.1.21(a): Two-pin socket Fig.1.21(b): Three-pin socket Fig.1.21(c): Multiple-pin socket

Figure 1.21 Various sockets

1.6 Iron connector:

The term "iron connector" includes a wide range of connectors or components constructed of iron or used in conjunction with iron materials. With further elucidation, it is easier to furnish exact details. Here are a few potential options:

a) **Iron Pipe Connectors:** In plumbing and construction, the term "iron connectors" typically refers to fittings used to join iron pipes together. These fittings encompass couplings, elbows, tees, and other components that facilitate the connection and alignment of iron pipes.

b) **Iron Wire Connectors:** Iron connectors or connectors utilized with iron wires may be referenced in electrical wiring. Nevertheless, using materials such as copper or aluminum for electrical connectors is more prevalent owing to their superior conductivity characteristics.

c) **Iron Hose Connectors:** Iron connectors typically denote fittings employed to join hoses together in gardening or industrial applications. These connectors may consist of threaded fittings, couplings, and adapters.

d) **Iron Framing Connectors:** Connectors utilized in iron or steel framing systems might be called iron connectors in the building field. These connectors enhance the joining process and ensure the strength of iron or steel frameworks.

e) **Iron Joint Connectors:** This term encompasses a range of connectors utilized to join iron components, including brackets, plates, or fasteners that secure iron pieces together.

1.7 Adapter:

An adapter is a device or component that enables the compatibility of different systems or devices. Adapters serve the purpose of connecting connectors, interfaces, or specifications that are not compatible with each other. They exhibit diverse manifestations and fulfill distinct objectives in various sectors. Below are several prevalent varieties of adapters:

a) **Electrical Plug Adapter:** An electrical plug adapter is a device that converts one type of electrical plug to another, facilitating the operation of a device designed for one region's electrical system in a different area.

b) **Voltage Adapter:** Adapts the voltage of an electrical device to correspond with the voltage specifications of a particular location or region. This is especially crucial when traveling abroad.

c) **USB Adapter:** Enables devices with varying USB connections or physical designs. An instance is when a USB-C to USB-A adapter establishes a connection between a USB-C device and a USB-A port.

d) **Audio Adapter:** Enables the conversion of audio connectors from one kind to another. For example, a headphone jack adapter can transform a 3.5mm audio connection into a 1/4-inch jack.

e) **Network Adapter:** A network adapter, commonly referred to as a network interface card (NIC), enables a computer to establish a connection with a network, whether it be through a wired or wireless medium.

f) **Display Adapter:** Utilised in graphics cards to establish a connection between a computer and a display device. Some examples consist of HDMI adapters, DisplayPort adapters, or VGA adapters.

g) **Thread Adapter:** Thread adapters are used in plumbing or piping systems to establish connections between pipes or fittings with varying thread kinds or sizes.

h) **Lens Adapter:** Lens adapters facilitate the utilization of lenses with one mount type on cameras equipped with a different mount type in photography.

i) **Power Adapter:** Adapts the power output of a power source to meet the precise requirements of an electronic device. For instance, a laptop power adapter transforms alternating current (AC) electricity into direct current (DC) power, which the laptop requires.

j) **Fiber Optic Adapter:** Fiber optic adapters facilitate the interconnection of two fiber optic connectors, enabling the transmission of optical signals between them.

Adapters facilitate interaction and connection between devices or systems with different standards. It is crucial to select the appropriate type of adapters for the intended use and adhere to safety requirements and compatibility considerations.

1.8 Ceiling Rose:

A ceiling rose is an ornamental and utilitarian feature employed in lighting fixtures. Typically affixed to the roof, it functions as a junction box that links the electrical wire from the ceiling to a light fixture or pendant lamp. Ceiling roses commonly feature numerous terminals to connect live, neutral, and ground wires.

Key Features of a Ceiling Rose:

a) **Wiring Terminals:** The ceiling rose features connections that facilitate the connection of electrical wires from the lighting circuit. The terminals are commonly marked as live (L), neutral (N), and ground (earth).

b) **Cover Plate:** The ceiling rise often accompanies an ornamental cover plate. The cover plate can be adorned with many styles and finishes to augment the visual appeal of the lighting fixture and the surrounding space.

c) **Cable Grip:** The ceiling may incorporate a cable grip or clamping mechanism. This is employed to secure electrical lines' fastening, preventing unintentional disconnection or excessive tension.

d) **Mounting Bracket:** A mounting bracket commonly holds the rose to the ceiling. This bracket guarantees the stability and accurate placement of the ceiling rise.

e) **Ceiling Rose Pendant:** In some designs, a ceiling may incorporate a pendant or hanging mechanism to suspend a light fixture or chandelier from the ceiling. This can contribute a supplementary ornamental aspect to the entire illumination scheme.

Installation of a Ceiling Rose:

✓ **Shut Off Power:** Before installing or handling a ceiling rose, it is imperative to ensure that the power to the lighting circuit is deactivated at the primary electrical panel.

✓ **Remove Existing Fixture:** If there is an existing light fixture, remove it from the ceiling, exposing the electrical wires.

✓ **Mounting Bracket Installation:** Attach the mounting bracket to the ceiling structure according to the instructions provided by the manufacturer.

✓ **Wiring Connection:** Attach the electrical circuit's live, neutral, and ground wires to the matching terminals on the ceiling rose. Adhere to established wiring color regulations and rules.

✓ **Secure Cables:** Utilise the cable grip or clamping mechanism to fasten the electrical cables, preventing connection tension firmly.

✓ **Attach Cover Plate:** After finishing the wiring, connect the decorative cover plate to the ceiling rose.

✓ **Install Light Fixture:** If applicable, affix the light fixture or pendant to the ceiling rose, adhering to the directions provided by the manufacturer.

✓ **Power On** Restore electricity to the electrical panel and verify the functionality of the lighting installation.

Ceiling roses have both practical and aesthetic purposes, enhancing the appearance of the lighting fixture and the room. They are frequently found in residential and business environments, mainly where pendant lights or chandeliers are installed.

1.9 Distribution Box:

A distribution box, sometimes called a distribution board or electrical panel, is an essential element in an electrical system that allocates electrical power to different circuits within a building or facility. It functions as a central hub where electrical circuits are partitioned and regulated to guarantee the secure dissemination of electricity to various locations or devices.

Key Components and Features of a Distribution Box:

a) **Main Switch or Circuit Breaker:** The main switch or circuit breaker functions as the principal disconnect for the entire electrical infrastructure. It enables the user to deactivate the power supply to the whole building in the event of maintenance or emergency.

b) **Individual Circuit Breakers or MCBs:** Distribution boxes include many individual circuit breakers or tiny circuit breakers (MCBs), each assigned to a single circuit. Miniature

Circuit Breakers (MCBs) safeguard individual circuits by preventing excessive currents or short circuits.

c) **Busbars:** Busbars are conductive strips or bars inside the distribution box that transmit electrical power from the main switch to the individual circuit breakers. They serve as a common junction for the circuits.

d) **Neutral Bar and Grounding:** Distribution boxes commonly feature a neutral bar that facilitates the connection of neutral wires from different circuits. Grounding is achieved by connecting a ground bar, which guarantees the safety and optimal operation of the electrical system.

e) **Door or Cover:** A distribution box is commonly equipped with a door or cover to safeguard against inadvertent contact with energized components. The door may also be fitted with a locking mechanism to enhance safety.

f) **Labelling and Identification:** Accurate labeling and identification of circuits are crucial for facilitating the clear identification of the purpose of each circuit. Labels might provide details regarding the specific room or area a given circuit serves.

g) **Voltage and Current Ratings:** Distribution boxes are engineered with precise voltage and current ratings to effectively manage the interconnected circuits' electrical load. Strict adherence to these values is necessary to ensure safety and compliance with electrical rules.

h) **DIN Rail:** A DIN rail, which stands for Deutsches Institut für Normung, is utilized in certain distribution boxes to securely install circuit breakers and other modular components in a standardized manner.

i) **Ventilation and Cooling:** More expansive distribution boxes may incorporate ventilation or cooling mechanisms to disperse the heat produced by the electrical components while in use.

Types of Distribution Boxes:

✓ **Main Distribution Board (MDB):** The main distribution board is the central hub for electrical power distribution throughout a structure. The system derives power from the utility and disperses it to sub-panels or branch circuits.

✓ **Sub-Distribution Board (SDB):** Sub-distribution boards serve as intermediate distribution stations that take electrical power from the main distribution board and distribute it to specified locations or loads.

✓ **Consumer Unit:** A consumer unit is a distribution box commonly found in residential settings. It contains the main switch, circuit breakers, and other components for distributing power within a home.

A consumer unit, a distribution box, is a frequently encountered apparatus in residential environments. It comprises the primary switch, circuit breakers, and additional elements for electricity distribution within a residential residence.

1.10 Neutral Link:

A neutral link, commonly referred to as a neutral bar or neutral busbar, is a crucial element in an electrical distribution system. It is widely located in distribution panels, breaker boxes, or boards. The primary purpose of the neutral link is to serve as a central connection point for the neutral conductors from different circuits within a building.

Key features and functions of a neutral link:

✓ **Common Connection Point:** The neutral link functions as a central hub where the grounded conductors, also known as neutral conductors, from various circuits are interconnected.

✓ **Busbar Structure:** A typical busbar is a metallic strip or bar composed of a conductive substance like copper or aluminum. The busbar facilitates the efficient flow of return currents to the power source by offering a channel with minimal resistance.

✓ **Connection to Neutral Conductor:** The neutral conductor of each circuit is connected to the neutral link. This connection facilitates the return current flow, which is responsible for completing the circuit, back to its source, usually the transformer or the main service panel.

✓ **Grounding Connection:** The neutral link is commonly interconnected with the grounding system of the electrical installation. This guarantees that the electrical system is adequately grounded, facilitating the safe transit of fault currents to the ground.

✓ **Isolation and Insulation:** The neutral link may occasionally be equipped with insulating barriers or insulation to avoid accidental contact with the energized components, guaranteeing safety during maintenance or alterations.

✓ **Provision for Neutral Wires:** The purpose of the neutral link is to facilitate the joining of several neutral wires. This is especially crucial in distribution panels that service several circuits.

✓ **Neutral-to-Ground Bonding:** Some electrical systems have a neutral link that allows for the connection of the neutral to the earth. The interconnection of the neutral and ground is crucial for safety purposes, particularly in systems where the neutral is grounded at its origin.

Importance of the Neutral Link:

✓ **Balancing Loads:** The neutral link enables the equalization of loads in a distribution system. It guarantees that the return currents from different circuits can circulate through a shared route, preventing any imbalances.

✓ **Safety and Grounding:** Ensuring proper grounding via the neutral link is essential for maintaining safety. It offers a route for fault currents to effectively trigger circuit breakers or cause fuses to blow, safeguarding the electrical system and users from potential dangers.

✓ **Code Compliance:** Adherence to electrical codes and standards necessitates correctly bonding the neutral and grounding systems. The neutral link is crucial in fulfilling these requirements.

Compliance with electrical rules and standards is essential for the safe and dependable functioning of the electrical distribution system while designing and installing the neutral link. Electrical experts adhere to particular protocols to accurately determine the appropriate dimensions and install and upkeep neutral linkages in different electrical panels and distribution boards.

1.11 Bus-bar Chamber:

A bus-bar chamber is a confined space within an electrical distribution system or switchgear that houses and interconnects bus-bars. Bus bars are metallic bars or strips that are a shared electrical connector for several electrical circuits. The bus bar chamber offers a protected and insulated space for these bars, facilitating electrical power transmission to different components within the system.

Key Features of a Bus-bar Chamber:

✓ **Bus bars:** The bus bars themselves are the main component within the bus bar chamber. These components are commonly constructed from copper or aluminum and serve as a path with little resistance to the flow of electrical current.

- ✓ **Insulation:** The chamber is constructed with insulation to limit the risk of accidental contact with energized components. Insulation encompasses various components, such as barriers, coverings, or enclosures, which safeguard against electrical shock.
- ✓ **Supports and Mounting:** The bus bars are positioned and secured inside the chamber with insulating materials. The design guarantees the stable positioning of the bus bars, preventing any contact with any conductive materials.
- ✓ **Accessibility:** The chamber is precisely engineered to facilitate convenient entry for maintenance and inspection. Removable panels or doors allow technicians to easily access the bus bars, facilitating their work on the system.
- ✓ **Temperature Control:** Bus-bar chambers can include temperature management elements, such as ventilation or cooling systems, that distribute the heat produced during operation.
- ✓ **Connection Points:** Bus-bar chambers have connection terminals where other electrical components, such as circuit breakers, transformers, or disconnect switches, can be attached to the bus bars.
- ✓ **Bus-bar Supports and Insulators:** Insulating materials or insulators are commonly used to support the bus bars to ensure electrical isolation from the structure of the chamber. This aids in the prevention of accidental electrical connections.
- ✓ **Grounding:** Proper grounding is a fundamental requirement for a bus-bar chamber. Grounding guarantees the system's security by establishing a route for fault currents to disperse safely into the ground.
- ✓ **Safety Interlocks:** Certain bus-bar chambers may incorporate safety interlocks to prevent the opening of access doors or panels while the system is powered, hence improving safety during repairs.

Applications of Bus-bar Chambers:

- ✓ **Power Distribution Systems:** Bus-bar chambers are frequently employed in power distribution systems to disseminate electrical power from the primary source to other sub-systems or loads.
- ✓ **Switchgear and Switchboards:** Bus-bar chambers of electrical switchgear and switchboards contain the bus-bars, which serve as the connections between different components such as circuit breakers, transformers, and disconnect switches.
- ✓ **Industrial Facilities:** Industrial establishments commonly employ bus-bar chambers to manage and distribute power to various sections inside the plant.

✓ **Commercial Buildings:** Busbar chambers are integral components of electrical distribution systems in commercial buildings, facilitating the effective distribution of electricity to different floors and sections.

Bus-bar chambers are essential electrical system components responsible for the reliable and safe electrical power distribution. They are crucial in various industrial, commercial, and infrastructure sectors.

1.12 Wooden / Mica Boards:

Wooden and mica boards are distinct materials utilized for various applications, each with distinctive attributes and purposes. Let us examine these materials:

a) **Wooden Boards:**

- **Materials:** Wooden planks can be crafted from various wood species, each possessing distinct characteristics. *Typical varieties comprise plywood, particleboard, MDF (Medium-*Density Fibreboard), and solid wood.
- **Characteristics:**

✓ **Strength and Durability:** Solid wood boards are renowned for their robustness and longevity, while engineered wood materials such as plywood and MDF provide stability and consistency.

✓ **Versatility:** Wooden boards possess versatility and can be employed in various applications, such as furniture making, cabinetry, construction, and decorative additions.

✓ **Finishing Options:** Wood can undergo staining, painting, or varnishing to attain the desired appearance. The inherent texture of wood enhances its visual attractiveness.

✓ **Insulation:** Wood has inherent insulating characteristics, offering thermal and acoustic insulation.

- **Applications:**

✓ **Furniture Construction:** Wooden planks are often utilized in the construction of furniture pieces such as tables, chairs, cabinets, and bed frames.

✓ **Construction:** Plywood and other timber panels are used for sheathing, subflooring, and wall panelling.

✓ **Crafting and DIY Projects:** Artisanal and Do-It-Yourself Endeavours: Wooden boards are widely favoured for many crafting and do-it-yourself (DIY) endeavors.

b) **Mica Boards:**

- **Materials:** Mica boards consist of layers of mica substance, frequently reinforced with other materials such as glass, cloth, or ceramic.
- **Characteristics:**

✓ **Thermal Resistance:** Mica boards possess exceptional thermal resistance, rendering them well-suited for applications that involve elevated temperatures.

✓ **Electrical Insulation:** Mica, known for its exceptional electrical insulating properties, is used in electrical applications to isolate components from heat and electricity effectively.

✓ **Dielectric Strength:** Mica boards are notable dielectric, rendering them appropriate for electrical insulation in various applications.

✓ **Chemical Resistance:** Mica boards provide chemical resistance, augmenting their longevity and appropriateness for specific industrial settings.

- **Applications:**

✓ **Electrical Insulation:** Mica boards are frequently employed as insulation in electrical apparatus, including heating elements, electric motors, and electrical panels.

✓ **Thermal Insulation:** Because of their thermal resilience, Mica boards are utilized in high-temperature settings, such as industrial furnaces and ovens.

✓ **Gaskets and Seals:** Mica boards are suitable for cutting and utilizing as gaskets or seals in applications that demand resilience to high temperatures and chemicals.

- **Selection Considerations:**

✓ **Application Requirements:** Select the appropriate material based on the precise specifications of your application. Mica boards are preferable for electrical insulation, although wooden boards are adaptable for various construction and furniture uses.

✓ **Environment:** Consider the environmental circumstances under which the item will be utilized. For instance, opting for mica boards may be more advantageous when dealing with elevated temperatures.

✓ **Aesthetic Preferences:** Wooden boards offer a visually pleasing and cozy look, whereas mica boards are primarily selected for their utilitarian capabilities rather than their beauty.

Wooden and mica boards possess unique characteristics and are utilized in specific applications. The selection depends on the intended application, prevailing environmental circumstances, and precise project specifications.

1.13 Moulded / MS Concealed Boxes of Different Sizes:

Moulded or MS hidden boxes are electrical enclosures utilized in construction for housing and concealing wire and electrical components within walls. The boxes are commonly constructed from moulded plastic or mild steel and are intended to be hidden behind walls or ceilings. They offer a polished appearance and safeguard electrical connections.

- **Key Features:**

a) **Material:**

- ✓ **Moulded Boxes:** These are frequently built from plastic materials, such as polycarbonate or ABS, because of their insulating characteristics. The plastic composition offers insulation and safeguards against electrical shocks.
- ✓ **MS Boxes:** These objects are constructed from mild steel, imparting robustness and long-lasting qualities. MS boxes are commonly employed in situations that necessitate enhanced structural durability.

b) **Mounting:** Concealed boxes are designed to be installed inside walls or ceilings, concealing electrical wiring from sight. Their orientation can be horizontal or vertical, depending on the use case.

c) **Sizes and Dimensions:** These boxes are available in various sizes and dimensions to handle diverse electrical components, switches, outlets, or junctions. Typical sizes consist of single gang, double gang, and bigger combinations.

d) **Concealed Installation:** The main objective of these boxes is to obscure electrical connections and wiring. They are embedded into the building's framework and concealed by finishing materials like plaster, plasterboard, or other wall coverings.

e) **Ease of Installation:** Moulded and MS hidden boxes are specifically engineered to facilitate effortless installation during construction. They commonly possess attributes like knockouts or pre-punched holes to facilitate the routing of electrical conduits.

f) **Safety:** Concealed boxes often feature a surface that can be painted for aesthetics. This enables them to be coated with paint that matches the color of the adjacent wall or ceiling.

g) **Paintable Surface:** Concealed boxes often feature a surface that can be painted for aesthetics. This enables them to be painted to correspond with the color of the adjacent wall or ceiling.

h) **Covers and Accessories:** Concealed boxes typically include removable coverings for maintenance or future changes. Additional accessories, such as extension rings or adjustable brackets, may be provided to meet special installation needs.

- **Applications:**

✓ **Residential Wiring:** Commonly employed in residential settings to facilitate the installation of electrical switches, outlets, and junctions while ensuring a visually pleasing and uncluttered aesthetic.

✓ **Commercial Buildings:** Utilised in business environments to hide electrical elements within walls or ceilings.

✓ **Industrial Installations:** Commonly employed in industrial environments to facilitate hidden wiring and connections, particularly when conditions require sturdy enclosures.

✓ **Institutional Buildings:** Widely employed in educational institutions, medical facilities, and other establishments to conceal electrical equipment.

✓ **Renovation and Retrofitting:** Hidden enclosures are also utilized in rehabilitation projects to modernize or alter electrical infrastructure.

Acknowledging that building requirements and electrical regulations differ depending on the region is crucial. Therefore, it is imperative to comply with local standards while choosing and installing concealed boxes. Furthermore, the decision between molded and MS concealed boxes may be influenced by cost, material preferences, and unique application needs.

1.14 Modular Accessories:

Modular accessories are electrical and electronic components designed to be modular, enabling effortless installation, replacement, or upgrade within electrical systems. These accessories are frequently employed in residential, commercial, and industrial environments to augment the functionality and adaptability of electrical infrastructure. Below are many prevalent instances of modular accessories:

a) **Modular Switches:** Modular switches are simply installable and replaceable components designed for switch panels. The switches comprise a base plate and a detachable switch module, allowing for easy modification of the switch's look or functioning without modifying the complete wiring system.

b) **Modular Outlets:** Modular outlets provide effortless installation and replacement, akin to modular switches. The amenities encompass power outlets, USB outlets, and other data or communication ports that can be customized to meet unique requirements.

c) **Modular Jacks:** Modular jacks establish connections between different communication lines, specifically Ethernet cables, in the context of networking applications. These objects are commonly located on wall plates and patch panels.

d) **Modular Data Connectors:** These connectors are used for data communication and are commonly used in networking systems. Illustrations encompass RJ45 connectors utilized for Ethernet cables.

e) **Modular Frames and Bezels:** Frames and bezels are the external framework for modular accessories, enabling their integration into switchboards, wall boxes, or other mounting structures. They are available in a variety of designs and finishes.

f) **Modular Dimmers:** Dimmer modules enable the regulation of lighting brightness. Modular switch panels readily accommodate their integration, allowing for convenient adjustment of light brightness levels.

g) **Modular USB Chargers:** These modules can be installed in electrical outlets or incorporated into modular switch panels to offer USB charging ports for charging electronic gadgets.

h) **Modular Fan Regulators:** Fan regulator modules enable the precise adjustment of fan velocity. Similar to dimmer modules, they can be included in modular switch panels.

i) **Modular Timer Switches:** Timer switches facilitate the automation of electrical devices by allowing the user to establish precise time intervals for their functioning. They can be incorporated into modular switch panels.

j) **Modular Blanking Plates:** Blanking plates conceal unoccupied spaces in modular switch panels. They offer a clean and polished look.

k) **Modular Indicator Lights:** Switch panels can be enhanced with indicator light modules to display the current state of electrical devices or circuits.

l) **Modular Fuse Holders:** Fuse holder modules facilitate the convenient installation and substitution of fuses in electrical circuits. These devices are specifically engineered to be seamlessly incorporated into modular switchboards.

Modular attachments provide benefits such as adaptability, effortless installation, and customization. They offer the capacity to modify electrical installations to accommodate evolving requirements and technology without requiring substantial rewiring. The modular nature of these accessories simplifies maintenance and upgrades, keeping them highly desirable in contemporary electrical systems.

1.15 BIS Symbols of Electrical Accessories

The Bureau of Indian Standards (BIS) offers standardized symbols for diverse electrical accessories and components to guarantee uniformity and lucidity in electrical diagrams, schematics, and designs. In electrical engineering diagrams, these symbols depict electrical devices, switches, outlets, and other constituents. Below are several prevalent BIS symbols representing electrical accessories:

1. **Switches:**

Type	Symbol
Single Pole Switch	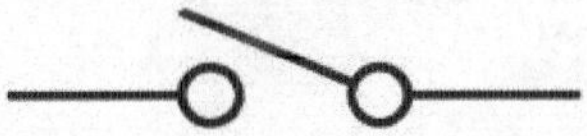
Double Pole Switch	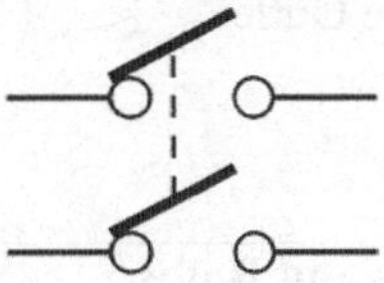
Three Pole Switch	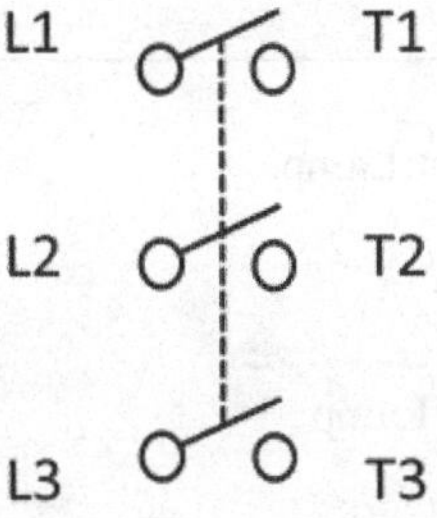 L1 T1 L2 T2 L3 T3
Selector Switch	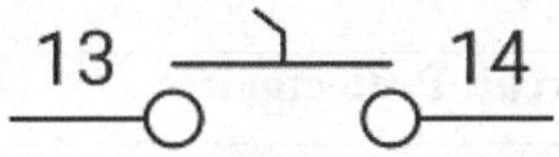 13 14
Push Button Switch	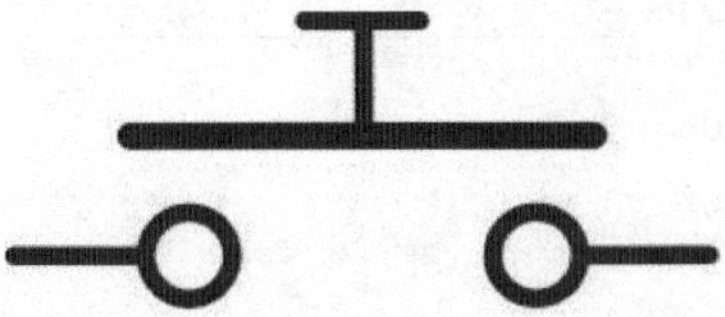

Type	Symbol
Rotary Switch	

2. **Sockets and Outlets:**

Type	Symbol
Socket Outlet	
Telephone Outlet	

3. **Lamps and Bulbs:**

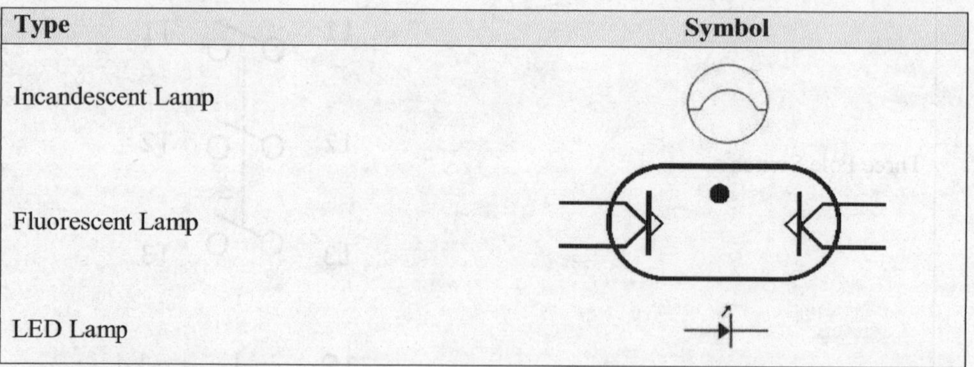

Type	Symbol
Incandescent Lamp	
Fluorescent Lamp	
LED Lamp	

4. **Circuit Protection:**

Type	Symbol
Fuse	
Circuit Breaker	1 2

5. **Others:**

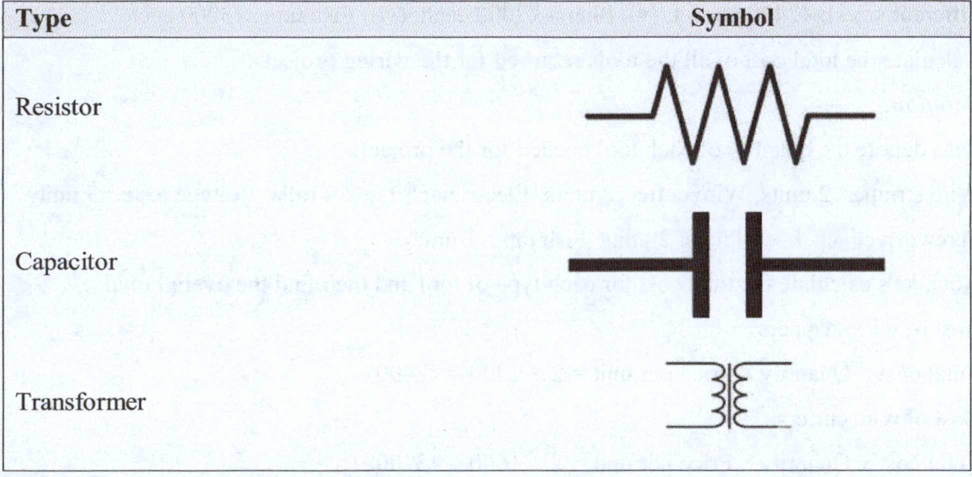

Type	Symbol
Resistor	
Capacitor	
Transformer	

These symbols are universally accepted and extensively employed in electrical engineering drawings, circuit diagrams, and construction plans to depict different electrical components precisely and consistently. They assist in facilitating better communication and enhancing the comprehension of electrical systems and installations for electrical engineers, technicians, and electricians.

More on BIS Symbols

Solved Numerical Problems

Example 1.1:

You are tasked with wiring a residential building. You have a list of tools required for the wiring job. Calculate the total cost of the tools needed for the project based on their prices. List of tools and their prices: (i) Wire stripper: ₹1200 each, (ii) Wire cutter: ₹1600 each, (iii)

Electrician's tape: ₹400 per roll, (iv) Voltage tester: ₹800 each, (v) Screwdriver set (containing different sizes): ₹2400 per set, (vi) Pliers: ₹2000 each, (vii) Fish tape: ₹3200 each

Calculate the total cost of all the tools required for the wiring project.

Solution:

Let's denote the quantity of each tool needed for the project:

Wire stripper: 2 units, Wire cutter: 2 units, Electrician's tape: 4 rolls, Voltage tester: 3 units

Screwdriver set: 1 set, Pliers: 2 units, Fish tape: 1 unit

Now, let's calculate the total cost for each type of tool and then find the overall total.

Cost of wire strippers:

Total cost = Quantity × Price per unit = 2 × 1200 = ₹2400

Cost of wire cutters:

Total cost = Quantity × Price per unit = 2 × 1600 = ₹3200

Cost of electrician's tape:

Total cost = Quantity × Price per unit = 4 × 400 = ₹1600

Cost of voltage testers:

Total cost = Quantity × Price per unit = 3 × 800 = ₹2400

Cost of screwdriver set:

Total cost = Quantity × Price per unit = 1 × 2400 = ₹2400

Cost of pliers:

Total cost = Quantity × Price per unit = 2 × 2000 = ₹4000

Cost of fish tape:

Total cost = Quantity × Price per unit = 1 × 3200 = ₹3200

Now, let's calculate the overall total cost by adding up the costs of all the tools:

Total cost = Cost of wire strippers + Cost of wire cutters + Cost of electrician's tape + Cost of voltage testers + Cost of screwdriver set + Cost of pliers + Cost of fish tape

= ₹(2400 + 3200 + 1600 + 2400+ 2400+ 4000+ 3200) = ₹19200

Therefore, the total cost of all the tools required for the wiring project is ₹19200.

Example 1.2:

You are designing the electrical layout for a commercial building. The building requires various electrical accessories, including switches, sockets, and circuit breakers. The specifications for each accessory are as follows:

Switches: Single-pole switches: ₹5 each, Double-pole switches: ₹8 each

Sockets: Single socket: ₹3 each, Double socket: ₹5 each

Circuit breakers: Single-pole circuit breaker: ₹10 each, Double-pole circuit breaker: ₹15 each

Calculate the total cost of electrical accessories needed for the building based on the following requirements:

20 single-pole switches, 10 double-pole switches, 30 single sockets, 15 double sockets, 25 single-pole circuit breakers, 12 double-pole circuit breakers

Solution:

Let's calculate the total cost for each type of electrical accessory and then find the overall total.

Cost of switches:

Total cost for single-pole switches:

Total cost = Quantity × Price per unit = 20 × ₹400 = ₹8000

Total cost for double-pole switches:

Total cost = Quantity × Price per unit = 10 × ₹640 = ₹6400

Cost of sockets:

Total cost for single sockets:

Total cost = Quantity × Price per unit = 30 × ₹240 = ₹7200

Total cost for double sockets:

Total cost = Quantity × Price per unit = 15 × ₹400 = ₹6000

Cost of circuit breakers:

Total cost for single-pole circuit breakers:

Total cost = Quantity × Price per unit = 25 × ₹800 = ₹20000

Total cost for double-pole circuit breakers:

Total cost = Quantity × Price per unit = 12 × ₹1200 = ₹14400

Now, let's calculate the overall total cost by adding up the costs of all the electrical accessories:

Total cost = Cost of switches + Cost of sockets + Cost of circuit breakers

= ₹8000 + ₹6400 + ₹7200 + ₹6000 + ₹20000 + ₹14400 = ₹62000

Therefore, the total cost of all the electrical accessories needed for the commercial building is ₹62000.

Example 1.3:

A technician needs to tighten a loose screw in an electrical panel. The screw requires a Phillips-head screwdriver with a diameter of 3.5mm. The technician selects the appropriate screwdriver and tightens the screw. What torque is applied if the screwdriver handle is 15cm long, and the technician applies a force of 8N perpendicular to the handle?

Solution:

Torque, $\tau = r \times F$, where r is the length of the handle and F is the force applied.

Given r=15cm=0.15m, and F=8N,

τ=0.15m×8N=1.2Nm

Example 1.4:

A technician uses tongue-and-groove pliers to grip a wire with a force of 50N. The length of the handle where the force is applied is 20cm, and the length of the gripping jaws is 5cm. What is the mechanical advantage of the pliers?

Solution:

MA(Mechanical Advantage)= Length of handle/Length of jaws.

Given that the length of the handle is 20cm = 0.2m and the length of the jaws is 5cm = 0.05m,

MA=0.2m/0.05m =4

Example 1.5:

A technician is testing a circuit using a digital multimeter. The technician sets the multimeter to measure voltage and places the probes across a 12V battery. The multimeter displays a reading of 11.8V. What is the percentage error in the measurement?

Solution:

Error%=$\frac{MV-AV}{AV} \times 100$.

Given the measured value (MV) is 11.8V and the actual value (AV) is 12V,

rro%=$\frac{11.8-12}{12} \times 100\%=-1.67\%$

Example 1.6:

A circuit consists of two switches connected in series. Each switch has a resistance of 5 ohms. What is the equivalent resistance of the circuit?

Solution:

In a series circuit, the equivalent resistance is the sum of individual resistances. Given that each switch has a resistance of 5 ohms, and they are in series,

Equivalent Resistance=5ohms+5ohms=10ohms

Exercise:

Multiple Choice Questions:

1. Which tool is commonly used to strip insulation from electrical wires?

   a) Pliers b) Wire cutter c) Wire stripper d) Screwdriver

2. What is the primary purpose of a cable tester?

   a) Measure voltage b) Test the continuity of wires c) Cut wires accurately d) Tighten nuts and bolts

3. Which of the following is NOT a type of screwdriver commonly used in electrical work?

   a) Flathead screwdriver b) Phillips screwdriver c) Hex screwdriver d) Torx screwdriver

4. Which tool is used to crimp connectors onto the end of wires?

   a) Pliers b) Wire stripper c) Crimper d) Screwdriver

5. A multimeter is primarily used for:

   a) Measuring voltage, current, and resistance b) Cutting wires accurately c) Stripping insulation from wires d) Testing the continuity of wires.

6. Which accessory is used to secure cables and wires to surfaces?

   a) Cable tie b) Wire nut c) Terminal block d) Conduit

7. Which of the following is a safety accessory used to protect against electrical shock? a) Voltage tester b) Insulated gloves c) Wire brush d) Fish tape

8. What is the purpose of a fish tape in electrical work?

a) To strip insulation from wires b) To test the continuity of wires c) To pull wires through conduit or walls d) To measure the voltage of electrical circuits.

9. Which tool is used to bend and shape wires or electrical components?

   a) Pliers b) Wire cutter c) Crimper d) Voltage tester.

10. What is the primary function of a conduit in electrical installations?

    a) To protect wires from mechanical damage b) To strip insulation from wires c) To measure voltage d) To test the continuity of wires.

11. Which tool is commonly used to tighten electrical connections?

    a) Pliers b) Adjustable wrench c) Hammer d) Screwdriver

12. What is the primary function of a junction box in electrical installations?

    a) To protect wires from mechanical damage b) To provide a safe enclosure for wire connections c) To measure voltage d) To strip insulation from wires.

13. Which tool creates a clean and precise hole in electrical panels or boxes for switches and outlets?

    a) Screwdriver b) Hammer c) Conduit reamer d) Knockout punch.

14. What is the purpose of a wire nut in electrical work?

    a) To strip insulation from wires b) To test the continuity of wires c) To securely connect two or more wires d) To measure voltage.

15. Which accessory is used to connect multiple wires securely and insulated?

    a) Wire nut b) Cable tie c) Conduit d) Fish tape.

16. What is the primary purpose of an electrical tape?

    a) To cut wires accurately b) To measure voltage c) To insulate electrical connections and prevent shorts d) To strip insulation from wires.

17. Which tool measures the thickness of wires or electrical components?

    a) Wire cutter b) Vernier caliper c) Wire stripper d) Pliers.

18. Which of the following is a tool used to thread wires through conduits or walls?

    a) Wire nut b) Fish tape c) Conduit bender d) Voltage tester.

19. What is the primary function of a voltage tester?

    a) To measure the thickness of wires b) To test the continuity of wires c) To measure voltage in electrical circuits d) To secure cables and wires to surfaces.

20. Which tool is used to remove burrs and rough edges from conduits or pipes?

    a) Wire brush b) Pliers c) Conduit reamer d) Hammer

Answers:

1. c) Wire stripper,
2. b) Test the continuity of wires
3. c) Hex screwdriver
4. c) Crimper
5. a) Measuring voltage, current, and resistance
6. a) Cable tie
7. b) Insulated gloves
8. c) To pull wires through conduit or walls
9. a) Pliers
10. a) To protect wires from mechanical damage
11. b) Adjustable wrench
12. b) To provide a safe enclosure for wire connections
13. d) Knockout punch
14. c) To securely connect two or more wires
15. a) Wire nut
16. c) To insulate electrical connections and prevent shorts
17. b) Vernier caliper
18. b) Fish tape
19. c) To measure voltage in electrical circuits
20. c) Conduit reamer

Short Answer Type Questions:

1) What is the purpose of a wire stripper?
2) What tool is commonly used to crimp connectors onto wires?
3) Define the function of a multimeter.
4) What is the primary use of a cable tie?
5) Name an accessory used to secure cables and wires to surfaces.
6) Why are insulated gloves essential in electrical work?
7) Explain the purpose of a fish tape.

8) What is a conduit used for in electrical installations?
9) Define the function of a junction box.
10) What is the primary purpose of a wire nut?
11) Why is electrical tape commonly used in wiring installations?
12) What is the function of a knockout punch?
13) Name a tool used to thread wires through conduits or walls.
14) Define the purpose of a voltage tester.
15) What is the function of a conduit reamer in electrical work?

Long Answer Type Questions:

1) Discuss the functions and features of a wire stripper. How does it contribute to efficient electrical work?
2) Explain the different types of screwdrivers commonly used in electrical work. Discuss their specific applications and advantages.
3) Describe the primary functions of a multimeter. How does it measure voltage, current, and resistance in electrical circuits?
4) Discuss the importance of cable ties in managing cables and wires. Provide examples of situations where cable ties are beneficial.
5) Explain the significance of insulated gloves in electrical work. How do they protect workers from electrical hazards, and what are the safety precautions associated with their use?
6) Describe the purpose of a fish tape in electrical installations. Provide examples of scenarios where fish tapes are indispensable tools.
7) Discuss the role of conduits in electrical wiring. What different conduits are available, and how are they used to protect and organize wires?
8) Explain the functions and benefits of junction boxes in electrical installations. Provide examples of situations where junction boxes are necessary for safety and efficiency.
9) Describe the primary purposes of wire nuts in electrical work. How are they used to connect and secure wires, and what are the best practices for their application?
10) Discuss the significance of a voltage tester in electrical maintenance and troubleshooting. Explain how it identifies live circuits, detects voltage, and ensures safety during electrical work.

Numerical Problems:

1) A junction box contains 5 wires that must be securely connected using wire nuts. Each wire nut can hold up to 3 wires. How many wire nuts are needed to connect all the cables securely? (Ans.2)
2) A cable tie reel contains 100 cable ties. If each cable tie can secure 2 wires together, how many wires can be secured in total using the entire reel of cable ties? (Ans.200)
3) A switch controls a light bulb rated at 60 watts. If the switch has an efficiency rating of 90%, how much power is wasted when the light is turned on? (Ans.6 watts)
4) A conduit reamer is used to remove burrs and rough edges from conduits. If a conduit reamer is used to smooth a conduit with a diameter of 2 inches, what is the circumference of the pipe after smoothing? (Ans.6.28inch)
5) fish tape is used to thread wires through conduits. If fish tape is 50 feet long and needs to be used to thread wires through a conduit that is 30 feet long, what length of fish tape will remain outside the conduit? (Ans.20 feet)
6) If a switch has a resistance of 0.5 ohms and carries a current of 5 amps, what is the voltage drop across the switch? (Ans.2.5 volts)
7) A multimeter measures voltage with a resolution of 0.1 volts. If the measured voltage is 12.5 volts, what is the absolute error in the measurement? (Ans.0 volts)
8) If a switch is rated for a maximum load of 10 amps and operates at 120 volts, what is the maximum power (in watts) it can control? (Ans.1200 watts)
9) A pack of wire nuts contains 50 pieces. If each wire nut costs 20, how much will it cost to purchase the entire pack? (Ans.1000rupees)
10) A voltage tester indicates a voltage of 120 volts in an electrical circuit. If the actual voltage is 117 volts, what is the percentage error in the voltage measurement? (Ans.2.56%)

Unit Summary

Electrical wiring tools and accessories cover a range of equipment used in electrical installations and maintenance. Here's a unit summary highlighting key points:

- Introduction to electrical wiring
- Essential tools for electrical wiring
- Advanced tools and equipment
- Accessories for wiring installations
- Maintenance and troubleshooting tools
- Specialized tools for specific tasks
- Proper handling and storage

This unit provides a comprehensive overview of the tools and accessories essential for electrical wiring projects, emphasizing safety, functionality, and efficiency in installations and maintenance tasks.

Practical

1. Demonstrate power tools and practice utility items.
2. Demonstration of advanced power tools, pneumatic tools, electrical wiring tools, and accessories. An electrical switchboard with two sockets, piano buttons, and electrical wiring is made. Any other item as per the requirement of the college/department, with safety precautions.
3. Practice on simple lamp circuits (i) One lamp controlled by one switch by surface conduit wiring, (ii) Lamp circuits- connection of lamp and socket by separate switches, (iii) Connection of Fluorescent lamp/tube light, (iv) Simple lamp circuits-install bedroom lighting. (v) Simple lamp circuits- install staircase wiring.

Know More

Sl No	Topic	QR Code
1	Standards for Plugs-Socket-Outlets and Switches	
2	Standards List	
3	Standards in Category: Electrical appliances	
4	IS 1356-1 (1972)	
5	Electrical Measurement and Instrumentation	
6	Utilization of Electrical Energy	

References and Suggested Readings

1. M. Brown, J. Rawtani, and D. Patil, *Practical troubleshooting of electrical equipment and control circuits*, Elsevier, 2004.
2. T. Candela, *Automotive wiring and electrical systems*, Vol. 160, CarTech Inc, 2009.
3. G. Stokes, and J. Bradley, *A practical guide to the wiring regulations: IEE wiring regulations (BS 7671: 2008)*. John Wiley & Sons, 2009.
4. E.L. Donnelly, *Electrical Installation-Theory and Practice Third Edition*. Nelson Thornes, 2014.
5. W.T. Grondzik, and A.G. Kwok, *Mechanical and electrical equipment for buildings*. John Wiley & Sons, 2019.
6. B.K. Reddy, *Electrical Equipment: A Field Guide*. John Wiley & Sons, 2021.
7. J.W. Vincoli, *Basic guide to system safety*. John Wiley & Sons, 2024.

UNIT 2

ELECTRICAL WIRES AND UNDERGROUND CABLES

UNIT SPECIFICS

This unit elaborately discusses the following topics:

- ✓ Introduction to electrical wires and underground cables
- ✓ Exploring the different types of conductors
- ✓ Classification of electrical cables
- ✓ Exploring the diverse cable insulation materials
- ✓ Various types of cable laying and cable jointing methods
- ✓ Exploring various factors in determining the selection of electric cables

This module provides a comprehensive examination of the significance of wire, cable, bus bar, stranded conductor, cable, armoured & flexible cable, solid conductor, PVC wires, CTS wire, LC wire, FR (fire retardant) wire and wire size as per the BIS (British Institute of Standards). This paper examines the instruments employed for wire size measurement and wire jointing techniques. This unit provides an overview of the categorization of cables into low-tension, high-tension, and extra-high-tension varieties and solid, oil-filled, and gas-filled variations. The text examines several cable insulation materials, namely vulcanised rubber (VIR), polyvinyl chloride (PVC), cross-linked polyethylene (XLPE), and impregnated paper. It also explores selecting an appropriate cable size and type based on established standard data. Practical illustrations and demonstrations exemplify various cable jointing techniques, cable laying methodologies, and the determinants that influence the selection of electric cables.

RATIONALE

This fundamental unit on electrical wires and underground cables stems from these components' critical role in modern electrical systems and infrastructure. As society continues to rely heavily on electricity for various applications, understanding the design, installation, and maintenance of electrical wires and underground cables is essential for ensuring safe, reliable, and efficient power distribution. Electricity transmission and distribution from power-producing sources to end-users are facilitated by electrical wires and subterranean cables, which are the fundamental components of electrical infrastructure. By studying these components, engineers and technicians can contribute to the planning, design, and construction of robust electrical networks that meet the growing demand for energy in urban, suburban, and rural areas. Compliance with national and international standards and regulations is essential for ensuring electrical wires and underground cables' quality, safety, and interoperability. Professionals can ensure that electrical installations meet regulatory requirements and industry best practices through study and adherence to relevant standards, such as those established by organizations like the International Electrotechnical Commission (IEC) and the National Electrical Code (NEC).

PRE-REQUISITES

Basics of Electrical Engineering

UNIT OUTCOMES

The list of outcomes of this unit is as follows:

U2-O1: Describe various types of electrical wires and underground cables for wiring.

U2-O2: Explain the classification of electrical cables.

U2-O3: Exploring the diverse cable insulation materials.

U2-O4: Explain diverse types of cable laying and cable jointing methods.

U2-O5: Describe various factors that determine the selection of electric cables.

Unit-2 Outcomes	Expected Mapping with Programme Outcomes *(1- Weak Correlation; 2- Medium correlation; 3- Strong Correlation)*					
	CO-1	CO-2	CO-3	CO-4	CO-5	CO-6
U2-O1	3	3	1	2	2	2
U2-O2	3	3	1	2	2	2
U2-O3	3	3	1	2	2	2
U2-O4	3	3	1	3	2	2
U2-O5	3	3	1	3	2	1

2.1 Introduction:

Electrical wires and underground cables are critical components of electrical networks that enable the transmission and distribution of electrical power. They are used for various purposes, including domestic electrical wiring and power distribution in industrial settings. Underground cables and wires are critical components of electrical power distribution networks that transfer electrical energy between places. These components are essential in various situations, including commercial, industrial, and residential areas.

2.1.1 Electrical Wires:

Purpose:

- ✓ Conductive materials, such as electrical cables, transmit electrical current between different locations. They are applied to connect electrical equipment, devices, and parts within buildings or between different structures.
- ✓ Electrical wire transfers energy from a transformer or another source to outlets, appliances, devices, distribution boards, sockets, and light fittings. They provide a path of minimal resistance for the flow of electricity, facilitating energy conduction.

Types:

- ✓ **Copper Wires:** Copper is a prevalent choice for electrical wiring because of its exceptional conductivity.
- ✓ **Aluminum Wires:** Aluminum is utilized for electrical wire, particularly in high-voltage transmission lines.

Sizes:

- ✓ The wire size or gauge to be used is determined by the current load that the application needs to carry.

Insulation:

- ✓ Insulation is commonly applied to wires to protect them from the environment and prevent electrical shocks.

Types of Electrical Wires:

- ✓ **Single Conductor Wires:** These wires are composed of a solitary conductor, usually crafted from copper or aluminium, and are employed for diverse purposes such as electrical wiring in buildings and internal connections within the equipment.

✓ **Multiconductor Cables:** These cables consist of numerous insulated conductors enclosed in a single protective sheath. They are frequently employed for power distribution and control systems.

✓ **Coaxial Cables:** Employed in communication systems for signal transmission. They comprise a centre conductor, insulation, a metallic shield, and an outer insulating layer.

Insulation Material:

✓ Insulation materials for wires can include PVC (Polyvinyl Chloride), XLPE (Cross-Linked Polyethylene), rubber, and others. The selection depends on variables such as the specific use case, prevailing ambient circumstances, and the necessary voltage specifications.

Applications:

✓ **Building Wiring:** Electrical cables are frequently utilised in residential, commercial, and industrial wiring systems to facilitate the transmission of electricity to various electrical devices such as lighting fixtures, outlets, appliances, and other related equipment.

✓ **Appliances and Equipment:** Wires are employed internally in appliances and electronic devices to connect different components.

✓ **Power Transmission:** High-voltage wires transport electrical power across large distances in overhead power lines.

2.1.2 Underground Cables:

Purpose:

✓ Electrical power can be conveyed subterraneous over long distances or within a singular establishment using subterranean cables. They are commonly employed in urban areas where overhead cables are not feasible or aesthetically unappealing.

✓ An underground cable refers to a cable that is buried beneath the surface. They distribute telecommunications or power. These cables can replace the overhead wires positioned significantly above the ground. Subterranean cables often serve as substitutes for aboveground wires.

Types:

✓ **Power Cables:**

These cables facilitate electricity transmission at high voltages from substations to distribution locations.

✓ **Communication Cables:** Underground installations may incorporate communication cables for data transmission, telephony, and power connections.

➧ **Components:**

✓ **Conductors:**

Underground cables, like electrical wires, contain conductors (often composed of copper or aluminum) that transport the electrical current.

✓ **Insulation:**

Insulation is applied to cables to minimize electrical leakage and safeguard the conductors from moisture and other environmental influences.

✓ **Sheath:**

The cable is safeguarded from physical harm and environmental factors by an external layer known as a sheath.

➧ **Advantages of Underground Cables:**

✓ **Aesthetics:** Underground cables are concealed, enhancing their visual appeal in metropolitan environments.

✓ **Reduced Environmental Impact:** Underground installations have a diminished environmental impact, particularly in environmentally sensitive or heavily populated regions.

✓ **Less Susceptible to Weather:** Underground wires are more resistant to weather-induced harm, such as storms or ice.

➧ **Challenges of Underground Cables:**

✓ **Installation Cost:** The cost of installing subterranean cables is higher than the above lines because of the need for excavation and installation.

✓ **Maintenance Challenges:** Identifying and fixing defects in underground cables might pose more significant difficulties and require more time than overhead lines.

✓ **Heat Dissipation:** Underground cables necessitate careful design considerations to regulate heat development efficiently.

➧ **Construction:**

✓ Underground cables comprise several insulated conductors grouped and surrounded by protective layers. The layers commonly comprise conductor insulation, metallic shielding, and an outer insulating sheath.

Types of Underground Cables:

- ✓ **Low Voltage (LV) Cables:** Employed for local distribution and underground wiring in residential and commercial regions.
- ✓ **Medium Voltage (MV) Cables:** Frequently employed for subterranean power distribution in urban regions.
- ✓ **High Voltage (HV) Cables:** Carry large amounts of electricity across extended distances.
- ✓ **Fiber Optic Cables:** Underground installations may use fiber optic cables alongside electrical lines to facilitate communication.

Advantages of Underground Cables:

- ✓ **Aesthetics:** Underground cables are imperceptible, enhancing urban spaces' cleanliness and visual appeal.
- ✓ **Reduced Vulnerability:** Underground cables are more resistant to weather-induced harm, such as storms or strong winds.
- ✓ **Less Maintenance:** Underground cables, due to their protection from environmental factors, typically necessitate less maintenance than overhead lines.

Installation and Challenges:

- ✓ Installing underground cables necessitates the excavation of trenches, a process that can be more intricate and costly compared to installing overhead lines.
- ✓ Special precautions must be taken to safeguard subsurface cables from harm during construction operations.
- ✓ Locating faults and performing repairs can be more difficult due to limited sight.

Applications:

- ✓ Underground cables are extensively utilized in metropolitan settings, where overhead lines may be impractical due to limitations in space and concerns regarding aesthetics.
- ✓ They are frequently utilized in residential, commercial, and industrial areas.
- ✓ It is employed for transmitting power between substations and in different utility distribution networks.

Electrical wires and underground cables are vital in ensuring the secure and effective transmission of electrical power. The selection between overhead lines and underground cables is contingent upon various factors, including environmental concerns, visual appeal, expenses, and the particular demands of the electrical infrastructure.

2.2 Conductors:

Conductors are substances that facilitate the flow of electric electricity. They possess specific characteristics that render them appropriate for the conduction and propagation of electric charge. Conductors are essential components in electrical and electronic systems as they facilitate the flow of electric charge. Below are several fundamental attributes and classifications of conductors:

2.2.1 Characteristics of Conductors:

a) **High Electrical Conductivity:**

Conductors possess a notable level of electrical conductivity, indicating that they offer minimal resistance to the passage of electric current. Materials possessing high conductivity facilitate the effective transmission of electrical information.

b) **Low Resistance:**

Conductors have low electrical resistance, enabling them to minimize electrical energy dissipation as heat. Highly conductive materials are essential for optimizing power transmission efficiency.

c) **Free Electron Movement:**

Conductors are materials in which electrons are weakly linked to atoms, allowing them to travel unrestrictedly inside the substance. The unrestricted movement of electrons enables the smooth passage of electric current.

d) **Malleability and Ductility:**

Several conductors possess the property of malleability, meaning they may be quickly hammered into thin sheets, as well as ductility, allowing them to be dragged into thin wires. These qualities render them adaptable for many applications, such as wiring and cabling.

e) **Thermal Conductivity:**

Conductors possess high thermal conductivity, enabling them to transfer heat effectively. This characteristic is crucial in situations where heat dispersion is a factor to be considered.

f) **Reflectivity:**

Certain conductors, particularly metals, demonstrate a notable ability to reflect light and other electromagnetic waves. This characteristic is crucial in applications such as electrical connections and optical coatings.

2.2.2 Types of Conductors:

a) **Metal Conductors:**

✓ **Copper:** Copper is well recognized as one of the most prevalent conductors due to its exceptional conductivity, malleability, and resistance to corrosion.

✓ **Aluminium:** Commonly utilized in power distribution networks and specific electrical applications.

✓ **Silver:** It possesses the best electrical conductivity compared to other metals but is costly and prone to tarnishing.

✓ **Gold:** Renowned for its exceptional corrosion resistance, this material finds application in specialized electronic contexts.

b) **Semiconductor Conductors:**

Semiconductors are materials with conductivity that fall between conductors and insulators. Illustrative instances encompass silicon and germanium, frequently employed in electronic apparatuses such as transistors.

c) **Electrolytic Conductors:**

Electrolytes, such as solutions containing salts or acids, can conduct electricity when ionizing them. Batteries and electrochemical cells often have electrolytic conductors.

d) **Carbon Conductors:**

✓ **Graphite:** Graphite is a highly conductive type of carbon that finds applications in pencils and specific electronic uses.

✓ **Carbon Composites:** Materials that blend carbon with other substances to produce conductive composites utilized in various applications, such as resistors.

2.2.3 Applications of Conductors:

✓ **Electrical Wiring:** Copper and Aluminium conductors are widely utilized in electrical wiring for home, commercial, and industrial purposes.

✓ **Power Transmission:** Overhead power lines and underground cables carry electrical power across extensive distances.

✓ **Electronic Components:** Conductors are crucial in fabricating electronic components, including resistors, capacitors, and inductors.

✓ **Electric Motors and Generators:** Copper conductors are frequently employed in electric motor and generator windings.

✓ **Telecommunications:** Copper and fiber optic conductors are employed in telecommunications to transport electrical signals and optical data.

✓ **Heating Elements:** Conductive materials, specifically nichrome, are utilized in heating elements for various purposes, such as electric stoves and water heater

Conductors play a crucial role in the operation of electrical and electronic systems. The choice of conductors is determined by criteria such as their ability to conduct electricity, cost, environmental conditions, and special application needs.

2.2.4 Wire:

A wire (Fig.2.1) is a thin, pliable, and elongated conductor composed of a conductive substance, usually metal, employed for transferring electrical signals or conducting electric current. Wires play a crucial role in electrical and electronic systems by facilitating the movement of electric charge. An individual metallic conductor is usually cylindrical. Earth, Live, and Neutral wires are the three types of wires commonly encountered in residential circuits. The earth wire is a precautionary measure, specifically for devices with metallic enclosures. The wire was coated with green insulation.

Figure 2.1 Wire

2.2.5 Cable:

A cable (Fig.2.2) is a collection of one or more conductors, usually isolated from each other and sometimes collectively insulated or shielded, which is utilized for transferring electrical signals or power. Cables are more intricate than individual wires, which frequently incorporate numerous conductors, shielding, and insulation materials. Cables have multiple functions in electrical and electronic systems, serving as a medium for carrying power, data, or signals across short and long distances. Below are a few fundamental characteristics of cables:

- **Components of a Cable:**

✓ **Conductors:**

The conductors within a cable refer to the individual wires or strands responsible for transporting electric current. Typically, these conductors are composed of metals such as copper or aluminum.

✓ **Insulation:**

Each conductor is provided with insulation to prevent the occurrence of short circuits and accidental contact with other conductors or surfaces. The insulation materials comprise PVC, XLPE, rubber, or non-conductive materials.

✓ **Shielding:**

Some cables are shielded to protect signals or power transfer from external interference or avoid signal leakage. Typical materials used for shielding are metallic foils or braiding.

✓ **Sheath or Jacket:**

The outer layer of the cable, known as the sheath or jacket, offers general protection and insulation for the complete cable assembly. It protects the internal components against external elements such as dampness, chemicals, and physical harm.

- **Types of Cables:**

a) **Power Cables:** These are intended to transmit electrical power between two locations. Illustrative instances encompass subterranean power cables and aerial power lines.

b) **Control Cables:** They are utilized to convey control signals in industrial automation and machinery. They frequently possess several insulated conductors.

c) **Communication Cables:** The transmission of data or signals between different devices. Illustrative instances encompass Ethernet cables, coaxial cables, and fiber optic cables.

d) **Instrumentation Cables:** They are utilized in instrumentation and process control applications, like control cables.

e) **Coaxial Cables:** It is composed of a center conductor, insulation, metallic shielding, and an outer insulating layer, frequently employed in the transmission of cable television signals and networking applications.

f) **Fiber Optic Cables:** Utilise optical fibers composed of glass or plastic to transmit data using light signals. It is commonly employed for rapid transmission of data in telecommunications.

g) **Submarine Cables:** She was deployed on the ocean floor to establish connections between continents, enabling international telecommunications and internet connectivity.

h) **Flat Cables:** They possess a planar, ribbon-like configuration and are frequently employed in situations with a constraint on available space, such as within computer systems and electrical apparatus.

- **Applications of Cables:**

✓ **Power Distribution:** Electricity cables transmit electricity from power plants to residential areas, industrial facilities, and other sites.

✓ **Telecommunications:** Communication cables, such as fiber optics, transmit voice, data, and video information in telephone and internet networks.

✓ **Industrial Automation:** It involves the utilization of control and instrumentation cables in industrial environments to convey signals that regulate machines and processes.

✓ **Networking:** Ethernet cables connect computers, routers, and other devices in local and wide area networks (LANs) networks (WANs).

✓ **Audio and Video:** Cables are transmitted between devices using cables, such as HDMI cables, which are specifically designed for high-definition video and audio.

✓ **Medical Applications:** Involve using specialized cables to transfer signals and power to devices such as ultrasound machines and patient monitoring.

Various cables are available, each designed to meet unique needs and purposes. The appropriate choice of cable is contingent upon various criteria, including the intended application, surrounding conditions, signal characteristics, and transmission span.

Figure 2.2 Cable

2.2.6 Bus Bar:

An electric conductor made of copper or aluminum, typically a large bar or strip, is used to distribute power. A busbar (Fig.2.3), often called a bus bar, is a conductive strip or bar utilized in localized high-current power distribution, typically within switchgear, panel boards, and busway enclosures. A bus bar, also known as a busbar, is a conductive strip or bar made of copper or aluminum that is utilized to transmit and distribute electrical power inside an electrical system. Bus bars function as a central electrical junction for several electrical circuits, facilitating the effective dispersion of electric current to different components. These devices are frequently encountered in power distribution panels, switchgear, and other electrical enclosures.

Figure 2.3 Bus bar

Functions of Busbars:

- ✓ **Power Distribution:** Busbars serve as a central conductor for distributing electrical power from a single source, like a transformer or distribution panel, to various loads or circuits.
- ✓ **Connection Point:** They are a central connection point for electrical equipment, facilitating simple and effective connectivity of circuit breakers, switches, relays, and distribution boards.
- ✓ **Redundancy:** Busbars in systems with redundant power supplies or backup generators enable the smooth power flow between sources to provide an uninterrupted electrical supply.
- ✓ **Fault Current Path:** Busbars offer a low-resistance route for fault currents to pass through, guaranteeing that high currents caused by short circuits or overloads are safely directed away from delicate equipment.
- ✓ **Voltage Regulation:** Busbars distribute power evenly across many circuits to maintain stable voltage levels in the electrical system, reducing voltage dips and maintaining consistent functioning of linked devices.

Types of Busbars:

a) **Main Busbar:** The main busbar is located in distribution panels or switchgear and is responsible for transmitting high-current feeds from the power source to distribution points within the system.

b) **Distribution Busbar:** Secondary busbars carry power from the main busbar to different circuits or loads as a connection point for circuit breakers or fuses.

c) **Neutral Busbar:** This is specifically designed to connect the neutral conductors of circuits to provide appropriate grounding and balance of electrical loads in a system.

d) **Grounding Busbar:** Committed to connecting conductors to the ground, creating a conduit with low resistance for fault currents to flow to the earth, improving electrical safety, and preventing electric shocks.

- **Construction and Materials:**

✓ **Material:** Busbars are commonly constructed from highly conductive metals like copper or aluminium because of their superior electrical conductivity and mechanical characteristics.

✓ **Shape:** Busbars can vary in shape and form, such as flat strips, rectangular bars, or tubular conductors, according to the specific application and needed current-carrying capability.

✓ **Insulation:** Insulating materials or coatings can be used on busbars to prevent unintended contact and ensure electrical safety.

✓ **Mounting:** Busbars are commonly installed on insulated or busbar supports inside electrical enclosures or switchgear assemblies.

- **Applications:**

✓ **Power Distribution Systems:** Busbars are frequently utilized in electrical distribution panels, switchgear, substations, and industrial power systems to distribute electricity to different loads and circuits.

✓ **Data Centers:** Busbars are utilized in data center architecture to efficiently and reliably transfer power to server racks and networking devices.

✓ **Industrial Applications:** Busbars are essential for industrial facilities, refineries, and process plants. They are crucial in distributing power to motors, machinery, and control systems.

Busbars are essential components in electrical distribution systems, as they efficiently and reliably distribute electricity while maintaining electrical safety and system integrity.

2.2.7 Stranded Conductor:

A stranded conductor (Fig.2.4) is an electrical conductor of numerous smaller wires or strands twisted or braided together. This design improves the flexibility, durability, and conductivity of the conductor. Stranded conductors are frequently utilized in situations requiring flexibility, such as cables and wires that must bend and move without fracturing.

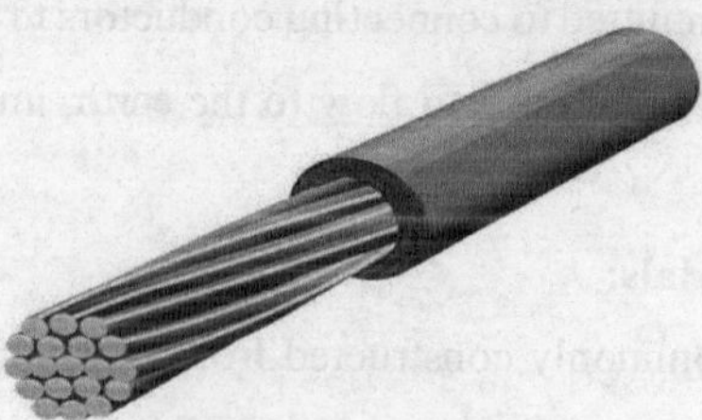

Figure 2.4 Stranded conductor

- **Characteristics of Stranded Conductors:**
  - ✓ **Flexibility:** Each wire strand in a stranded conductor may move separately, enabling the conductor to bend and flex more easily. Stranded conductors are well-suited for regular movement or vibration applications, like portable cords or cables.
  - ✓ **Durability:** Strands in a stranded conductor transfer mechanical stress uniformly, decreasing the chances of breakage or metal fatigue. Improved durability increases the longevity and dependability of stranded conductors in challenging working conditions.
  - ✓ **Conductivity:** Stranded conductors have a high electrical conductivity, like solid conductors of the same size. The greater surface area created by the numerous wire strands guarantees effective conduction of electrical current with negligible impedance.
  - ✓ **Resistance to Corrosion:** Having numerous wire strands ensures backup in case of localized damage or corrosion, which helps sustain the overall integrity and conductivity of the conductor in the long run.
  - ✓ **Termination:** Stranded conductors are more pliable than solid conductors, making them easier to connect to terminal screws or lugs. This enables safe and dependable electrical connections for a variety of uses.
- **Construction of Stranded Conductors:**
  - ✓ **Wire Strands:** Stranded conductors usually consist of wire strands made of copper or aluminium, possibly using materials like tinned or copper-clad aluminium.
  - ✓ **Stranding Pattern:** Stranded conductors can be made using various stranding patterns such as concentric-lay, inlay, or rope-lay configurations, chosen according to the required flexibility and mechanical characteristics.
  - ✓ **Strand Size:** The gauge of a stranded conductor is determined by the size and total number of wire strands, which also affects its current-carrying capacity.

Applications of Stranded Conductors:

✓ **Flexible Cords and Cables:** Stranded conductors are frequently used in flexible cords and cables for various applications such as power extension, appliance, and electronic cables.

✓ **Building Wiring:** Stranded conductors are utilized in building wiring applications for branch circuits, fixture wires, and equipment grounding conductors when flexibility and durability are needed.

✓ **Automotive Wiring:** Stranded conductors are commonly utilized in automotive wiring harnesses for cars due to their flexibility and ability to withstand vibrations and mechanical stress.

The stranded conductors provide flexibility, endurance, and conductivity, making them suitable for many electrical and electronic applications that require consistent performance under changing situations.

2.2.8 Armored Cable:

Armored cable (Fig.2.5), sometimes called armored electrical cable or armored wire, is an electrical cable shielded by a metal armor covering. This armor enhances the cable's durability against physical damage, moisture, and environmental elements by providing extra mechanical protection. Armored cables are frequently utilized in diverse situations that necessitate additional protection.

Figure 2.5 Armored cable

Construction of Armoured Cable:

✓ **Conductors:** The armoured cable's core comprises one or more insulated electrical conductors, usually composed of copper or aluminium. These conductors transmit electrical current between two points.

✓ **Insulation:** Every conductor is coated with an insulating substance like PVC or XLPE to prevent electrical leakage and assure safety.

✓ **Armor:** Insulated conductors are encased in a protective layer of metal armour or sheathing. This armour is typically constructed from steel or aluminium and offers mechanical defence against physical harm, including crushing, abrasion, or puncture by sharp things.

✓ **Outer Jacket:** Some armoured cables may have an extra outer jacket for increased protection and insulation. The outer coat is usually constructed from PVC or other appropriate materials and can offer additional protection against moisture, chemicals, and UV radiation.

Types of Armoured Cable:

a) **Steel Armoured Cable (SWA):** This armoured cable includes a steel wire armour layer, often consisting of steel wires coiled in a helical pattern around the insulated conductors. SWA cable is frequently utilized in subterranean installations, outdoor settings, and locations that need mechanical protection.

b) **Steel Tape Armoured Cable (STA):** The armour layer in STA cable comprises a single strip of steel tape wrapped around the insulated conductors. STA cable provides comparable mechanical safeguarding to SWA cable but is more pliable and straightforward to install in specific scenarios.

c) **Aluminium Armoured Cable:** Certain armoured cables utilize aluminium armour instead of steel. Aluminium armoured cables are lightweight and may provide corrosion resistance in some conditions. Nevertheless, they might offer a different degree of mechanical safeguard than steel armoured cables.

Applications of Armoured Cable:

✓ **Industrial Installations:** Armoured cables are frequently utilized in industrial settings, factories, and manufacturing facilities to provide electricity to machinery, equipment, and control systems. The metal armour shields the cables against harm inflicted by heavy machinery, automobiles, and other dangers found in industrial settings.

✓ **Commercial Buildings:** Armoured cables are commonly utilized in commercial buildings, offices, and retail areas to wire power distribution systems, lighting circuits, and HVAC systems. The armoured construction offers robustness and dependability in high-traffic areas and locations prone to periodic repairs or alterations.

✓ **Outdoor and Underground Installations:** Armoured cables are appropriate for outdoor and underground installations when cables are subjected to severe weather conditions, dampness, and physical harm. The metal armour shields the wires from harm due to digging, trenching, and accidental strikes.

✓ **Critical Infrastructure:** Armoured cables are utilized in vital infrastructure settings like hospitals, data centers, telecommunications facilities, and military installations, where dependable electrical power delivery is crucial. Armoured cables offer mechanical protection to ensure continuous operation and reduce the likelihood of electrical malfunctions.

2.2.9 Flexible Cable:

Flexible cable (Fig.2.6), sometimes called flexible cord or flex cable, is an electrical wire specifically created to be pliable and capable of bending, making it ideal for situations involving frequent movement or bending. Flexible cables can endure multiple bends and flexes without causing harm to the conductors or insulation, unlike rigid cables intended for stationary installations.

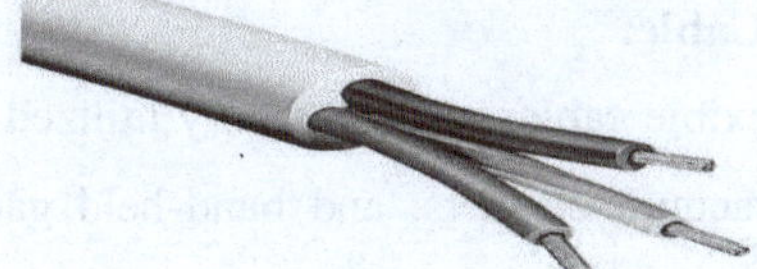

Figure 2.6 Flexible cable

Construction of Flexible Cable:

✓ **Conductors:** Flexible cables comprise one or more electrical conductors crafted from copper or aluminium. The conductors transmit electrical current within the cable.

✓ **Insulation:** Every conductor is coated with an insulating material like PVC or rubber to prevent electrical leakage and assure safety.

✓ **Stranding:** Flexible cables use stranded conductors of numerous tiny wire strands twisted or braided together, unlike the solid conductors of rigid cables. This stranding enables the cable to stretch and bend without causing damage to the conductors.

✓ **Fillers:** Flexible cables may include filler materials like cotton or synthetic fibers between the conductors for added mechanical support and insulation.

✓ **Sheathing:** The insulated conductors are covered with a protective outer layer of flexible and resilient materials like PVC, rubber, or thermoplastic elastomer (TPE). The sheathing shields the conductors from external elements, abrasion, and mechanical harm.

✓ **Colour Coding:** Flexible cables may feature color-coded insulation or labels on the outside sheath to distinguish the conductors and aid in installation and maintenance.

Types of Flexible Cables:

a) **Flexible Power Cables:** Utilised to provide electrical power to portable devices, appliances, tools, and machinery. These wires are frequently on building sites, workshops, and industrial settings.

b) **Flexible Control Cables:** Intended for transferring control signals and data among devices, like sensors, actuators, and automation systems. They are utilized in machinery, robotics, and industrial control applications.

c) **Flexible Instrumentation Cables:** Utilised to link instruments, meters, and sensors in measurement and monitoring systems. These cables offer precise signal transfer and are commonly utilized in laboratories, testing facilities, and scientific equipment.

d) **Flexible Multicore Cables:** Comprising numerous conductors in a single cable assembly, offering flexibility and adaptability for different uses. They are frequently utilized in audiovisual systems, telecommunications, and electrical equipment.

Applications of Flexible Cable:

✓ **Portable Equipment:** Flexible cables are commonly utilized in portable equipment and tools like power drills, vacuum cleaners, and hand-held gadgets, where mobility and flexibility are crucial.

✓ **Machinery and Robotics:** Flexible cables in industrial machinery and robotic systems ensure dependable power and data transfer, enduring constant movement and operation.

✓ **Medical Devices:** Flexible cables are utilized in medical equipment such as patient monitoring systems, diagnostic tools, and imaging devices to ensure patient safety and comfort due to their crucial flexibility and durability.

✓ **Automotive and Transportation:** Flexible cables are utilized in automotive settings, such as car wire harnesses, battery cables, and sensor connections, to endure vibrations, temperature changes, and mechanical strain.

✓ **Entertainment and Events:** Flexible cables are utilized in stage lighting, audio systems, and event production equipment to facilitate power distribution and signal transmission in dynamic settings, including concerts, theatres, and live events.

2.2.10 Solid Conductor:

A solid conductor (Fig.2.7) is an electrical conductor crafted from a single metal piece, usually copper or aluminum. Stranded conductors comprise several smaller wire strands twisted or braided together, while a solid conductor is a single continuous wire.

Figure 2.7 Solid conductor

- **Characteristics of Solid Conductors:**

✓ **Single-piece Construction:** Solid conductors are crafted from a singular metal piece, typically copper or aluminium, with a consistent cross-sectional area along the wire's length.

✓ **Rigid Structure:** Solid conductors are less flexible than stranded conductors due to their one-piece design. They are rigid and not easily bendable, making them most suitable for stationary installations with little movement or vibration anticipated.

✓ **Low Resistance:** Solid conductors have low electrical resistance, enabling efficient transfer of electrical current with minimal energy loss. They are ideal for situations that demand high conductivity, like power distribution systems and electrical wiring in buildings.

✓ **Easy Termination:** Solid conductors are simple to terminate and attach to electrical equipment, including switches, outlets, and terminal blocks. They offer a steady and dependable electrical connection when correctly installed and fastened.

✓ **Cost-effective:** Solid conductors are typically more economical than stranded conductors because of their uncomplicated production method and reduced material expenses. This quality makes them a favoured option for fundamental electrical wiring tasks in residential, commercial, and industrial environments.

- **Applications of Solid Conductors:**

✓ **Building Wiring:** Solid conductors are frequently utilized in building wire systems for electrical distribution, lighting circuits, and power outlets. They offer a steady and dependable electrical link between electrical equipment and the main distribution panel.

✓ **Electrical Appliances:** Solid conductors are utilized within electrical appliances and gadgets like refrigerators, stoves, and air conditioners to transmit electrical current to different components and circuits.

✓ **Telecommunications:** Solid conductors are used in telecommunications cables like Ethernet and telephone wires to transport data signals across short distances. They establish a reliable electrical connection between networking equipment and enable fast data transmission.

✓ **Automotive Wiring:** Solid conductors are utilized in automotive wire harnesses and electrical systems to supply electricity to car components like lights, sensors, and motors. They offer a dependable electrical connection under the challenging operating circumstances of vehicle settings.

✓ **Electronics:** Solid conductors are utilized in electrical circuits and PCBs to connect components and traces. They establish a reliable electrical connection between components and enhance the transmission of signals inside the circuit.

Although solid conductors offer benefits like low resistance and cost efficiency, their inflexibility and lack of pliability may restrict their appropriateness for tasks involving frequent bending or movement. Stranded conductors are commonly chosen because they offer increased flexibility and are more resistant to metal fatigue.

2.2.11 PVC Wires:

PVC wires (Fig.2.8), or PVC insulated wires, are electrical cables with insulation made of PVC (Polyvinyl Chloride), a plastic material. These wires are commonly utilized in various electrical applications because of their exceptional insulation, flexibility, durability, and cost-effectiveness.

Construction of PVC Wires:

✓ **Conductors:** PVC wires commonly use copper conductors, while aluminium conductors are utilized in some applications. Copper conductors are highly conductive and are the chosen choice for many electrical wiring applications.

✓ **Insulation:** The conductors are coated with PVC (Polyvinyl Chloride) for electrical insulation and protection from moisture, chemicals, and environmental elements. PVC insulation is robust, pliable, and resistant to abrasion, making it appropriate for indoor and outdoor use.

✓ **Colour Coding:** PVC wires may feature color-coded insulation to distinguish between various conductors and aid in correct installation and upkeep. Typical designations are red for live cables, black for neutral wires, and green or green/yellow for grounding wires.

✓ **Size and Gauge:** PVC wires are offered in various sizes and gauges, from thin gauges for low-voltage applications to thicker gauges for larger currents and power ratings.

✓ **Sheathing (Optional):** Some PVC wires may feature an extra outer sheath or jacket composed of PVC or other appropriate materials to provide increased protection and insulation. The outer layer protects against physical harm, UV radiation, and moisture penetration.

- **Characteristics of PVC Wires:**

✓ **Electrical Insulation:** PVC insulation offers superior electrical insulation, limiting current leakage and assuring electrical circuits' safe and dependable functioning.

✓ **Flexibility:** PVC wires are pliable and convenient for installation in confined areas and around corners due to their flexibility. PVC insulation's flexibility decreases the likelihood of conductor breakage and makes cable routing easier.

✓ **Durability:** PVC wires are robust and resistant to wear, tear, and abrasion damage. They can endure mechanical stress, bending, and twisting without affecting their electrical performance.

✓ **Chemical Resistance:** PVC insulation is chemically robust, making PVC wires ideal for industrial settings with frequent exposure to severe chemicals.

✓ **Temperature Rating:** PVC wires have a moderate temperature rating, usually between -15°C to 70°C or greater, according to the unique PVC composition and application needs.

- **Applications of PVC Wires:**

✓ **Building Wiring:** PVC wires are commonly utilized for electrical wiring in residential, commercial, and industrial structures. They are used for power distribution, lighting circuits, electrical outlets, switches, and other electrical systems.

✓ **Appliance Wiring:** PVC wires are utilized in electrical appliances like refrigerators, washing machines, air conditioners, and kitchen appliances for internal wiring and connections.

✓ **Automotive Wiring:** PVC wires are utilized in automobile wiring harnesses and electrical systems to supply electricity to car components such as lights, sensors, ignition systems, and entertainment systems.

✓ **Telecommunications:** PVC wires are used in telecommunications cables, such as telephone and data cables, to convey signals and data between devices.

✓ **Electronics:** PVC wires connect components and traces in electronic circuits, PCBs, and electronic devices. They offer a dependable electrical connection and insulation in electronic applications.

Figure 2.8 PVC wires

2.2.12 CTS Wire (Copper-Clad Steel):

CTS wire (Fig.2.9), short for Copper-Clad Steel wire, is a composite wire with a steel core covered by a layer of copper. CTS wire combines the advantages of steel and copper, making it excellent for applications requiring conductivity, strength, and cost-effectiveness.

Figure 2.9 CTS wire

✚ Construction of CTS Wire:

✓ **Steel Core:** The center of CTS wire consists of high-strength steel, offering mechanical strength, durability, and flexibility against bending and stretching. The steel core provides structural support for the wire, maintaining its integrity when subjected to strain and compression.

✓ **Copper Coating:** A copper layer is applied to the steel core using methods like electroplating or cladding. The copper coating improves electrical conductivity and corrosion resistance while enhancing the wire's look. The thickness of the copper plating may vary based on the specific application needs.

Characteristics of CTS Wire:

- ✓ **Conductivity:** CTS wire merges steel's robustness with copper's superior electrical conductivity. The steel core maintains structural integrity, while the copper coating enhances electrical current transmission with little resistance. CTS wire is appropriate for applications that demand both strength and conductivity.
- ✓ **Strength:** CTS wire's steel core offers greater mechanical and tensile strength than solid copper wire. CTS wire can endure high strain and mechanical stress without breaking or deforming, making it ideal for toughness and robustness applications.
- ✓ **Corrosion Resistance:** The copper layer on CTS wire provides corrosion protection, shielding the steel core from rust and oxidation. This guarantees CTS wire's enduring dependability and efficiency in outdoor and harsh settings where contact with moisture, chemicals, and other corrosive substances is a worry.
- ✓ **Cost-effectiveness:** CTS wire is typically more economical than solid copper wire since it utilizes steel as the core material. Utilizing steel and copper reduces costs while maintaining high electrical performance and mechanical robustness.

Applications of CTS Wire:

- ✓ **Grounding Systems:** Electric systems, phone networks, and lightning protection systems frequently employ CTS wire for grounding. Electrodes, rods, and wires for grounding can benefit from its high conductivity and mechanical strength.
- ✓ **Antenna Systems:** The transmission and reception of radio frequency (RF) signals through CTS wire is facilitated by antenna systems. Because of its high conductivity and durability, it is well-suited for use in constructing ground planes, radials, and antenna elements for radio, television, and wireless networks, among other communication systems.
- ✓ **Cable Television (CATV) Systems:** CTS wire is utilized in CATV systems to transmit television signals through coaxial cables. The conductivity and longevity of this product provide dependable signal transmission and reception in cable television networks.
- ✓ **Power Transmission and Distribution:** CTS wire is utilized in power transmission and distribution systems for overhead power lines, substations, and electrical grids. Its tremendous strength and conductivity make it ideal for transmitting electrical currents across vast distances.

- ✓ **Electrical Fencing:** CTS wire is utilized in electric fence systems for animal containment and perimeter protection. The fence's conductivity facilitates the effective transmission of electric pulses to repel animals and intruders, while its robustness guarantees the fence's longevity and dependability.

The CTS wire provides a flexible and economical option for applications that demand a blend of durability, electrical conductivity, and protection against corrosion.

2.2.13 LC Wire (Lead Covered Wire):

LC wire (Fig.2.10) stands for Lead Covered Wire, which is electrical wiring encased in a layer of lead. This wiring is frequently utilized in specialized applications where safeguarding against external elements like moisture, corrosion, or environmental pollutants is crucial.

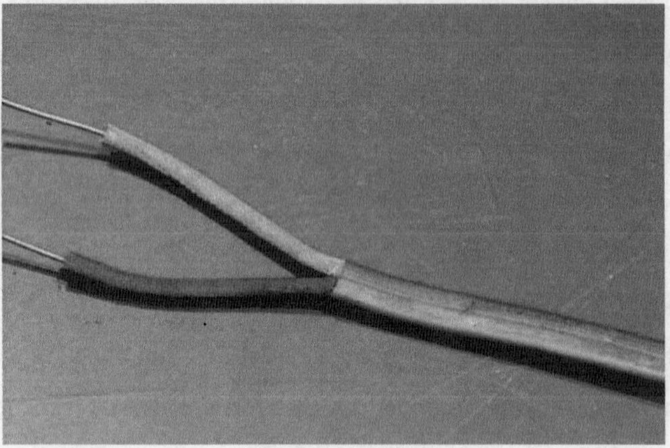

Figure 2.10 LC wire

Construction of LC Wire:

- ✓ **Conductors:** The LC wire conductors are usually composed of copper or aluminum, chosen for their superior conductivity characteristics, making them popular choices for electrical wiring.
- ✓ **Lead Sheathing:** The conductors are encased or coated with a lead layer, creating the wire's outer sheath. The lead sheathing protects from moisture, corrosion, chemical exposure, and mechanical harm.
- ✓ **Sealing:** Sometimes, the lead sheathing could be sealed or glued to the insulation or conductors to offer extra protection and guarantee a waterproof seal. This prevents water from entering and extends the wire's lifespan in challenging conditions.

- **Characteristics of LC Wire:**

✓ **Corrosion Resistance:** Lead is highly corrosion-resistant, making LC wire ideal for use in settings with high humidity, dampness, or chemical exposure. The lead sheathing shields the underlying conductors from oxidation and deterioration as time passes.

✓ **Chemical Resistance:** Lead sheathing offers strong resistance to various chemicals and solvents, making LC wire ideal for industrial settings with frequent contact with corrosive substances.

✓ **Mechanical Strength:** Lead is a pliable metal that can quickly adapt to the shape of conductors beneath it, offering protection from physical harm. LC wire has excellent tensile strength and durability, guaranteeing dependable operation in challenging environments.

✓ **Flexibility:** Lead sheathing, while protective, remains somewhat flexible, enabling LC wire to be bent, routed, and inserted in different configurations without losing its integrity.

- **Applications of LC Wire:**

✓ **Underground Installations:** LC wire is frequently utilized for subterranean electrical installations, including power distribution cables, grounding systems, and submersible pump cables. The lead sheathing shields the conductors from moisture and corrosion in underground conditions.

✓ **Marine and Offshore Applications:** LC wire is appropriate for maritime and offshore applications that involve exposure to saltwater, seawater, and severe weather conditions. The lead wrapping effectively prevents corrosion and prolongs the wire's lifespan in marine settings.

✓ **Chemical Plants and Refineries:** LC wire is utilized in chemical plants, refineries, and industrial facilities that frequently encounter corrosive chemicals and solvents. The lead sheathing offers dependable defence against chemical harm and maintains the integrity of electrical circuits.

✓ **High-Temperature Environments:** LC wire is utilized in chemical plants, refineries, and industrial facilities frequently encountering corrosive chemicals and solvents. The lead sheathing offers dependable defence against chemical harm and maintains the integrity of electrical circuits.

✓ **Historic Restoration:** LC wire is used in historic restoration projects to preserve ancient buildings and constructions' authenticity and aesthetic charm. The lead sheathing combines a classic look with a contemporary defence against environmental harm.

The LC wire provides efficient protection against moisture, corrosion, and chemical exposure in challenging conditions where traditional wiring materials may not be adequate.

2.2.14 FR (Fire Retardant) Wire:

FR wire, or Fire Retardant wire, is an electrical wiring product made with components that prevent fire propagation or combustion when exposed to heat or flames. The wires are designed to meet fire safety standards and regulations, making them appropriate for use in areas where fire protection is crucial.

Construction of FR Wire:

✓ **Conductors:** FR wires usually use conductors composed of copper or aluminium, which are popular materials in electrical wiring for their exceptional conductivity characteristics.

✓ **Insulation:** The conductors are covered with specialized fire-resistant insulation material, like PVC (Polyvinyl Chloride) or XLPE (Cross-Linked Polyethylene), designed to resist burning and stop the fire from spreading.

✓ **Outer Sheath (Optional):** Certain FR wires may feature an extra outer sheath or jacket composed of fire-retardant material to provide increased protection and insulation. The outer sheath improves the wire's fire resistance and offers mechanical protection from physical harm.

Characteristics of FR Wire:

✓ **Fire Retardant Properties:** FR wires are designed to withstand igniting and burning when in contact with heat or flames. FR wires employ fire-retardant insulating material to restrict fire spread and reduce property damage.

✓ **Low Smoke Emission:** FR wires resist fire and often have minimal smoke emission properties. The insulation emits low levels of smoke and harmful chemicals when in contact with fire, enhancing visibility and aiding safe evacuation in fire situations.

✓ **Thermal Stability:** FR wires demonstrate excellent thermal stability, preserving their electrical characteristics and structural strength even under high temperatures. This guarantees dependable operation in high-temperature and fire scenarios while maintaining safety standards.

- ✓ **Electrical Performance:** FR wires provide fire-retardant qualities while maintaining high electrical conductivity and insulation resistance, ensuring dependable power transmission and signal integrity in electrical circuits.
- ✓ **Compliance with Standards:** FR wires are produced by fire safety industry norms and regulations, including the National Electrical Code (NEC) and International Electrotechnical Commission (IEC) requirements. They are subjected to thorough testing to guarantee they fulfil the necessary fire performance standards.

Applications of FR Wire:

- ✓ **Residential Buildings:** FR wires are frequently utilized for electrical wiring in residential structures such as houses, apartments, and condominiums. They are put in electrical systems such as lighting circuits, power outlets, and switches to provide fire prevention and assure occupant safety.
- ✓ **Commercial and Industrial Facilities:** Using FR wires, commercial and industrial buildings prioritize fire safety. These facilities include offices, factories, warehouses, and hospitals. They are found in control panels, emergency lighting systems, electrical distribution systems, and machinery wiring.
- ✓ **Public Infrastructure:** Flame-retardant cables are used in public infrastructure projects such as schools, government buildings, transit hubs, and public utilities to adhere to fire safety rules and safeguard essential infrastructure from fire risks.
- ✓ **Fire Alarm and Detection Systems:** FR cables are utilized in fire alarm and detection systems to link sensors, detectors, control panels, and notification devices. Their fire-resistant characteristics guarantee the dependability and efficiency of fire detection and evacuation systems during crises.
- ✓ **Data Centers and Telecommunications:** FR wires are used in data centers, telecommunications facilities, and network infrastructure installations to guarantee the safety and dependability of electrical and communication systems. They supply power to servers, networking devices, and telecommunications cables.

FR wires are essential for fire prevention and safety since they offer dependable electrical wiring solutions that are resistant to ignition, reduce fire propagation, and safeguard lives and property during a fire.

2.2.15 Size of Wire According to BIS (Bureau of Indian Standards):

The Bureau of Indian Standards (BIS) determines the dimensions of electrical cables according to the criteria set in the Indian standards. Wire size is usually specified by its cross-sectional area, measured in square millimeters (mm^2) or square inches (AWG). BIS has set standards for many types of cables utilized in electrical installations, such as:

a) **Single-Core PVC Insulated Non-Sheathed Cable:** According to IS 694, BIS outlines the dimensions of single-core PVC insulated cables without sheathing for voltages up to 1100 volts. The standard includes wire diameters from 0.5 mm^2 to 400 mm^2.

b) **Multicore PVC Insulated Sheathed Cables:** Standard 1554 specifies the dimensions of multicore PVC insulated and PVC sheathed cables with copper conductors for up to 1100 volts. This standard covers wire diameters ranging from 0.5 mm^2 to 400 mm^2.

c) **Flexible Cords:** BIS specifies the dimensions of flexible cords, usually referred to as flexible cables or flex wires, that are utilized for portable appliances, tools, and equipment. The cords comply with IS 694 and come in diameters from 0.5 mm^2 to 6 mm^2.

d) **House Wiring Cables:** Standard IS 694 includes specifications for PVC insulated and PVC sheathed cables for permanent electrical installations in buildings and structures. The diameters vary from 1.5 square millimeters to 6 square millimeters.

e) **Armoured Cables:** Standard 1554 specify the dimensions of armoured cables containing copper conductors and PVC insulation for up to 1100 volts. The standard includes diameters ranging from 1.5 mm^2 to 400 mm^2.

f) **Flexible Cables for Submersible Pump Motors:** IS 694 outlines the dimensions of pliable cables designed for submersible pump motor purposes. The diameters vary from 1.5 mm^2 to 10 mm^2.

g) **Welding Cables:** BIS standards, like IS 9857, specify the dimensions of welding cables utilized in electric arc welding processes. The wires usually vary in size from 10 mm^2 to 120 mm^2.

Consult the precise BIS standards for complete information regarding wire diameters, conductor materials, insulation types, and other pertinent parameters. BIS standards guarantee the quality, safety, and dependability of electrical wire goods utilized in Indian marketplaces.

More on BIS Standards

2.2.16 Tools Used for Measurement of Wire Size:

Various equipment is frequently utilized to assess wire size in electrical and industrial environments. These tools measure the diameter or cross-sectional area of wires, which is essential for choosing the correct wire diameters for different purposes. Below are some tools utilized for measuring wire gauges:

a) **Wire Gauge (AWG Gauge):** The American Wire Gauge (AWG) system is a standardized way of measuring the diameter of electrical wires. AWG gauges can be obtained as handheld devices or as printed charts that feature notches aligned with wire diameters. The indicator is placed into the corresponding notch for the wire diameter to achieve a rapid and effortless measurement.

b) **Vernier Caliper:** A vernier caliper is a precise measurement instrument utilized to measure the diameter of wires and cables properly. The device comprises a primary and movable vernier scales, enabling accurate and exact measurements. The caliper's jaws are adjustable to hold the wire for measuring purposes securely.

c) **Micrometer:** A micrometer, or screw gauge, is a precision tool utilized for measuring wire diameter with exceptional accuracy. The device has a calibrated screw mechanism and a thimble or spindle that adjusts to measure the wire's diameter. Micrometers are frequently utilized in contexts of quality assurance and precision engineering.

d) **Wire Stripper with Gauge:** Certain wire stripping equipment includes built-in wire gauges or measurement indicators. These gadgets enable users to remove insulation from wires and measure the wire diameter simultaneously. Users may adequately assess the wire size by choosing the correct notch or gauge setting.

e) **Cable Measuring Device:** Cable measurement devices are utilized in industrial settings to measure the length and diameter of cables and wires swiftly. The devices usually include rollers or guides that enable the wire to move through to measure its diameter and length. They are frequently utilized in cable manufacturing and assembly procedures.

f) **Digital Caliper:** Digital calipers provide accurate wire diameter measurements with electronic sensors and a digital display. They offer prompt and precise readings, making them ideal for various wire measurement applications in electrical and manufacturing settings.

g) **Wire Size Gauge Tool:** Tools designed specifically for measuring wire diameter or cross-sectional area are available and known as wire size gauge tools. The tools have slots or holes designed for specific wire diameters, enabling users to identify the wire size by inserting it into the corresponding slot.

This equipment is crucial for precise wire size measurements and guaranteeing adherence to standards and specifications in electrical installations, manufacturing processes, and other applications that demand accurate wire measurements.

2.2.17 Wire Jointing Methods:

Wire jointing is joining two or more electrical wires to ensure electrical continuity and conductivity. Multiple techniques exist for wire jointing, each tailored to specific uses, conditions, and conductor kinds. Below are a few typical wire joining techniques:

✓ **Twist and Tape:** This is a straightforward and fundamental wire jointing procedure. The process entails intertwining the exposed tips of two or more wires and then encasing the twisted connection with electrical tape for insulation. Although simple to execute, twist and tape couplings may need more reliability and longevity for permanent installations, making them generally not advisable.

✓ **Wire Nut Connector:** Wire nut connections, sometimes called twist-on wire connectors, are frequently used to connect solid or stranded wires. The connectors have a threaded metal top and an insulating plastic casing. The cables are placed into the connector, and the cap is twisted onto the threads to establish a secure and insulated connection. Wire nut connections are available in different diameters to fit various wire gauges.

✓ **Soldering:** Soldering is a wire jointing technique where a low-melting-point metal alloy (solder) is melted to form a lasting connection between the wires. The wires are stripped of insulation, twisted together, and then heated with a soldering iron while solder is applied to the connection. Once the solder cools, it hardens, forming a robust and conductive connection. Soldered connections are frequently utilized in electronics and precision applications to establish a permanent and dependable connection.

- ✓ **Crimp Connectors:** Crimp connectors connect wires by crimping or squeezing them onto metal terminals or pins. The connector is positioned onto the exposed wire ends, and a crimping tool is employed to compress the connector, establishing a reliable mechanical and electrical link. Crimp connectors are available in many varieties, such as butt connectors, spade connectors, ring terminals, and bullet connectors, each designed for specific wire and terminal combinations.
- ✓ **Compression Connectors:** Compression connectors are akin to crimp connectors, except they utilize a compression tool to establish the connection rather than crimping. These connectors are frequently utilized for connecting more prominent conductors or aluminum wires due to their superior reliability and durability compared to only crimping.
- ✓ **Insulation Displacement Connectors (IDC):** Insulation displacement connectors (IDCs) are specifically engineered to connect wires without the need to remove the insulation. The connectors contain sharp metal blades or contacts that penetrate the insulation to establish contact with the conductors. Insulation Displacement Connectors (IDCs) are commonly utilized in telecommunications and data cabling for efficient and dependable wire terminations.
- ✓ **Mechanical Connectors:** Mechanical connectors like terminal blocks, lug connectors, and screw terminals establish a reliable and detachable link between wires. These connections usually have screws or clamps that hold the cables in place, simplifying installation and upkeep.
- ✓ **Heat Shrink Tubing:** Heat shrink tubing is utilised to insulate and safeguard wire joints by placing them over the connection and then using heat to shrink them securely around the joint. Heat shrink tubing offers superior insulation and mechanical defence against moisture, abrasion, and vibration.

The selection of the wire jointing method is contingent upon elements including the type of wires being connected, application specifications, environmental circumstances, and the desired level of reliability and durability. Adhering to industry standards and best practices is crucial when choosing and executing wire jointing procedures to guarantee secure and dependable electrical connections.

2.3 Classification of Cables:

Cables are categorized according to their structure, insulation materials, conductor materials, and intended uses. Below are some typical categorizations of wires:

More on Cables

2.3.1 Based on Construction:

- ✓ **Single Core Cable:** A cable consisting of a solitary conductor surrounded by an insulator. Single-core wires are utilized in applications where electromagnetic interference (EMI) is not a factor.
- ✓ **Multicore Cable:** A cable containing numerous conductors surrounded by insulation that are typically grouped in a single protective covering. Multicore cables are utilized for tasks that necessitate multiple connections, like power distribution and signal transmission.
- ✓ **Twisted Pair Cable:** A multicore cable with insulated wire pairs twisted together. Twisted pair cables are frequently utilized in telecommunications and networking for their capacity to minimize electromagnetic interference.
- ✓ **Coaxial Cable:** A cable composed of a centre conductor, an insulating layer, a metallic shield, and an outer insulating sheath. Coaxial cables transmit high-frequency signals and are commonly found in cable television and data connections.
- ✓ **Triaxial Cable:** This resembles a coaxial cable but with an extra outside conductor layer. Triaxial cables offer superior shielding and are utilized in applications that demand high degrees of noise tolerance.
- ✓ **Shielded Cable:** A cable with one or more conductors enclosed by a metallic shield to guard against electromagnetic interference. Shielded cables are utilized in areas with elevated electromagnetic interference (EMI) levels, such as industrial automation and control systems.
- ✓ **Unshielded Cable:** An unshielded cable. Unshielded cables are utilized when electromagnetic interference (EMI) is not a significant factor or affordability is the main priority.

2.3.2 Based on Insulation Material:

- ✓ **PVC Insulated Cable:** A cable insulated with polyvinyl chloride (PVC). PVC insulated cables are commonly utilized for many applications because of their flexibility, durability, and cost efficiency.
- ✓ **XLPE Insulated Cable:** A cable insulated with cross-linked polyethylene (XLPE). XLPE insulated cables have more excellent electrical and thermal qualities than PVC, making them appropriate for high-voltage and high-temperature applications.
- ✓ **Rubber Insulated Cable:** A cable with synthetic or natural rubber insulation. Rubber-insulated cables are pliable and durable, resistant to abrasion, oil, and chemicals, making them ideal for challenging conditions and mobile devices.
- ✓ **Silicone Insulated Cable:** A cable with silicone rubber insulation. Due to their exceptional temperature resistance and flexibility, silicone insulated cables are well-suited for high-temperature applications in industries including aerospace, automotive, and industrial heating. **Based on Conductor Material:**

a) **Copper Conductor Cable:** A cable featuring copper conductors known for their strong conductivity and resilience to corrosion. Copper conductor cables are commonly utilized for the transmission and distribution of power.

b) **Aluminium Conductor Cable:** An aluminium cable with conductors that provide cost efficiency and reduced weight compared to copper. Aluminium conductor cables are frequently utilized for overhead power lines and extensive electrical systems.

2.3.3 Based on Applications:

- ✓ **Power Cable:** The conveyance of electrical current through a specially constructed cable. Power cables are essential for distributing electricity in homes, businesses, and factories.
- ✓ **Control Cable:** A cable utilized in automation and control systems for transmitting control signals or low-power electrical signals. Machines, robots, and building automation all make use of control cables.
- ✓ **Instrumentation Cable:** A cable engineered explicitly for transferring analog or digital signals in measurement and instrumentation contexts. Instrumentation cables are utilized in oil and gas, petrochemical, and process control industries.
- ✓ **Communication Cable:** A wire that carries electrical impulses in a network or telecommunications system, such as data, voice, or video. The three main types of communication cables are fiber optic, coaxial, and twisted pair.

- **Fire-resistant Cable:** A cable engineered to preserve circuit integrity and functionality in the event of a fire. Fire-resistant cables are utilized in vital infrastructure like hospitals, high-rise buildings, and transportation systems.

These are some of the standard classifications of cables, each suited to specific applications and requirements.

2.3.4 Low Tension Cables:

Low-tension cables are cables designed to carry electrical power at low voltages. In electrical engineering, "low tension" is commonly used synonymously with "low voltage." The cables are intended for conducting electrical currents at voltages, usually ranging up to 1,000 volts AC or 1,500 volts DC. Here are some essential aspects of low-tension cables:

- **Voltage Range:** Low tension cables are intended for use at voltages under 1,000 volts AC or 1,500 volts DC. These voltages are categorized as low compared to medium voltage systems (1 kV to 35 kV) and high voltage systems (above 35 kV).
- ***Applications:*** *Low voltage cables* are frequently utilized in various applications, such as residential, commercial, and industrial environments. They are used for power distribution, lighting, control circuits, and other low-voltage electrical systems.
- **Conductor Material:** Low-tension cables usually have conductors made of either copper or aluminium, chosen based on the application's specific needs and financial constraints.
- **Insulation:** Low-tension cables are insulated to guard against risks, such as short circuits and electrical leaks. Rubber, XLPE (cross-linked polyethylene), and PVC (polyvinyl chloride) are common insulation materials. These materials vary according to the application and environmental factors.
- **Types of Low-Tension Cables:** There are several varieties of low-tension cables, such as single-core, multicore, armored, and unarmored cables. The application, installation environment, and necessary protection level all influence the selected cable type.
- **Standards and Regulations:** Low-tension cables must adhere to all applicable standards and laws to guarantee performance, safety, and dependability. These cables are produced in India per guidelines established by agencies like the Central Electricity Authority (CEA) and the Bureau of Indian Standards (BIS).

- ✓ **Installation:** Low-tension cables can be installed directly underground, in trenches, cable trays, conduits, or anywhere else, depending on the needs of the particular application and local laws. A cable system's durability and integrity depend on proper installation techniques.

Low-tension cables are essential to lighting and power distribution systems because they offer a dependable and effective way to carry electricity at lower voltages.

2.3.5 High Tension Cables:

High-voltage electrical power transmission is the purpose of high-tension cables. In electrical engineering applications, "high voltage" and "high tension" are frequently synonymously. The higher voltages required in medium- and high-voltage electrical transmission and distribution systems are specially handled by these cables. Here are some essential details regarding high-tension cables:

- ✓ **Voltage Range:** Typically, high-tension cables are intended to function at voltages of more than 1,500 volts DC or 1,000 volts AC. These voltages are classified as high voltage (beyond 35 kV) and medium voltage (1 kV to 35 kV) systems, respectively, and are regarded as high compared to low voltage systems.
- ✓ **Applications:** In electrical transmission and distribution networks, high-tension cables are mainly used to transfer power over large distances between power plants and substations and between substations and residential, commercial, and industrial regions. They are also utilized in the interconnection lines connecting several nations or areas.
- ✓ **Conductor Material:** High-tension cables commonly use Copper or aluminium as conductor materials. Because they are less expensive and heavier, aluminium conductors are frequently utilized for overhead transmission lines. On the other hand, copper conductors are used for subterranean cables and high-performance applications that demand more excellent conductivity.
- ✓ **Insulation:** Insulation shields high-tension cables from electrical leaks, short circuits, and other risks related to high voltages. Depending on the particular needs of the application and the surrounding conditions, XLPE (cross-linked polyethylene), EPR (ethylene propylene rubber), and oil-impregnated paper are employed as insulation materials in high-tension cables.
- ✓ **Types of High-Tension Cables:** There are several kinds of high-tension cables, such as subterranean, overhead insulated, overhead transmission lines, and underwater cables (also

known as covered conductors). Every type of cable has a particular purpose and set of operating parameters in mind.

✓ **Standards and Regulations:** For high-tension cables to be reliable, safe, and function well, they must adhere to all applicable standards and laws. Depending on the nation and area, these standards may differ. However, they frequently contain regulations issued by national regulatory authorities, the Institute of Electrical and Electronics Engineers (IEEE), and the International Electrotechnical Commission (IEC).

✓ **Installation:** Installing high-tension cables requires specific tools and methods to handle high voltages and heavy conductors. Installation techniques might differ based on the kind of cable and the application's particular needs, but they usually involve techniques for installations above ground, below ground, and underwater.

High-tension cables are essential for distributing and transmitting electricity because they allow power to be reliably and efficiently transferred across large distances and between diverse geographical areas.

2.3.6 Extra High Tension Cables:

Specialized Extra High Tension (EHT) cables are made to transfer electricity at very high voltages, usually above 230 kV (kilovolts) and up to 1,200 kV or higher. These cables are used in ultra-high voltage (UHV) transmission systems to move bulk electricity from generating stations to load centers, connect massive power grids, and transmit power over vast distances. Essential information regarding extra high-tension wires is as follows:

✓ **Voltage Range:** Compared to standard high-tension cables, EHT cables are made to function at voltages far greater. Typically, the voltages are between 230 to well over 1,000 kV. In the language of electrical engineering, these voltage levels are referred to as "ultra-high voltage" or "extra-high tension."

✓ **Applications:** Ultra-high voltage transmission systems are the main application for EHT cables in long-distance power transfer. They connect large power grids, bulk power is sent from distant generating stations (such as thermal or hydroelectric power plants) to urban load centers, and the accessibility of integrating renewable energy sources situated at a considerable distance from demand centres is enhanced.

✓ **Conductor Material:** High-strength aluminium or aluminium alloy is commonly used for the conductors in EHT cables. This material offers superior mechanical strength and conductivity at a lower weight and cost than copper conductors. In EHT systems, aluminium conductors are frequently utilized for overhead transmission lines.

- ✓ **Insulation:** To avoid electrical breakdown and preserve the integrity of the transmission system at extremely high voltages, insulation in EHT cables is essential. Cross-linked polyethylene (XLPE), oil-impregnated paper (OIP), and gas-insulated designs (such as SF6 gas-insulated cables), which provide excellent dielectric qualities and insulation strength, are common insulation materials used in EHT cables.
- ✓ **Types of EHT Cables:** Depending on the needs of the transmission system and the surrounding environment, EHT cables are available in various forms and combinations. These could include buried cables, submerged cables, overhead transmission lines with bundled conductors, and cutting-edge technology like high-temperature superconducting (HTS) cables that lower losses and boost transmission capacity.
- ✓ **Standards and Regulations:** EHT cables must follow strict criteria and laws to guarantee performance, dependability, and safety. International organizations like the Institute of Electrical and Electronics Engineers (IEEE), the International Electrotechnical Commission (IEC), and national regulatory authorities that oversee power transmission frequently develop these standards.
- ✓ **Installation and Maintenance:** Because installing and maintaining EHT wires involves high voltages, specific tools, knowledge, and safety measures are needed. Depending on the kind of cable and the particular needs of the transmission system, installation procedures might vary, although they usually involve methods for overhead, subterranean, and underwater installations.

In ultra-high voltage transmission systems, the EHT cables are essential because they allow massive amounts of electrical power to be efficiently and reliably transmitted over great distances while minimizing losses and maximizing grid stability.

More on Indian Standard Details

2.3.7 Solid Cables:

In contrast to stranded cables, which have numerous tiny strands twisted or braided together to form the conductor, solid cables usually refer to wires that have a single solid conductor core. Solid cables are frequently utilized in many electrical and telecommunications applications where flexibility is not a critical need. Key details of solid wires are as follows:

- **Conductor Construction:** A single, continuous conductor core, such as one composed of copper or aluminium, is the feature of solid cables. Usually, a single wire with a constant cross-sectional area is used to make up the conductor.
- **Applications:** Electrical wiring for residential, commercial, and industrial buildings frequently uses solid cables. Applications, including data transmission, illumination, and power distribution, make use of them. Voice, data, and video signals are also sent over solid cables in telecommunications.
- **Advantages:**
  - ✓ **Simplicity:** Solid cables are easy to make and install because of their straightforward design, which features a single conductor core.
  - ✓ **Cost-Effectiveness:** Solid cables are frequently more affordable than stranded cables because of their more straightforward design.
  - ✓ **Stability:** Besides offering consistent electrical performance, solid cables are less likely than stranded cables to have problems like corrosion or strand breakage.
- **Disadvantages:**
  - ✓ **Flexibility:** Solid cables are less suited for applications requiring regular bending or flexing because they are less flexible than stranded cables.
  - ✓ **Susceptibility to Breakage:** Solid wires may break more easily if they are bent or mechanically stressed repeatedly.
  - ✓ **Limited Ampacity:** Solid cables have less surface area for heat dissipation than stranded cables of the same gauge, so they may have a lower ampacity (current-carrying capability).
- **Sizes and Gauges:** Solid cables come in a range of diameters and gauges to meet diverse voltage and current needs. Their conductor diameter, insulation type, and current-carrying capacity are usually used to classify them.
- **Installation:** Standard wiring procedures are followed while installing solid cables, such as passing the wires through raceways, conduits, or cable trays and terminating them at

switches, outlets, or other equipment. Proper installation methods are necessary to guarantee the electrical system's functionality and safety.

Solid cables are frequently utilized in electrical and telecommunications applications where cost-effectiveness and consistent electrical performance are crucial factors but flexibility is not the main issue.

2.3.8 Oil-Filled Cables:

Oil-filled cables, sometimes called oil-impregnated paper (OIP) cables, are a high-voltage power cable frequently utilized for transmitting power underground and underwater. When submerged in oil, these cables protect the conductors from high voltages and offer insulation. Essential details regarding oil-filled wires are as follows:

- **Construction:** Oil-filled cables comprise layers of paper insulation that have been impregnated with oil around one or more insulated conductor cores. Usually composed of copper or aluminium, the conductors are stranded to lower electrical losses and increase flexibility. To improve the dielectric strength and thermal conductivity of the paper insulation, a unique insulating oil, such as mineral oil or synthetic ester oil, is infused into it.
- **Insulating Oil:** The insulating oil used in oil-filled cables serves multiple purposes:

✓ **Dielectric Strength:** The cable can tolerate high voltages without breaking because of the oil's superior dielectric strength.

✓ **Heat Dissipation:** By aiding in the dissipation of heat produced during operation, the oil keeps the cable operating within safe temperature ranges.

✓ **Pressure Equalization:** The oil keeps the cable's internal pressure constant by adjusting for temperature-related expansion and contraction of the cable's constituent parts.

- **Applications:**

✓ **Underground Transmission:** Underneath power transmission in urban settings with limited space and impracticability of overhead lines is often accomplished with oil-filled cables.

✓ **Submarine Transmission:** Oil-filled cables are also utilized for underwater power transmission to connect offshore wind farms, oil platforms, and island grids to mainland power grids.

✓ **HVDC Transmission:** High-voltage direct current (HVDC) transmission systems can use oil-filled cables to transmit power over great distances with low losses.

- **Advantages:**
- ✓ **High Voltage Capability:** Oil-filled cables are appropriate for high-power, long-distance transmission because they can tolerate extraordinarily high voltages.
- ✓ **Reliability:** Oil insulation offers superior defense against electrical failure, moisture seepage, and external harm, guaranteeing dependable performance throughout an extended lifespan.
- ✓ **Environmental Compatibility:** When it leaks or spills, the oil used in oil-filled cables is usually non-toxic and biodegradable, reducing the environmental impact.
- **Disadvantages:**
- ✓ **Maintenance Requirements:** To guarantee sustained dependability and performance, oil-filled cables need to undergo routine maintenance, which includes oil samples, testing, and replacement.
- ✓ **Cost:** Oil-filled cables can be more costly to produce and install than other high-voltage cable types because they require specific jointing methods and oil-filled terminations.
- ✓ **Environmental Considerations:** Even if the oil utilized in oil-filled cables is non-toxic and biodegradable, there is still a chance of spilling or leakage. Thus, containment and careful handling are necessary to prevent environmental hazards.

A tried-and-true method for transmitting high voltage power in harsh settings like subterranean and underground applications is the oil-filled cables.

2.3.9 Gas-Filled Cables:

As-filled cables, sometimes called gas-insulated cables (GICs), are a class of high-voltage power cables that insulates using a gaseous medium instead of conventional solid or liquid dielectric materials. These cables are made to carry electrical power over great distances with low loss and little effect on the environment. Essential details regarding gas-filled cables are as follows:

- **Construction:** One or more conductor cores encased in a gas-insulated casing make up gas-filled cables. To provide electrical insulation, a high dielectric strength gas, such as sulfur hexafluoride (SF6), or a mixture of gases, is pumped into the enclosure. The gas medium insulates the conductors, normally composed of copper or aluminum, from the enclosure and one another.

- **Insulating Gas:** The insulating gas used in gas-filled cables serves multiple purposes:
  - ✓ **Dielectric Strength:** Because of the gas's superior dielectric strength, the cable can tolerate high voltages without malfunctioning.
  - ✓ **Thermal Conductivity:** The gas aids in dissipation heat produced during operation, guaranteeing that the cable stays within safe temperature ranges.
  - ✓ **Chemical Stability:** Because of the gas's chemical inertness and lack of reactivity, stability and dependability are maintained over time.
- **Applications:**
  - ✓ **High-Voltage Transmission:** Gas-filled cables are frequently utilized for high-voltage power transmission over long distances, such as connecting massive power grids, transferring power from distant producing stations, and integrating renewable energy sources into the grid.
  - ✓ **Urban Areas:** In densely populated metropolitan areas, when space is limited and overhead lines are either unfeasible or unappealing, gas-filled cables can be a suitable solution.
  - ✓ **Underground and Submarine Transmission:** For uses including offshore wind farms, underwater power transmission, and urban distribution networks, gas-filled cables can be buried or erected.
- **Advantages:**
  - ✓ **High Voltage Capability:** Gas-filled cables are appropriate for high-power, long-distance transmission because they can tolerate extraordinarily high voltages.
  - ✓ **Low Losses:** Gas insulation lowers the electrical losses in the cable, increasing transmission efficiency and using less energy.
  - ✓ **Compact Design:** Compared to standard cables, gas-filled cables are smaller, making them easier to install in tight locations and requiring less expensive land acquisition.
- **Disadvantages:**
  - ✓ **Complex Design:** Because gas-filled cables require high-pressure gas handling equipment and gas-tight enclosures, they require specific design and production processes.
  - ✓ **Cost:** Compared to other high-voltage cable types, gas-filled cables can be more expensive to produce and install because they require specific tools and materials.
  - ✓ **Environmental Impact:** SF6 gas has a great potential for global warming even though it is non-toxic and non-flammable. Alternative insulating gases with less adverse effect on the environment are being developed.

Gas-filled cables have several benefits for transmitting high-voltage power, such as high efficiency, compact construction, and minimal environmental impact.

2.4 Cable Insulation Materials:

Cable insulation materials are essential to give conductors in different types of cables both mechanical and electrical protection. These materials are chosen with the care of safety, environmental issues, voltage rating, and flexibility requirements in mind. Here are a few typical materials used for cable insulation:

2.4.1 Vulcanized Rubber (VIR):

Vulcanized rubber (Fig.2.11), usually referred to as Vulcanised India Rubber, is an insulation substance frequently used in electrical cables in previous times. The vulcanization process includes heating natural rubber with sulphur or other chemicals to enhance its mechanical qualities and durability.

- **Composition:** Vulcanised rubber consists mainly of natural rubber that has been through the vulcanization process. To improve its characteristics, sulfur is commonly included in the rubber compound during vulcanization, along with accelerators and antioxidants.
- **Properties:** Vulcanised rubber insulation provides excellent electrical insulation, flexibility, and resistance to moisture, heat, and chemicals. It is sturdy and resistant to mechanical stress and abrasion.
- **Applications:** Power distribution and electrical wiring in buildings, factories, and other outdoor locations sometimes used vulcanized rubber-insulated wires. They worked well for uses involving low to medium voltages.
- **Advantages:**
  - ✓ Good Electrical Insulation: To prevent electrical leakage and short circuits, vulcanized rubber is a dependable insulator.
  - ✓ Flexibility: Vulcanised rubber's adaptability made it a good fit for uses requiring cables to flex or bend.
  - ✓ Durability: The vulcanized rubber insulation helped the cable last longer by protecting it from elements like heat, chemicals, and moisture.

- **Disadvantages:**
  - ✓ Aging: Insulation of vulcanized rubber can deteriorate with time if subjected to heat, oxygen, or mechanical stress. The insulating characteristics may be compromised, and the likelihood of electrical failure may increase.
  - ✓ Limited Temperature Range: Due to its narrow temperature range, vulcanized rubber is not an appropriate insulating material for use in environments with high temperatures.
  - ✓ Environmental Concerns: The production and disposal of vulcanized rubber, composed of natural rubber and additives, can harm the environment.
- **Replacement:** There has been a decrease in the use of vulcanized rubber-insulated cables due to the emergence of synthetic insulation materials, including PVC, XLPE, and EPR. These synthetic materials are more eco-friendly than vulcanized rubber, perform better, and are more reliable.

Modern synthetic insulation materials provide more excellent performance and environmental qualities, primarily replacing vulcanized rubber-insulated wires, which were once widely employed in electrical installations.

Figure 2.11 Vulcanized rubber

2.4.2 Polyvinyl Chloride (PVC):

Polyvinyl chloride (Fig.2.12) is a commonly utilized synthetic polymer known for its exceptional electrical, mechanical, and chemical qualities, making it ideal for insulation and sheathing in electrical cables. Here are some essential aspects of PVC regarding electrical wires:

- **Composition:** PVC is a thermoplastic polymer produced through the polymerization of vinyl chloride monomers. PVC is created via polymerization, in which vinyl chloride molecules (CH2=CHCl) are linked together chemically to create extended chains of PVC molecules.

- **Properties:**
  - ✓ **Electrical Insulation:** PVC possesses outstanding electrical insulation characteristics, making it ideal for application in electrical cables to avoid current leakage and provide secure power transfer.
  - ✓ **Mechanical Strength:** PVC provides mechanical solid strength and durability, shielding the conductors from external impact, abrasion, and mechanical stress.
  - ✓ **Chemical Resistance:** PVC has resistance to a broad spectrum of substances such as acids, alkalis, oils, and solvents, rendering it appropriate for usage in severe conditions.
  - ✓ **Fire Resistance:** PVC can be modified to comply with fire safety standards by including flame retardant chemicals, rendering it appropriate for fire resistance situations.
  - ✓ **Weather Resistance:** PVC is resistant to UV radiation and weathering, making it excellent for outdoor applications exposed to sunshine and environmental conditions.
- **Applications:** PVC is widely used in the insulation and sheathing of electrical cables for various applications, including:
  - ✓ Building Wiring: PVC-insulated cables are utilized for power distribution, lighting, and wiring in residential, commercial, and industrial structures.
  - ✓ Underground Cables: PVC sheathing shields cables buried underground from moisture and mechanical harm.
  - ✓ Outdoor Cables: PVC wrapping shields outside cables from sunshine, moisture, and environmental elements.
  - ✓ Flexible Cables: PVC can be tailored to provide pliable cables for applications that necessitate flexibility, like portable tools, appliances, and automotive wiring.
  - ✓ Low-Voltage and Medium-Voltage Cables: PVC is appropriate for use in low-voltage (LV) and medium-voltage (MV) cables, usually up to 33 kV.
- **Manufacturing Process:** PVC cables are produced by extruding PVC compounds through an extrusion process, in which the molten PVC material is pushed through a die to create the appropriate shape of the cable insulation or sheathing.
- **Standards and Regulations:** PVC cables must adhere to applicable norms and laws for electrical installations, safety, and environmental sustainability. Standards can differ based

on the country and location, typically established by organizations like the International Electrotechnical Commission (IEC) and national regulatory authorities.

PVC is a flexible and cost-effective material used for insulation and sheathing in electrical cables. It provides a mix of electrical performance, mechanical strength, chemical resistance, and fire safety features that make it acceptable for various applications.

Figure 2.12 Polyvinyl chloride

2.4.3 Cross-Linked Polyethylene (XLPE):

Cross-linked polyethylene (Fig.2.13) is a thermosetting polymer commonly utilized as insulation in electrical cables because of its exceptional electrical, mechanical, and thermal characteristics. Here are some essential aspects of XLPE insulation:

- **Composition:** XLPE is created by crosslinking polyethylene molecules to create a three-dimensional network structure. This is commonly accomplished using a process known as chemical cross-linking, in which the polyethylene molecules are subjected to high-energy radiation or chemical agents like peroxide compounds.
- **Properties:**
  - ✓ **Electrical Insulation:** XLPE provides exceptional electrical insulating characteristics, including a solid dielectric property and minimal dielectric losses, making it ideal for high-voltage uses.
  - ✓ **Mechanical Strength:** XLPE insulation offers strong mechanical resilience and durability, withstanding deformation, abrasion, and mechanical stress during installation and operation.
  - ✓ **Thermal Stability:** XLPE exhibits excellent thermal stability, allowing it to endure high temperatures without notable deterioration, making it appropriate for use in low and high-temperature scenarios.

- ✓ **Chemical Resistance:** XLPE is impervious to moisture, chemicals, oils, and solvents, safeguarding the conductors from environmental elements and chemical contact.
- ✓ **Weather Resistance:** XLPE insulation is UV radiation and weathering-resistant, making it ideal for outdoor use in areas exposed to sunshine and environmental elements.

- **Applications:** XLPE insulation is widely used in various types of electrical cables for different applications, including:
  - ✓ Power Cables: XLPE-insulated power cables transmit and distribute medium to high-voltage electricity, whether underground or overhead.
  - ✓ Submarine Cables: XLPE insulation offers superior water resistance, making it ideal for undersea power cables utilized in offshore wind farms, island systems, and interconnections.
  - ✓ Instrumentation Cables: XLPE-insulated cables are utilized for signal transmission and control, offering dependable insulation and shielding characteristics.
  - ✓ Communication Cables: XLPE insulation is suitable for communication cables for data transfer due to its minimal signal attenuation and outstanding electrical performance.
- **Manufacturing Process:** XLPE cables are produced by extruding the XLPE compound onto the conductor core to create the insulation layer. The XLPE insulation is cross-linked after extrusion to attain the appropriate characteristics.
- **Standards and Regulations:** XLPE cables need to adhere to applicable norms and regulations for electrical installations, safety, and performance. Standards may differ based on the country and location, typically encompassing criteria established by organizations like the International Electrotechnical Commission (IEC) and national regulatory authorities.

XLPE insulation is used for electrical cables in different applications because of its exceptional electrical qualities, mechanical strength, thermal stability, and resistance to external influences.

Figure 2.13 Cross-linked polyethylene

2.4.4 Ethylene Propylene Rubber (EPR):

Ethylene propylene rubber (Fig.2.14) is a synthetic rubber frequently utilized as insulation in electrical cables because of its outstanding electrical, mechanical, and thermal characteristics. Here are some essential aspects of EPR insulation:

- **Composition:** EPR is a copolymer composed of ethylene and propylene, plus a minor quantity of diene monomer to aid cross-linking. Cross-linking generates a three-dimensional network structure, enhancing the material's mechanical and thermal properties.
- **Properties:**
  - ✓ **Electrical Insulation:** EPR provides outstanding electrical insulating characteristics, including high dielectric strength and low dielectric losses, making it ideal for medium and high-voltage uses.
  - ✓ **Mechanical Strength:** EPR insulation offers mechanical solid strength and flexibility, which helps it withstand deformation, abrasion, and mechanical stress during installation and operation.
  - ✓ **Thermal Stability:** EPR exhibits excellent thermal stability, maintaining its integrity at high temperatures without notable deterioration, rendering it appropriate for use in both low and high-temperature settings.
  - ✓ **Chemical Resistance:** EPR is impervious to moisture, chemicals, oils, and solvents, shielding the conductors from ambient elements and chemical contact.
  - ✓ **Weather Resistance:** EPR insulation is UV radiation and weathering-resistant, making it ideal for outdoor applications exposed to sunlight and environmental conditions.

- **Applications:** EPR insulation is widely used in various types of electrical cables for different applications, including:
  - ✓ Power Cables: EPR-insulated power cables transmit and distribute medium to high-voltage power, whether underground or overhead.
  - ✓ Submarine Cables: EPR insulation has exceptional water resistance, making it ideal for undersea power cables utilized in offshore wind farms, island systems, and interconnections.
  - ✓ Portable Cables: EPR-insulated cables are utilized in portable power distribution, temporary installations, and industrial applications that necessitate flexibility and durability.
  - ✓ Mining Cables: EPR insulation is utilized in underground mining cables to protect against abrasion, chemicals, and mechanical strain.
- **Manufacturing Process:** EPR cables are produced by extruding the EPR compound onto the conductor core to create the insulation layer. The EPR insulation is cross-linked after extrusion to attain the appropriate characteristics.
- **Standards and Regulations:** EPR cables must adhere to applicable norms and regulations for electrical installations, safety, and performance. These standards may differ based on the country and location, typically encompassing regulations established by organizations like the International Electrotechnical Commission (IEC) and national regulatory authorities.

EPR insulation is commonly used for electrical cables because of its exceptional electrical qualities, muscular mechanical strength, thermal stability, and resistance to external influences.

Figure 2.14 Ethylene propylene rubber

2.4.5 Silicone Rubber:

Silicone rubber (Fig.2.15) is an artificial elastic material recognized for its superb electrical insulating characteristics, elevated thermal durability, and capacity to withstand extreme temperatures and environmental factors. Here are some essential aspects of silicone rubber:

- **Composition:** Sulphur, carbon, hydrogen, and oxygen atoms come together to form the synthetic polymer known as silicone rubber. Polymerization of siloxane monomers, which typically consist of silicon and oxygen atoms bound to organic groups, is the common process for creating silicone.
- **Properties:**
  - ✓ **Electrical Insulation:** Silicone rubber provides exceptional electrical insulating characteristics, including high dielectric strength and minimal dielectric losses, making it ideal for high-voltage uses.
  - ✓ **Thermal Stability:** Silicone rubber can withstand a broad range of temperatures, usually from -50°C to 200°C or above, with minimal degradation. It maintains its flexibility and mechanical characteristics across a wide temperature range.
  - ✓ **Flexibility:** Silicone rubber is pliable and durable, enabling it to endure mechanical strain, vibration, and motion without fracturing or shattering.
  - ✓ **Chemical Resistance:** Silicone rubber is impervious to moisture, chemicals, oils, and solvents, making it ideal for challenging and outdoor situations.
  - ✓ **Weather Resistance:** Silicone rubber strongly resists UV radiation, ozone, and weathering, making it ideal for long-term exposure to sunshine and outdoor environments.
  - ✓ **Flame Retardancy:** Silicone rubber can be designed to fulfill flame retardant standards, providing fire resistance and self-extinguishing characteristics.
- **Applications:** Silicone rubber is used in a wide range of electrical and electronic applications, including:
  - ✓ High-Voltage Insulation: Many electrical components, including high-voltage power cables, transformers, and switchgear, employ silicone rubber as an insulating material.
  - ✓ Insulating Sleeves and Boots: Cable terminations, connectors, and splices are mechanically protected and electrically insulated with silicone rubber boots and sleeves.

✓ Electrical Encapsulation: Silicone rubber offers environmental protection, moisture resistance, and insulation when used to pot and encapsulate electrical components and circuits.

✓ Thermal Management: For electronic device and system thermal management, silicone rubber finds utility in gaskets, seals, and thermal interface materials.

✓ Automotive and Aerospace: Silicone rubber's excellent temperature resistance and dependability make it a popular choice for various automotive and aerospace components, gaskets, and wiring harnesses.

- **Manufacturing Process:** Compression moulding, extrusion, or liquid injection moulding are the most common methods for producing silicone rubber goods, though they might vary according to the needs of the product and its intended use.
- **Standards and Regulations:** Regarding electrical insulation, safety, and performance, silicone rubber goods and materials must adhere to all applicable rules and regulations. Organizations like the International Electrotechnical Commission (IEC) and Underwriters Laboratories (UL) establish requirements that are typically included in these standards. However, they may differ by application and industry.

Silicone rubber is the way to go because of its exceptional electrical, thermal, and mechanical qualities when it comes to electrical and electronic applications that need reliability, long-term performance, and high performance.

Figure 2.15 Silicone rubber

2.4.6 Mineral-Insulated Cable (MIC):

Mineral-Insulated Cable (MIC), commonly referred to as "MI cable" or "Pyro cable," is an electrical cable recognized for its robust build and outstanding fire resistance characteristics. Here are some essential aspects of mineral-insulated cable:

- **Construction:** MIC comprises one or more conductors, usually copper, enclosed by a densely packed mineral insulator, typically magnesium oxide (MgO), inside a continuous metal covering, either copper or stainless steel. The conductors and the metal sheath are isolated from each other by the mineral insulation.

- **Properties:**
  - ✓ **Fire Resistance:** Mineral-insulated cables are fire-resistant because the mineral insulation and metal sheath are non-combustible. They can endure very high temperatures without burning or releasing harmful gases, making them ideal for crucial applications where fire safety is essential.
  - ✓ **High-Temperature Rating:** MICs can function at elevated temperatures, usually reaching 250°C or above, depending on the type of insulation material, without experiencing a decline in electrical characteristics.
  - ✓ **Mechanical Strength:** The metal sheath offers mechanical protection to the conductors, making MICs highly resilient against mechanical damage, abrasion, impact, and crushing.
  - ✓ **Chemical Resistance:** Mineral insulation is impervious to moisture, chemicals, oils, and solvents, shielding the conductors from ambient elements and chemical contact.
  - ✓ **Electrical Performance:** MICs provide exceptional electrical insulation characteristics, including high dielectric strength and minimal dielectric losses, which make them ideal for high-voltage uses.
  - ✓ **Longevity:** Mineral-insulated cables are recognized for their longevity, reliability, and resilience, especially in challenging environments.
- **Applications:** Mineral-insulated cables are utilized in diverse sectors and applications that require fire resistance, high-temperature tolerance, and durability, including:
  - ✓ Fire Alarm Systems: Microphone-based Intelligent Control Systems (MICs) are frequently utilized in buildings, tunnels, and other facilities to ensure fire safety through alarm and detection systems.
  - ✓ Emergency Lighting: MICs offer dependable power distribution for emergency lighting systems that must stay functional during fires or other emergencies.
  - ✓ High-Temperature Environments: Microprocessor-based integrated circuits (MICs) are utilized in industrial applications, ovens, furnaces, and kilns operating at temperatures above the capabilities of standard cables.
  - ✓ Hazardous Locations: MICs are appropriate for use in dangerous conditions with flammable gases, vapours, or dust since they do not promote fire spread.

- **Installation:** Specialised tools and techniques are necessary for installing mineral-insulated cables because of their stiff design and metal wrapping. Manipulating MICs is usually more complex than working with flexible wires, necessitating meticulous preparation and adequate training for installation.
- **Standards and Regulations:** Mineral-insulated cables need to adhere to applicable norms and regulations for electrical installations, fire safety, and performance. These standards might differ based on the application and industry, typically encompassing criteria established by organizations like the International Electrotechnical Commission (IEC), Underwriters Laboratories (UL), and local regulatory bodies.

Mineral-insulated cables provide exceptional fire resistance, high heat resistance, and robust mechanical properties, making them well-suited for crucial applications that require safety, dependability, and durability.

2.4.7 Oil-Impregnated Paper (OIP):

Oil-impregnated paper (Fig.2.16) is an insulating material in high-voltage electrical cables and power transformers. Here are some essential aspects of oil-impregnated paper insulation:

- **Composition:** Oil-impregnated paper is made of cellulose sheets soaked in a carefully designed insulating oil, usually mineral or synthetic ester oil. The paper is a stable dielectric, while the oil offers further insulation and thermal conductivity.
- **Properties:**
  - ✓ **Dielectric Strength:** Oil-impregnated paper has a high dielectric strength, enabling it to endure muscular electric field strains without breaking down.
  - ✓ **Thermal Conductivity:** The impregnating oil improves the thermal conductivity of the insulation, aiding in heat dissipation and enhancing the cable's or transformer's thermal efficiency.
  - ✓ **Moisture Resistance:** The impregnating oil serves as a barrier to keep moisture from entering, therefore protecting the cellulose paper from collecting moisture and preserving the insulation's dielectric qualities.
  - ✓ **Chemical Stability:** Oil-impregnated paper is chemically stable and resistant to oxidation, ensuring long-term stability and reliability of the insulation system.
  - ✓ **Flexibility:** Oil-impregnated paper is pliable and may be readily shaped to match cable or transformer parts' precise form and structure.

- ✓ **Self-Healing:** Impregnating oil can move to the location of a breakdown or partial discharge and repair the insulation, decreasing the chance of additional damage.

- **Applications:** Oil-impregnated paper insulation is frequently utilized in high-voltage power lines and transformers for:

- ✓ Power Transmission: OIP-insulated cables transmit power over long distances at high voltages, usually between 66 kV to 400 kV or more.
- ✓ Power Distribution: OIP-insulated transformers increase or decrease voltage levels in electrical distribution networks, substations, and industrial facilities.
- ✓ Submarine Cables: Submarine cables insulated with oil-impregnated paper (OIP) are utilized for undersea power transmission, connecting islands, and offshore wind farms because of their exceptional resistance to moisture and high reliability.

- **Manufacturing Process:** Oil-impregnated paper insulation is created by saturating cellulose paper sheets with insulating oil using vacuum or pressure impregnation methods. The paper infused with oil is dried to eliminate surplus moisture and guarantee even oil dispersion.

- **Standards and Regulations:** Oil-impregnated paper insulation must meet all electrical insulation, safety, and performance requirements. While specific standards may differ by use case and sector, they frequently incorporate criteria established by groups like the IEEE, the International Electrotechnical Commission (IEC), and national regulatory bodies.

The oil-impregnated paper insulation has exceptional electrical and thermal characteristics, making it ideal for high-voltage applications that require reliability, durability, and long-term performance.

2.4.8 Gas (e.g., SF_6):

Sulphur hexafluoride (SF_6) is frequently used in gas-insulated cables (GICs) because of its exceptional dielectric properties and insulating solid capabilities. Here are some essential aspects regarding SF_6 in gas-insulated cables:

- **Dielectric Properties:** SF_6 gas possesses outstanding dielectric characteristics, enabling it to endure high electric field pressures without breaking down. SF_6's properties make it a perfect insulating medium for gas-insulated cables, allowing for compact designs and efficient electrical power transmission.

- **Insulating Strength:** SF_6 gas offers superior insulation properties, allowing for the creation of gas-insulated cables that are smaller and lighter than traditional cables. This reduces material expenses and simplifies installation, particularly in metropolitan regions with limited space.
- **Environmental Considerations:** SF_6 gas has excellent insulating characteristics but is also a potent greenhouse gas with a high global warming potential (GWP). Efforts are being made to reduce SF_6 emissions and provide alternative insulation solutions with less environmental impact. Recycling and recovering SF6 gas from defunct equipment is done to address environmental issues.
- **Applications:** Gas-insulated cables (GICs) are utilized for transmitting and distributing high-voltage electrical power, particularly in highly populated urban areas, subterranean installations, and locations with restricted space. SF_6 gas-insulated cables provide benefits, including decreased electromagnetic interference, reduced transmission losses, and enhanced dependability compared to traditional overhead lines or air-insulated cables.
- **Manufacturing Process:** Gas-insulated cables are composed of one or several conductors in a sealed metal tube filled with SF_6 gas. The conductors are commonly composed of copper or aluminium, with the outside metal tube offering mechanical protection and functioning as an earth conductor. SF_6 gas functions as an insulator and a coolant, dispersing heat produced during operation.
- **Maintenance and Monitoring:** Gas-insulated wires need regular maintenance and monitoring to guarantee optimal performance and dependability. This may involve conducting inspections to detect leaks, verifying the quality of seals and connections, and monitoring gas pressure and purity levels in the cable system. Gas chromatography and infrared spectroscopy are advanced diagnostic methods to identify and analyze SF_6 gas leaks.

SF_6 gas-insulated cables provide notable benefits in electrical performance, reliability, and compact design, making them appropriate for various high-voltage transmission and distribution uses. Efforts to reduce SF6 emissions and deal with environmental issues related to its use are now underway.

Figure 2.16 Oil-impregnated paper

2.4.9 Polyethylene (PE):

Polyethylene (Fig.2.17) is a flexible thermoplastic polymer commonly utilized in various applications, such as providing electrical insulation for cables. Below are some essential aspects of polyethylene:

- **Chemical Structure:** Polyethylene is a polymer made up of extended chains of ethylene monomers. It is categorized into many varieties according to its density and molecular structure, such as low-density polyethylene (LDPE), high-density polyethylene (HDPE), and linear low-density polyethylene (LLDPE).
- **Properties:**

✓ **Electrical Insulation:** Polyethylene has excellent dielectric characteristics, making it appropriate for electrical insulation applications. It possesses a high dielectric strength and low dielectric constant, enabling it to endure high voltages without breakdown.

✓ **Mechanical Strength:** Polyethylene is recognized for its exceptional tensile strength, toughness, and flexibility. It is capable of enduring mechanical stress and deformation, making it appropriate for use in bent and handled cables.

✓ **Chemical Resistance:** Polyethylene is impervious to moisture, chemicals, acids, and alkalis, shielding the conductors from ambient elements and chemical contact.

✓ **Thermal Stability:** Polyethylene exhibits excellent thermal stability and can function throughout a broad temperature span, usually ranging from -50°C to 80°C or above, without notable deterioration of its electrical characteristics.

- ✓ **Low Water Absorption:** Polyethylene's low water absorption qualities enhance its electrical stability and resistance to breakdown caused by moisture.
- ✓ **UV Resistance:** Some forms of polyethylene, including HDPE, have strong resistance to UV radiation, making them appropriate for outdoor use in areas where exposure to sunshine is a factor.

- **Applications:**

- ✓ **Electrical Cables:** Polyethylene is commonly used as insulation in power, telecommunications, and control cables. It offers dependable electrical insulation, mechanical safeguarding, and longevity for conductors.
- ✓ **Pipe and Tubing:** Polyethylene produces pipes and tubing for water distribution, gas distribution, irrigation, and other fluid transport purposes. HDPE and MDPE are frequently utilized for these applications because of their exceptional chemical resistance and long-lasting properties.
- ✓ **Packaging:** LDPE and LLDPE are utilized in packaging films, bags, and containers for their flexibility, durability, and resistance to moisture.
- ✓ **Construction:** HDPE is used for geomembranes, drainage pipes, and building envelope insulation because of its durability, chemical resistance, and UV stability.
- ✓ **Consumer Products:** Polyethylene is utilized in a wide range of consumer goods such as bottles, toys, household items, and automobile parts because of its versatility, affordability, and ease of manufacturing.

- **Manufacturing Process:** Polyethylene is commonly manufactured by polymerizing ethylene monomers through high-pressure or low-pressure polymerization methods. The polymerization process can be adjusted to regulate the molecular weight, density, and other characteristics of the resultant polyethylene.
- **Standards and Regulations:** Polyethylene materials in electrical cables must adhere to applicable norms and regulations for electrical insulation, safety, and performance. The standards might differ based on the application and locality, typically established by organizations like the International Electrotechnical Commission (IEC), Underwriters Laboratories (UL), and national regulatory bodies.

Polyethylene is a commonly used insulating material in electrical cables due to its blend of electrical qualities, mechanical strength, chemical resistance, and cost-effectiveness, making it appropriate for various applications. The choice of cable insulation material is based on application requirements, voltage rating, environmental conditions, regulatory standards, and cost concerns.

Figure 2.17 Polyethylene

2.4.10 Selection of Suitable Cable Size and Type From Standard Data:

Choosing the appropriate cable size and type requires evaluating factors such as current-carrying capability, voltage rating, installation conditions, and regulatory standards. Here is a standard procedure for choosing the size and kind of cable based on standard data.

More on Indian Standard

- **Determine Load Characteristics:**
  - ✓ Identify the electrical load parameters such as rated voltage, current, power factor, and duty cycle. This information aids in determining the cable's maximum current capacity.
- **Calculate Current-Carrying Capacity:**
  - ✓ Consult standard tables or calculation methods to establish the cable's current-carrying capacity considering installation variables such as ambient temperature, cable installation method (e.g., buried, ducts, overhead), and grouping factors.
  - ✓ Consider issues such as derating for ambient temperature, burial depth (subterranean cables), and cable bundling (for many cables connected).

- **Select Cable Size:**
  - ✓ Choose a cable size that meets or surpasses the calculated current-carrying capacity, considering aspects like voltage drop, short-circuit withstand capability, and potential future load growth.
  - ✓ Consult standard cable sizing tables from regulatory organizations or cable manufacturers to choose the correct cable size according to the calculated current needs.
- **Consider Voltage Rating:**
  - ✓ Make sure the chosen cable has a sufficient voltage rating to endure the working voltage of the electrical system. Consider elements such as phase-to-phase voltage, phase-to-ground voltage, and transient overvoltage circumstances.
- **Evaluate Environmental Conditions:**
  - ✓ Consider the installation environment, considering aspects such as temperature, humidity, moisture exposure, chemical exposure, mechanical stress, and UV radiation.
  - ✓ Choose a cable type with insulation and sheathing materials that offer protection from environmental dangers and assure long-term reliability.
- **Compliance with Standards and Regulations:**
  - ✓ Make sure the chosen cable type and size adhere to the appropriate standards and laws for electrical installations, including national electrical codes, international standards like IEC and IEEE, and local legislation.
  - ✓ Ensure that the chosen cable complies with the criteria for fire resistance, flame retardancy, smoke emission, and toxicity (where relevant).
- **Consult Manufacturer's Data:**
  - ✓ Consult manufacturer's datasheets, product catalogs, and technical specifications for in-depth details regarding cable construction, electrical properties, performance characteristics, and installation guidelines.
  - ✓ When choosing a cable type, consider elements including cable flexibility, bending radius, pulling tension, and terminating methods.
- **Review Project Requirements:**
  - ✓ Consider project-specific criteria such as budget limitations, installation schedules, spare parts availability, and compatibility with current infrastructure.
  - ✓ Consult with electrical experts, contractors, and stakeholders to confirm that the chosen cable aligns with the project's technical and operational needs.

By adhering to these procedures and utilizing standardized data, individuals may accurately choose the appropriate cable size and type for a particular application, guaranteeing dependable and secure electrical setups.

2.5 Cable Jointing Methods:

Cable jointing is connecting two cable ends or splicing cables to form an uninterrupted electrical connection. Ensuring the integrity and safety of the electrical system necessitates correct cable jointing. Various procedures are used for cable jointing, with the specific method chosen according to the type of cable, its intended application, and the environmental circumstances. Here are some standard methods for cable splicing:

- **Compression Joint:**
  - ✓ **Description:** Compression joints link cable ends by compressing a connector onto the cable with specialized crimping tools.
  - ✓ **Applications:** Typically utilized for medium and low voltage wires.
- **Soldered Joint:**
  - ✓ **Description:** Soldering is melting solder and applying it to the cable joint to establish a reliable connection.
  - ✓ **Applications:** Designed for use with low voltage and signal cables.
- **Cold Shrink Joint:**
  - ✓ **Description:** Cold shrink couplings utilize pre-expanded rubber or silicone tubes that shrink around the cable joint when the supporting core is removed, ensuring a tight fit.
  - ✓ **Applications:** Frequently used for medium-voltage wires.
- **Heat Shrink Joint:**
  - ✓ **Description:** Heat shrink joints utilize heat shrink tubing that contracts when exposed to heat, offering insulation and safeguarding the cable joint.
  - ✓ **Applications:** Appropriate for cables ranging from low to high voltage.
- **Resin/Potting Joint:**
  - ✓ **Description:** The cable joint is encapsulated with a resin or potting material that solidifies for insulation and safeguarding.
  - ✓ **Applications:** It can be utilized for low and high-voltage cables, suitable for outdoor and underground applications.

- **Slip-On Joint (Prefabricated Joint):**

✓ **Description:** Prefabricated joints are pre-made joints placed on cable ends and secured with mechanical connectors.

✓ **Applications:** Typically utilized for medium and high voltage cables.

- **Taped Joint:**

✓ **Description:** Taped junctions entail using insulating tape, typically with glue or mastic, to fasten and insulate the cable joint.

✓ **Applications:** It is ideal for use with low and medium-voltage cables.

- **Stress Cone Joint:**

✓ **Description:** Stress cone joints are utilized in high-voltage cables to spread electrical field loads and prevent breakdowns.

✓ **Applications:** Suitable for high voltage applications.

- **Oil-Filled Joint:**

✓ **Description:** Within oil-filled junctions, cable ends are linked in an enclosure filled with insulating oil to offer electrical insulation and cooling.

✓ **Applications:** Typically utilized in high voltage and highly high voltage scenarios.

- **Gas-Insulated Joint:**

✓ **Description:** Gas-insulated joints utilize pressurized gases like sulfur hexafluoride (SF6) for insulating and safeguarding cable joints.

✓ **Applications:** High voltage and ultra-high voltage applications.

- **Crimp Connectors:**

✓ **Description:** Crimp connections are utilized to physically join and fasten wires together with the help of a crimping tool.

✓ **Applications:** Typically used for cables with low to medium voltage.

- **Welding Joint:**

✓ **Description:** Welding is melting and fusing the cable ends to create a seamless junction.

✓ **Applications:** Appropriate for high current applications and specific scenarios.

Criteria including the cable type, voltage level, environmental circumstances, and installation needs determine the choice of a particular jointing method. Ensuring the safety and dependability of cable joints is vital and depends on proper installation and respect for norms and regulations.

2.6 Cable Laying Method:

The cable laying process is crucial in electrical installations, and the method selection is based on criteria including the cable type, the surroundings, and the project's particular needs. Here are some typical cable installation techniques:

- **Direct Burial:**
  - ✓ **Description:** Cables are buried underground without conduits or ducts.
  - ✓ **Applications:** Typical for power distribution and communication cables in regions with appropriate soil conditions.
- **Conduit or Duct Installation:**
  - ✓ **Description:** Cables are enclosed within conduits or ducts for protection and convenient maintenance.
  - ✓ **Applications:** They are utilized in structures, industrial sites, and locations requiring cable safeguarding.
- **Trenching:**
  - ✓ **Description:** Cables are placed in trenches and then covered with soil or other materials.
  - ✓ **Applications:** Appropriate for underground and outdoor installations, particularly in metropolitan settings.
- **Cable Tray Installation:**
  - ✓ **Description:** Cables are contained in cable trays that are mounted on walls or ceilings.
  - ✓ **Applications:** Typically found in industrial and commercial facilities that require an organized cable management system.
- **Aerial Installation:**
  - ✓ **Description:** Cables are suspended above the ground, typically on poles or support structures.
  - ✓ **Applications:** They are typically used for overhead power distribution and telecommunication lines in rural and urban regions.
- **Cable Trenching:**
  - ✓ **Description:** Excavated trenches were created for the specific purpose of installing cables, serving as a dedicated channel.
  - ✓ **Applications:** Power transmission and distribution in expansive outdoor spaces.
- **Ploughing:**

- ✓ **Description:** A plow digs a narrow trench for laying cables, burying the cable as it moves forward.
- ✓ **Applications:** It is ideal for burying cables at shallow depths in agricultural or open-field settings.

Submersible Installation:

- ✓ **Description:** Cables are submerged underwater in rivers, lakes, or oceans and might be buried in the waterbed.
- ✓ **Applications:** They are typically used in undersea power lines to connect offshore wind farms or islands.

Horizontal Directional Drilling (HDD):

- ✓ **Description:** Underground cables are installed by employing drilling equipment to create a horizontal borehole.
- ✓ **Applications:** It is ideal for urban locations where conventional trenching methods may not be feasible.

Jetting or Blowing:

- ✓ **Description:** Installation involves passing cables via pre-existing ducts or conduits using compressed air or water.
- ✓ **Applications:** Used for fiber optic cables and other communication lines.

Cable Pulling:

- ✓ **Description:** Cables are drawn through conduits or ducts with the help of cable-pulling equipment.
- ✓ **Applications:** Typical for indoor installations and structured cabling systems.

Cable Ladder Installation:

- ✓ **Description:** Cables are held in place within cable ladders, creating an elevated route.
- ✓ **Applications:** They are utilized in industrial environments for cable organization and reinforcement.

Project requirements, such as cable type, environmental circumstances, geography, and regulations determine the selection of cable laying method. Adhering to industry standards and best practices is crucial for ensuring cable installations' safety, dependability, and lifespan.

2.7 Factors Determining the Selection of Electric Cables:

Choosing electric cables for a particular use requires evaluating multiple criteria to guarantee the electrical setup's safety, dependability, and effectiveness. These are the primary elements that influence the choice of electric cables:

- ✓ **Voltage Rating:** The cable's voltage rating should be equal to or higher than the voltage of the electrical system where it will be utilized. Cables are built explicitly for low voltage (LV), medium voltage (MV), and high voltage (HV) applications.
- ✓ **Current Carrying Capacity:** The cable must handle the highest current anticipated in the system without surpassing its thermal thresholds. Factors like conductor size, insulation material, and installation conditions determine the current carrying capacity.
- ✓ **Conductor Material:** The selection of conductor material, such as copper or aluminum, is based on conductivity, cost, weight, and corrosion resistance considerations. Copper is a superior conductor but comes at a higher price compared to aluminum.
- ✓ **Insulation Type:** Insulation material is crucial for preventing electrical leakage and safeguarding the cable from environmental elements. Various insulating materials such as PVC, XLPE, EPR, and others possess distinct qualities tailored for specific uses.
- ✓ **Temperature Rating:** Cables need a temperature rating that matches the ambient temperature and the temperature produced during operation. Increased temperature ratings are typically required for applications in high ambient temperatures or with electrical solid loads.
- ✓ **Environmental Conditions:** Factors including moisture, chemicals, sunshine, and abrasion are essential in deciding the appropriate cable type and sheathing material. Cables with improved durability in extreme conditions may be necessary for challenging settings.
- ✓ **Mechanical Strength:** Cables must possess adequate mechanical strength to endure physical loads encountered during installation, operation, and maintenance. Armored wires or cables with reinforced sheathing can be utilized for increased protection.
- ✓ **Flexibility:** Cable flexibility is essential, particularly in situations requiring bending or flexing. Flexible cables are utilized in scenarios where rigidity is not feasible.
- ✓ **Fire Performance:** Fire-resistant or fire-retardant cables, such as buildings and essential infrastructure, are necessary in some situations to meet safety rules and inhibit the spread of fire.

- ✓ **Cable Size and Type:** Considering the cable's cross-sectional area and current-carrying capacity dictates its size. The purpose of the cable determines whether it is a power cable, control cable, or communication cable.
- ✓ **Regulatory Compliance:** Following all applicable rules and regulations, both domestic and foreign, is paramount. Groups like the National Electrical Code (NEC) and the International Electrotechnical Commission (IEC) may establish regionally distinct standards.
- ✓ **Installation Method:** The mode of cable installation, such as underground, overhead, conduits, or within buildings, impacts the choice of cables. Various installation methods may require insulation, sheathing, and protection.
- ✓ **Budgetary Constraints:** Cost is a crucial factor in choosing a cable. Balancing performance and cost to adhere to project budget limitations is crucial.
- ✓ **Expected Lifespan:** Considering the cable's estimated lifespan is necessary, especially in sensitive applications where long-term reliability is essential.
- ✓ **Maintenance Requirements:** Specific cables may necessitate minor upkeep, while others can require regular checks or replacements. Maintainability is a consideration in choosing cables.
- ✓ **Corrosion Resistance:** Choosing cables with suitable corrosion-resistant materials is crucial in corrosive situations to guarantee sustained performance over time.
- ✓ **Future Expansion and Upgrades:** It is considering possible future expansions or system upgrades aids in choosing cables that can handle higher loads or modifications in the electrical system.
- ✓ **Technical Support and Documentation:** Access to technical help from the cable manufacturer and detailed documentation, such as specs and installation guides, is essential for selecting and installing cables correctly.

A thorough assessment of these elements is crucial for making well-informed selections when choosing cables. Collaborating with electrical engineers, evaluating project specifications, and following relevant standards are essential to guarantee that the selected cables satisfy the precise requirements of the electrical system.

Solved Numerical Problem

Example 2.1:

Find the voltage drop in a wire

Given:

Current flowing through the wire, I=10 A

Resistance of the wire, R=2 ohms

Length of the wire, L=50 meters

Solution:

Using Ohm's Law, $V=I\times R=10\times2=20$ volts

So, the voltage drop across the wire is 20volts

Example 2.2:

Calculate the power loss in a wire

Given:

Current flowing through the wire, I=5 A

Resistance of the wire, R=4 ohms

Time, t=10 hours

Solution:

Using the formula for power loss, $P=I^2R=(5)^2\times4=100$ watts

So, the power loss in the wire is 100 watts.

Example 2.3:

Determining Voltage Drop in Underground Cable

Given:

Current flowing through the underground cable, I=20 A

Resistance of the cable, R=0.5 ohms per kilometer

Length of the cable, L=2 kilometers

Solution:

First, calculate the total resistance of the cable: $R_{total}=R\times L=0.5\times2=1$ohm

Then, Applying Ohm's Law

$$V=I\times R_{total}=20\times1=20\text{volts}$$

So, the voltage drop across the underground cable is 20 volts.

Example 2.4:

Calculate cable size

Given:

Maximum current the cable should carry, I=30 A

Maximum allowable voltage drop, V_d =3 volts

Resistance of the cable per meter, R=0.1 ohms per meter

To find the minimum cross-sectional area of the cable.

Solution:

First, calculate the maximum allowable resistance of the cable: $R\text{max}=V_d/I=3/30=0.1$ohms

Then, the minimum cross-sectional area can be calculated using the formula

A=R / ($\rho\times L$), where ρ is the material's resistivity and L is the length.

$A=R\text{max} / (\rho\times L)$

$=0.1/L\times0.1=1/L$

So, the minimum cross-sectional area of the cable is 1/L.

Example 2.5:

Calculate the resistance of a copper wire with a diameter of 1.5 mm and a length of 100 meters. The resistivity of copper is $1.68\times10^{-8}\Omega\cdot m$.

Solution:

Given: Diameter (d) = 1.5 mm = 0.0015 m

Length (L) = 100 m

Resistivity (ρ) = $1.68\times10^{-8}\ \Omega\cdot m$

The cross-sectional area (A) of the wire can be calculated using the formula for the area of a circle:

$A=\pi d^2/4$

$A=\pi\ (0.0015)^2/4\approx1.767\times10^{-6}m^2$

Now, we can use the formula for resistance:

$R/A=\frac{1.68\times10^{-8}\times100}{1.767\times10^{-6}}=0.951\Omega$

Therefore, the resistance of the copper wire is approximately 0.951 ohms.

Example 2.6:

A 3-phase cable has a line voltage of 440 volts. Calculate the line current if the power factor is 0.8 and the power transmitted is 50 kW.

Solution:

Given: Line voltage (V) = 440 volts

Power factor (PF) = 0.8

Power transmitted (P) = 50 kW = 50000 watts

We know that:

$P=\sqrt{3}\times V\times I\times PF$, where I is the line current.

Therefore: $I=\frac{P}{\sqrt{3}\times V\times PF}$

$=\frac{50000}{\sqrt{3}\times 440\times 0.8}$

$\approx$85.36 amps

The line current is approximately 85.36 amps.

Example 2.7:

Determine the voltage rating of a cable if its insulation is designed to withstand maximum electric field strength of $20\times10^6 V/m$ and the thickness of the insulation is 0.03m.

Solution:

Given: Maximum electric field strength ($Emax$) = $20\times10^6 V/m$

Insulation thickness (t) = 0.03m

The voltage rating (V) can be calculated using the formula:

$V=Emax\times t$

$V= (20\times10^6) \times 0.03=600{,}000 volts$

Therefore, the voltage rating of the cable is 600,000 volts.

More on Problem

Exercise

Multiple Choice Questions:

1. What is the primary purpose of insulation in electrical wires?

   a) To reduce electrical resistance b) To increase electrical conductivity c) To prevent electric shock and short circuits d) To increase the current-carrying capacity.

2. Which material is commonly used for insulation in electrical wires?

   a) Copper b) Aluminum c) PVC (Polyvinyl Chloride) d) Steel.

3. The resistance of a wire depends on the following:

   a) Length and temperature b) Voltage and temperature c) Current and voltage d) Diameter and material

4. Which of the following factors affects the voltage drop in a wire?

   a) Length of the wire b) Temperature of the wire c) Current flowing through the wired) All of the above

5. Underground cables are preferred over overhead lines for transmission because:

   a) They are cheaper to install, b) They are less affected by weather conditions c) They are easier to maintain d) They have lower resistance.

6. The resistance of a conductor is inversely proportional to its:

   a) Cross-sectional area b) Length c) Temperature d) Voltage.

7. What is the SI unit of resistance?

   a) Ampere (A) b) Ohm (Ω) c) Volt (V) d) Watt (W).

8. In electrical transmission, which parameter is most important to minimize for efficient power delivery?

   a) Voltage b) Current c) Resistance d) Power.

9. Which cable type is commonly used for transmitting high-voltage electricity over long distances?

   a) Twisted pair cable b) Coaxial cable c) Submarine cable d) High-voltage power cable.

10. What is the primary function of an earth wire in electrical circuits?

    a) To carry the load current b) To provide a return path for fault currents c) To regulate the voltage d) To increase the resistance.

11. The formula V=IR represents which law in electricity?

    a) Ampere's Law b) Ohm's Law c) Faraday's Law d) Kirchhoff's Law.

12. Which cable type is most commonly used for low-voltage applications like household wiring?

    a) Twisted pair cable b) Coaxial cable c) Optical fiber cable d) Non-metallic sheathed cable.

13. What happens to the resistance of a wire when its length is doubled?

    a) It doubles b) It becomes half c) It quadruples d) It remains the same.

14. Superconductors are characterized by:

    a) High resistance b) Low resistance c) High voltage d) Low voltage.

15. In which scenario would a larger wire gauge be preferred?

    a) Low current applications, b) High current applications, c) Short distance transmission, and d) High voltage applications.

16. The resistance of a conductor increases with an increase in:

    a) Current b) Voltage c) Temperature d) Cross-sectional area.

17. The purpose of cable ducts in underground cable installations is to:

    a) Protect the cables from moisture, b) Increase the resistance of the cables, c) Provide ventilation for the cables, and d) Facilitate easy maintenance and repair.

18. What is the primary disadvantage of using overhead power lines compared to underground cables?

    a) They are more expensive, b) They are less susceptible to damage, c) They require more maintenance, and d) They cause visual pollution.

19. Which of the following materials is a good conductor of electricity?

    a) Rubber b) Glass c) Copper d) Plastic.

20. What is the purpose of armor in some types of underground cables?

    a) To increase flexibility b) To provide mechanical protection c) To reduce resistance d) To improve insulation

Answer:

1. c) To prevent electric shock and short circuits
2. c) PVC (Polyvinyl Chloride)
3. a) Length and temperature
4. d) All of the above
5. b) They are less affected by weather conditions
6. a) Cross-sectional area

7. b) Ohm (Ω)
8. c) Resistance
9. d) High-voltage power cable
10. b) To provide a return path for fault currents
11. b) Ohm's Law
12. d) Non-metallic sheathed cable
13. a) It doubles
14. b) Low resistance
15. b) High-current applications
16. c) Temperature
17. d) Facilitate easy maintenance and repair
18. c) They require more maintenance
19. c) Copper
20. b) To provide mechanical protection

Short Answer Type Questions:

1) Define electrical resistance and state its SI unit.
2) Explain the concept of voltage drop in electrical wires.
3) What factors affect the resistance of a conductor?
4) Differentiate between overhead power lines and underground cables.
5) Why is insulation important in electrical wiring?
6) Describe the function of an earth wire in electrical circuits.
7) What is the role of armour in underground cables?
8) How does the cross-sectional area of a wire affect its resistance?
9) Define the term "conductor" in the context of electricity.
10) What are the advantages of using underground cables over overhead lines?
11) Explain why high-voltage cables are preferred for long-distance transmission.
12) What is the primary function of cable ducts in underground cable installations?
13) Describe the purpose of sheathing in electrical cables.
14) Explain the significance of wire gauges in electrical wiring.
15) How does temperature affect the resistance of a conductor?

Long Answer Type Questions:

1) Discuss the factors that influence wire size selection and gauge selection in electrical wiring installations. How does the choice of wire size impact the efficiency and safety of the electrical system?
2) Explain the significance of insulation materials in electrical wires. Compare and contrast the properties of commonly used insulation materials such as PVC, rubber, and thermosetting plastics. How does the choice of insulation material affect the performance and longevity of electrical wiring systems?
3) Describe the process of calculating voltage drop in electrical wires. Discuss the factors contributing to voltage drop and explain why it is essential to minimize voltage drop in electrical systems. Provide examples of practical applications where voltage drop calculations are crucial.
4) Underground cables are often preferred for electrical transmission in urban areas. Discuss the advantages and disadvantages of underground cables compared to overhead power lines. Consider cost, installation complexity, maintenance requirements, and environmental impact.
5) Underground cables are subject to various environmental and operational challenges. Discuss the methods and techniques to mitigate these challenges, including cable burial depth, cable protection, and fault detection systems. How do these strategies contribute to the reliability and longevity of underground cable systems?
6) Explain the concept of cable resistance and its impact on electrical transmission efficiency. Describe how cable resistance is influenced by factors such as conductor material, cross-sectional area, length, and temperature. Provide examples to illustrate the calculation of cable resistance in practical scenarios.
7) Faults in underground cables can pose significant challenges for maintenance and repair. Discuss the common types of cable faults, including insulation breakdown, conductor damage, and ground faults. Describe the diagnostic techniques and equipment for detecting and locating cable faults in underground electrical systems.
8) Describe the construction and design features of high-voltage underground cables for long-distance transmission. Discuss the materials, insulation systems, and shielding techniques to ensure the safe and reliable operation of high-voltage cable systems. What are the key considerations in the design of high-voltage cable installations?

9) Cable joints and terminations play a critical role in the integrity and performance of underground cable systems. Discuss the importance of proper cable jointing and termination techniques in maintaining electrical continuity, preventing insulation failures, and ensuring operational safety. Describe the types of cable joints and terminations used in underground cable installations.

10) Discuss the environmental and regulatory considerations associated with the installation and operation of underground cable systems. How do factors such as soil conditions, groundwater levels, and environmental regulations influence underground cable networks' design, installation, and maintenance? Provide examples of sustainable practices and technologies used to minimize the environmental impact of underground cable installations.

Numerical Problems:

1) Current flowing through a wire I=15 A, Resistance of the wire, R=0.2 ohms. Calculate the voltage drop if the wire length is 30 meters.
2) Current flowing through a wire I=8 A, Resistance of the wire R=0.5 ohms. Calculate the power loss in the wire if used for 6 hours.
3) A copper wire has a length of 100 meters and a cross-sectional area of $2mm^2$. Calculate the wire's resistance if copper's resistivity is 1.68×10^{-8} ohm-meter.
4) A current of 25 A flows through an underground cable with a resistance of 0.3 ohms per kilometer. If the cable length is 3 kilometers, calculate the voltage drop.
5) A circuit requires a maximum current of 20 A, and the allowable voltage drop is 22 volts. If the resistance of the cable is 0.1 ohms per meter, what minimum cross-sectional area of cable should be used for a 50-meter length?
6) An aluminum wire has a resistance of 0.05 ohms and a cross-sectional area of 4mm^2. If the resistance per meter is 0.01 ohms, what is the length of the wire?
7) An overhead power line has a resistance of 0.2 ohms per kilometer. If the current flowing through the line is 30 A and the line length is 2 kilometers, calculate the voltage drop.
8) Two wires with 44 ohms and 66 ohms resistances are connected in parallel. Calculate the total resistance of the combination.
9) A copper wire with a resistance of 2 ohms and a length of 50 meters is used in a circuit. If the resistivity of copper is 1.68×10^{-8} ohm-meter, what should be the cross-sectional area of the wire?

10) A cable with a resistance of 0.1 ohms per meter carries a current of 15 A for 8 hours. Calculate the total power loss in the cable.

Unit Summary

The following unit summary of electrical wires and underground cables is discussed here:

- Introduction to electrical wires and cables
- Types of electrical wires and cables
- Insulation and jacketing materials
- Installation of electrical wires and cables
- Underground cable systems
- Cable testing and maintenance
- Safety practices and regulations

This unit provides a comprehensive overview of electrical wires, cables, and underground cable systems, covering their types, installation techniques, maintenance procedures, safety considerations, and future electrical engineering and construction trends.

Practical

1. Practice stripping insulation from wires and terminating them with appropriate connectors (such as terminal blocks or wire nuts).
2. Demonstrate proper techniques for crimping connectors onto wires and ensuring secure connections.
3. Create mock-up scenarios for routing cables in various configurations (e.g., in conduits, cable trays, or along walls).
4. Teach students how to use cable ties, clips, and clamps for organizing and securing cables neatly.
5. Demonstrate insulation resistance testing to ensure the integrity of insulation materials.
6. Simulate underground cable installations using appropriate tools and techniques.
7. Discuss considerations such as trenching, conduit sizing, and backfilling for underground cable systems.

Know More

Sl No	Topic	QR Code
1	Electrical Wires and Underground Cables	
2	Electrical Wires and Underground Cables- Poly Notes Hub	
3	IS 1255 (1983)	
4	Bureau of Indian Standards	
5	Guidelines for Use of Under Ground Cable System and Overhead Conductor System Along with Cost Benefit Analysis	

6	Electrical Measurement and Instrumentation	
7	Utilization of Electrical Energy	

References and Suggested Readings

1. R.M. Black, *The history of electric wires and cables*, No. 4, IET, 1983.
2. M.A. Laughton, and M.G. Say, eds. *Electrical engineer's reference book*, Elsevier, 2013.
3. T. Edition, Electrical Power Cable Engineering.
4. A.J. Pansini, *Guide to electrical power distribution systems*, River Publishers, 2020.
5. P. Argaut, Underground Cables. *Springer Handbook of Power Systems*, pp.759-866, 2021.

UNIT 3

WIRING METHODS AND WIRING LAYOUT

UNIT SPECIFICS

This unit elaborately discusses the following topics:

- ✓ Introduction to electrical wiring and factors required for wiring
- ✓ Classification of electrical wiring methods
- ✓ Exploring the various wiring systems and their comparisons
- ✓ General BIS rules for domestic installations and wiring
- ✓ Exploring various types of design, working, and drawing of electrical circuits

This unit gives an in-depth understanding of various wiring methods, techniques, and layout principles used in electrical installations. One can learn about safety standards, codes, and best practices for designing, installing, and maintaining electrical wiring systems in residential, commercial, and industrial settings. Here, the classification of wiring methods and different wiring casing are also reported for understanding. This unit also covers further aspects of conduit wiring, including various types of conduit, a comparison between metal and PVC conduit, the distinction between surface and concealed conduit wiring, conduit wiring accessories, and the BIS requirements for metal and PVC conduit wiring. This section discusses the general BIS guidelines that apply to home installations. This unit covers the design, operation, and drawing of various electrical circuits, including essential light and fan circuits; wiring for stairs and lifts; lighting for bedrooms and corridors; series-parallel circuits; circuits for controlling a master switch; design of wiring circuits according to user requirements; and various lighting circuits using an intermediate switch and a bell indicator.

RATIONALE

This fundamental unit on well-planned wiring layouts minimizes the risk of electrical shocks, fires, and short circuits by adhering to safety codes, using appropriate insulation, and implementing protective devices. Compact and well-organized wiring methods (e.g., conduit systems and cable trays) maximize usable space in electrical rooms, distribution panels, and equipment enclosures, enhancing overall building functionality. Concealed wiring methods and carefully planned layouts contribute to a clean, uncluttered appearance in residential, commercial, and industrial spaces, enhancing visual appeal and architectural aesthetics. Robust wiring methods and layouts reduce mechanical stresses, abrasion, and degradation of electrical insulation, extending the lifespan of wiring components and preventing premature failures.

PRE-REQUISITES

Basics of Electrical Engineering

UNIT OUTCOMES

The list of outcomes of this unit is as follows:

U3-O1: Understand the fundamentals of electrical wiring systems

U3-O2: Learn about different wiring methods and their applications

U3-O3: Explore safety measures, codes, and regulations related to wiring installations

U3-O4: Develop skills in designing and planning wiring layouts for buildings

Unit-1 Outcomes	Expected Mapping with Programme Outcomes (*1- Weak Correlation; 2- Medium correlation; 3- Strong Correlation*)					
	CO-1	**CO-2**	**CO-3**	**CO-4**	**CO-5**	**CO-6**
U3-O1	3	3	1	3	2	2
U3-O2	3	3	1	3	2	2
U3-O3	3	3	1	3	2	2
U3-O4	3	3	1	3	2	2

Introduction

Proper wiring techniques and architecture are essential for electrical systems to distribute electricity and signals efficiently and safely across a building or other facility. The following parts provide a high-level summary of the wiring techniques and layout.

3.1 Wiring Methods:

a) **Concealed Wiring:**

✓ Concealed wiring refers to running electrical cables or conductors through walls, ceilings, or floors to provide a tidy and inconspicuous appearance.

✓ Common techniques involve the installation of cables by threading them through conduits, ducts, or raceways embedded into structural features.

✓ Concealing wiring decreases the likelihood of harm and limits the vulnerability of electrical wires to external elements.

b) **Surface Wiring:**

✓ Surface wiring refers to putting electrical cables or conduits directly on the exterior of walls, ceilings, or floors. This commonly uses cable trays, surface-mounted conduits, or wire mouldings.

✓ Surface wiring is commonly employed in retrofit or temporary installations when it is impossible to use concealed wiring.

✓ It facilitates convenient access to wire for maintenance or alterations.

c) **Open Wiring:**

✓ Open wiring pertains to installing uncovered electrical lines in industrial or commercial environments, such as factories or warehouses.

✓ Conductors are upheld by insulators or cable hangers and are installed along walls, ceilings, or structural components.

✓ Open wiring is appropriate for locations that require frequent alterations to the wiring or where aesthetics are not a main priority.

d) **Overhead Wiring:**

✓ Overhead wiring refers to installing electrical lines above ground level, using poles, overhead racks, or cable trays for support.

✓ It is frequently employed to illuminate outdoor areas, distribute power in industrial settings, and transmit electricity over overhead lines.

e) **Underground Wiring:**

✓ Underground wiring refers to burying electrical cables or conduits beneath the ground, typically in trenches or conduits.

✓ It provides power, telecommunications, or data connections in locations where overhead wire is not feasible or visually unappealing.

3.2 Wiring Layout Considerations:

Load Distribution:

✓ Design the wiring arrangement to distribute the electrical loads among circuits equally, thus avoiding overloading and ensuring a balanced operation.

✓ Arrange loads with similar characteristics near streamline circuit design and reduce voltage loss.

Circuit Separation:

✓ Separate electrical circuits are used for lighting, power outlets, and specialized equipment to prevent interference and simplify troubleshooting.

✓ Utilise specialized circuits for high-power appliances or equipment to avoid frequent interruptions and maintain consistent and dependable performance.

Accessibility and Maintenance:

✓ Ensure that wiring and electrical components are easily reachable for inspection, maintenance, and repairs.

✓ Ensure the provision of access panels, junction boxes, or service conduits to provide convenient troubleshooting and adjustments.

Safety and Code Compliance:

✓ Follow the applicable local electrical codes and regulations on wiring procedures, conductor sizing, grounding, and protection against electrical hazards.

✓ Ensure safety by utilizing suitable wire materials, such as fire-rated cables, and implementing preventive measures like circuit breakers or fuses.

Future Expansion:

✓ Ensure future scalability and development by allocating sufficient space for supplementary circuits, outlets, or equipment.

✓ Ensure the installation of conduit or raceway systems with additional capacity to support future electrical upgrades or adjustments.

Aesthetics:

✓ Consider the visual effect of the wiring arrangement on the area's overall appearance.

✓ Hide electrical wiring whenever feasible or opt for surface-mounted conduits and fittings that harmonize with the architectural style and design.

National Electrical Code of India (NEC)

3.3 Factors Determining Wiring Method Selection:

Electrical wiring connects cables and other devices, such as fuses, switches, sockets, lights, and fans. These are connected to a primary distribution board, a framework attached to a utility pole, to ensure a consistent power supply. The procedure guarantees the secure and dependable transmission of power within a structure. Several criteria, such as safety regulations, environmental circumstances, aesthetic preferences, and practical limitations, determine the choice of wiring systems in electrical installations. The following are the primary elements that determine the selection of a wiring method:

Safety Requirements:

✓ Ensuring safety is of utmost importance in electrical installations. The selected wiring technique must adhere to applicable safety standards and regulations to mitigate the risks of electric shock, fire hazards, and other potential problems.

✓ Wiring procedures must ensure sufficient insulation, safeguard against mechanical harm, and establish adequate grounding or earthing to reduce the likelihood of electrical accidents.

Environmental Conditions:

✓ The selection of the proper wiring method is heavily influenced by the ambient circumstances at the installation site.

✓ Factors such as humidity, temperature variations, and exposure to corrosive substances should be considered while deciding between concealed, surface, or open wiring for indoor installations.

✓ Outdoor installations necessitate wiring techniques that resist moisture, UV radiation, and mechanical impacts, such as underground conduits, overhead wires, or weatherproof enclosures.

- **Building Structure and Layout:**

✓ The architectural design and layout of the structure influence the selection of wiring systems. Residential and commercial locations may opt for concealed wiring due to its aesthetic appeal, but industrial or warehouse environments may find surface or overhead wiring more practical.

✓ The building structure determines the possibility of concealing wiring within walls, ceilings, or floors, including wall materials, ceiling heights, and floor architecture.

- **Installation Accessibility:**

✓ Ensuring accessibility for the installation, maintenance, and future alterations is a crucial factor to consider. Wire systems should facilitate convenient access to wire, electrical components, junction points, and service conduits.

✓ Accessibility considerations impact the selection of concealed, surface, or open wiring and the utilization of accessible conduits, junction boxes, and access panels.

- **Load and Circuit Requirements:**

✓ The electrical load, circuit architecture, and distribution needs influence the selection of wiring methods.

✓ High-power equipment or appliances may have specialized circuits or specific wiring techniques to accommodate electrical requirements and avoid overloading.

✓ Separate wiring runs or conduits may be necessary to ensure circuit separation for lighting, power outlets, HVAC systems, and other loads.

- **Cost and Installation Complexity:**

✓ Cost concerns, such as the expenses associated with materials, labour, and the complexity of installation, influence the selection of wiring systems.

✓ Concealed wiring, although providing a neater appearance and better protection against damage, generally entails more extraordinary expenses for installation and requires more labour-intensive efforts than surface or exposed wiring techniques.

Regulatory Compliance:

- ✓ It is crucial to Complying with local electrical codes, standards, and regulations are crucial when choosing wiring solutions. Wiring installations must adhere to relevant norms and standards to guarantee safety, dependability, and conformity with the law.
- ✓ Standards encompass conductor sizing, insulation ratings, grounding, bonding, and protection against electrical dangers.

3.4 Classification of Wiring Methods:

Electrical wiring systems employ many techniques to establish connections, which can be classified according to the wire arrangement and how connections are made. Below are many frequently used techniques:

3.4.1 Loop-in Wiring:

This approach connects a single cable from the main distribution board to each outlet, creating a closed loop. Every outlet is interconnected across the loop, facilitating effortless distribution and connecting. Loop-in wiring, sometimes called loop-in circuit or loop-in method, is a prevalent wiring technique employed in electrical installations, specifically for lighting circuits. Loop-in wiring involves the creation of a circuit that creates a closed loop or ring, with many light fittings or outlets connected in parallel along the loop.

Basic Configuration:

- ✓ In loop-in wiring, the circuit originates from the consumer unit, where the supply cable, typically a twin and earth wire, is connected to the circuit breaker or fuse.
- ✓ From the consumer unit, a solitary cable (often twin and earth) extends to the circuit's initial light fixture or outlet.

Looping Configuration:

- ✓ From the initial light fitting or outlet, further cables are interconnected or linked in a series to connect to succeeding light fittings or outlets in the circuit.
- ✓ Each light fixture or electrical outlet is linked in parallel, indicating that the same supply wire powers them but does not rely on one another for functioning.

Connection Method:

- ✓ At every light fitting or outlet, the live (line) and neutral wires from the power source are linked to the terminals of the fitting or outlet.
- ✓ A brief segment of cable, known as loop wire, is subsequently linked from the live terminal of one fitting to the live terminal of the adjacent fitting, establishing a loop connection.

✓ Similarly, the neutral wires are interconnected, while the earth wires (if available) are linked to the earth terminal of each fixture.

- **Advantages of Loop-in Wiring:**

✓ Simplified Installation: Loop-in wiring streamlines the installation procedure by eliminating the requirement for separate wires to be routed back to the consumer unit for each light fitting or outlet.

✓ Flexibility: The loop-in wiring system enables convenient installation or removal of light fittings or outlets from the circuit without causing any disruption to the other parts of the circuit.

✓ Reduced Cable Usage: Instead of radial wiring, Loop-in wiring minimizes the wire needed by interconnecting the cables between light fittings. Each outlet requires a separate cable to run back to the consumer unit in radial wiring.

- **Applications:**

✓ Loop-in wiring is a frequently employed method for connecting lighting circuits in residential, commercial, and industrial structures.

✓ It is appropriate for circuits with numerous light fittings or outlets connected in a series arrangement throughout a corridor, hallway, or room.

Loop-in wiring (Fig.3.1) is a pragmatic and successful technique for wiring lighting circuits, offering versatility, simplicity of installation, and cost-efficiency in diverse electrical setups.

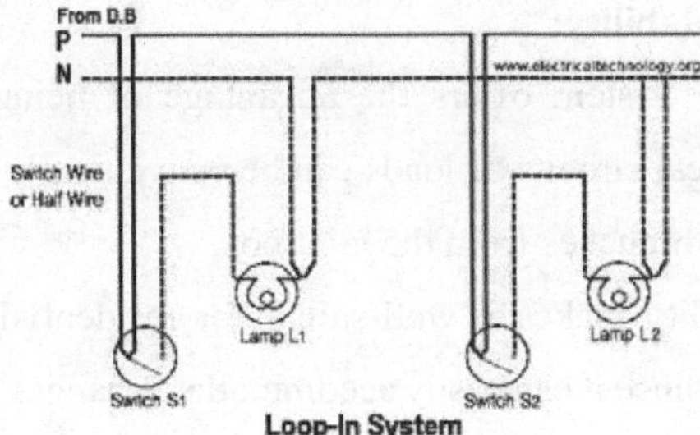

Figure 3.1 Loop-in wiring

3.4.2 Joint Box or Tee System:

This technique entails the utilization of joint boxes or tee connectors at various locations within the wiring system. It enables the division of the wiring into several paths that connect to multiple devices, offering versatility in the arrangement. The "Joint Box" or "Tee System" is a prevalent wiring configuration frequently employed in electrical installations, specifically in residential or small-scale business environments. This method is utilized to disseminate electrical power to various locations or divisions within a structure.

- **Basic Configuration:**
  - ✓ In the Joint Box or Tee System, the primary power supply line (usually from the distribution board or consumer unit) is linked to a central junction point, commonly known as the "joint box."
  - ✓ From this central distribution point, separate branch circuits or wires flow to different electrical outlets, fixtures, or appliances within the structure.
- **Connection Method:**
  - ✓ At the joint box, the incoming supply line is commonly split into several outgoing branches using terminal blocks, connectors, or busbars
  - ✓ Each outgoing branch is a conduit to various parts or sections of the building, such as different rooms or floors.
- **Distribution Approach:**
  - ✓ The Joint Box or Tee System utilizes a hierarchical distribution method. The primary supply line functions as the central axis and the outgoing branches serve as extensions or offshoots stemming from the axis.
  - ✓ Each branch can divide into sub-branches to handle more electrical loads or devices, resulting in a hierarchical distribution network.
- **Flexibility and Expandability:**
  - ✓ The Joint Box or Tee System offers the advantage of being flexible and capable of expansion. New electrical circuits or loads can be conveniently accommodated by adding or extending additional branches from the joint box.
  - ✓ This system's adaptability makes it well-suited for residential structures or small-scale commercial businesses since it can easily accommodate changes in the electrical layout that may occur over time due to repairs or extensions.
- **Applications:**
  - ✓ The Joint Box or Tee System is a frequently employed method for distributing power to lighting circuits, power outlets, and other electrical loads within buildings.
  - ✓ It is ideal when a single supply line must be divided and sent to various locations or areas, such as homes, small offices, stores, or flats.

Considerations:

- ✓ Properly identifying and documenting the wire connections at the joint box is crucial for facilitating maintenance, troubleshooting, and future upgrades.
- ✓ Adherence to electrical standards and regulations, encompassing appropriate conductor sizing, overcurrent protection, and grounding, guarantees safety and dependability.

The Joint Box or Tee System (Fig.3.2) offers a straightforward and efficient approach to distributing electrical power in buildings. It enables flexibility, expandability, and easy installation for various applications.

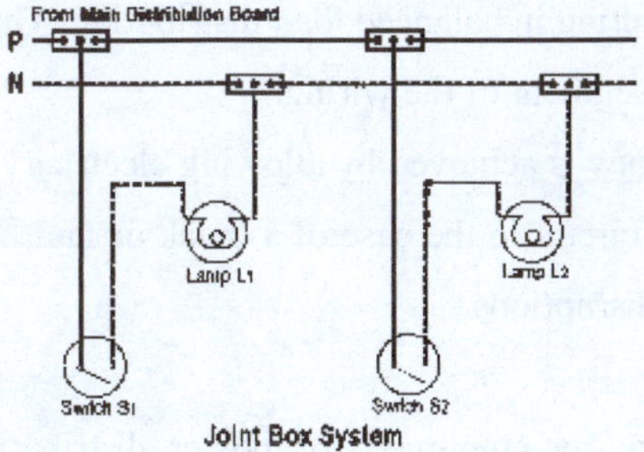

Figure 3.2 Joint box systems

3.4.3 Ring Circuit Wiring:

Ring circuit wiring (Fig.3.3) involves creating a cable ring from the main distribution board. Outlets and devices are then connected to the ring at different points. This method provides redundancy and even distribution of current. A ring circuit wiring system is commonly used in electrical installations, primarily residential buildings, to distribute power to multiple electrical outlets (sockets) from a single circuit. This system is prevalent in many countries, including the United Kingdom.

Basic Configuration:

- ✓ A ring circuit consists of a single wire, typically a twin and earth cable, that starts from the distribution board or consumer unit and forms an unbroken loop around the building.
- ✓ The cable commonly extends horizontally over walls or ceilings, interconnecting several electrical outlets (sockets) in its path.
- ✓ After the loop, the cable returns to the distribution board, thereby finalizing the ring

Connection Method:

- ✓ Each electrical outlet (socket) is connected in series with incoming and outgoing cables. Consequently, each socket is linked to the subsequent one sequentially.

✓ The last socket in the loop is linked back to the distribution board, creating a continuous loop or ring.

Advantages:

✓ **Reduced Cable Length:** One significant benefit of the ring circuit is that it necessitates less cable than standard radial circuits, which require a separate cable for each outlet that runs back to the distribution board. These results in a decrease in materials expenses and a reduction in the time required for installation.

✓ **Balanced Load Distribution:** The load in the circuit is evenly distributed due to the formation of a loop, resulting in balanced load distribution. This aids in averting excessive strain on any particular segment of the wiring.

✓ **Redundancy:** Redundancy is achieved by allowing electricity to continue flowing via the other component of the circuit in the case of a break or fault in one section. This ensures continuity and reduces disruptions.

Applications:

✓ Ring circuits are frequently employed for power distribution to electrical outlets in domestic buildings, such as living rooms, bedrooms, kitchens, and other locations that necessitate several outlets.

✓ They are appropriate for versatile power outlets for appliances, lamps, chargers, and other electrical devices.

Considerations:

✓ **Proper Sizing:** The cables in ring circuits must be adequately dimensioned to accommodate the highest anticipated load without experiencing excessive heat. This encompasses the assessment of both the ampacity and voltage drop.

✓ **Circuit Protection:** Ring circuits are commonly safeguarded by a circuit breaker or fuse at the distribution board. Ensuring enough overcurrent protection is crucial to safeguard the wire and maintain safety.

✓ **Testing and Inspection:** Regular testing and inspection of ring circuits are vital to guarantee optimal performance and safety. This encompasses continuity, insulation resistance, and earth fault loop impedance examinations.

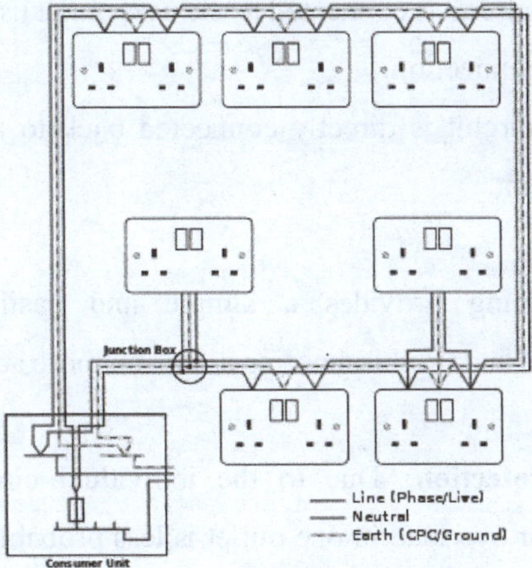

Figure 3.3 Ring circuit wiring

Ring circuit wiring is a cost-efficient and effective approach for distributing power to multiple electrical outlets within a building.

3.4.4 Radial Wiring:

The user's text is a checkmark symbol. Radial wiring is a direct method in which each outlet or device is linked individually to the main distribution board using a single cable. This is a straightforward and frequently employed method for wiring in residential settings. Radial wiring is a prevalent technique employed in electrical installations to disperse electricity from a solitary source to numerous electrical outlets or gadgets. In contrast to the ring circuit wiring system, radial wiring entails each outlet being linked to a separate wire that runs directly back to the distribution board or consumer unit.

Basic Configuration:

✓ In radial wiring, every electrical outlet (socket) is linked to the distribution board or consumer unit through an individual cable run.

✓ A solitary cable, usually a twin and earth cable, originates from the distribution board and directly connects to the initial outlet.

✓ Radial wiring refers to a configuration where cables extend from the initial outlet to link with successive outlets in a branching pattern.

Connection Method:

✓ At each electrical outlet, the cable from the distribution board is linked to the outlet terminal.

✓ Each outlet in the radial circuit is connected to the next outlet using outgoing wires, creating a sequential chain-like connection.

✓ The last socket in the circuit is directly connected back to the distribution board, thus finalizing the radial circuit.

Advantages:

✓ **Simplicity:** Radial wiring provides a simple and easily comprehensible wiring arrangement. Installation and troubleshooting are made more accessible by a separate cable run for each outlet.

✓ **Individual Circuit Protection:** Due to the individual circuitry of each outlet, the occurrence of a defect or overload in one outlet is less probable to impact the functioning of other outlets.

✓ **Flexibility:** Radial wire offers circuit design and construction versatility, rendering it appropriate for various applications and building types.

Applications:

✓ Radial wire is a frequently employed method for distributing electrical power to sockets in residential, commercial, and industrial structures.

✓ It is appropriate for systems where outlets are distributed across several rooms or locations, and a continuous loop (like a ring circuit) is not feasible or required.

Considerations:

✓ **Cable Sizing:** The cables utilized in radial wiring must be correctly dimensioned to withstand the maximum anticipated load without experiencing excessive heat. This encompasses the evaluation of both the ampacity and voltage drop.

✓ **Circuit Protection:** A circuit breaker or fuse at the distribution board must safeguard each radial circuit. Appropriate overcurrent protection is crucial for preserving the wiring and ensuring safety.

✓ **Circuit Length:** Maintaining a sufficient length for cable runs in radial wiring is essential to minimize voltage drop and ensure efficient power distribution.

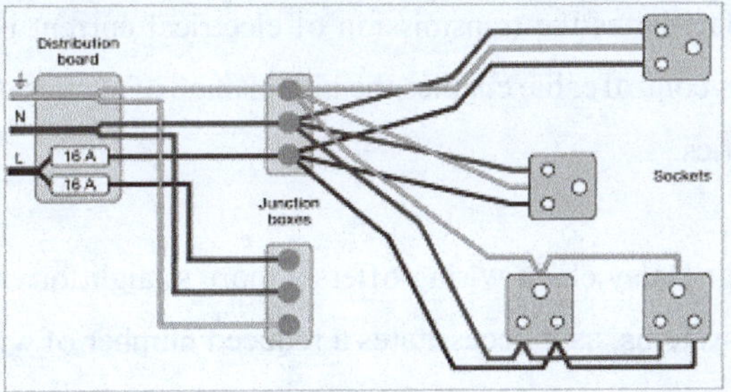

Figure 3.4 Radial wiring

Radial wiring (Fig.3.4) is a flexible and extensively employed technique for delivering power in electrical facilities. It provides a straightforward design, individualized circuit safeguarding, and adaptability, rendering it appropriate for various applications and building configurations.

3.4.5 Daisy Chain Wiring:

This approach entails establishing a connection between devices sequentially, akin to the arrangement of links in a chain. The wire extends linearly from one device to another, creating a sequential chain. It is frequently employed in the fields of networking and lighting systems. Daisy chain wiring, often called series wiring or linear wiring, is a technique for sequentially interconnecting electrical devices or components, allowing the current to pass from one device to the next in a consecutive fashion. Daisy chain wiring involves the direct connection of each item to the next one, creating a configuration that resembles a chain.

Basic Configuration:

- ✓ In daisy chain wiring, electrical devices or components are connected linearly, with each unit immediately connected to the next one in sequence.
- ✓ The wiring commences with either a power supply or a control unit and advances to the initial device in the sequence.
- ✓ Each subsequent device is connected to the previous one using a cable, and this sequence is repeated until all devices are connected.

Connection Method:

- ✓ Each item in the daisy chain is commonly linked using two cables: one for power (usually the live and neutral wires) and one for communication or control signals (if applicable).

✓ The power cable facilitates the transmission of electrical current between devices, while the communication/control cable enables the interchange of signals or data between devices with such capabilities.

Advantages:

✓ **Simplified Wiring:** Daisy chain wiring offers a more straightforward installation process than other wiring systems, as it necessitates a reduced number of wires and connections.

✓ **Space Efficiency:** Daisy chain is a space-saving method that minimizes the requirement for extra cable runs to a central control point, making it more efficient than other wiring systems.

✓ **Sequential Control:** Daisy chain wiring offers a simple means of controlling or activating components sequentially within a system.

Applications:

✓ **Lighting Systems:** Daisy chain wiring is frequently employed in lighting systems, wherein several light fixtures are interconnected linearly.

✓ **Sensor Networks:** Automation and control systems can be interconnected in a daisy chain configuration to enable efficient communication and control.

✓ **Audio and Video Systems:** Audio and video systems can be connected in a daisy chain topology to simplify wiring and control. This allows for the connection of devices such as speakers or monitors.

Considerations:

✓ **Voltage Drop:** In daisy chain wiring, voltage drop can occur over the entire chain length, mainly if the cable runs are lengthy or if the devices connected consume a substantial amount of current. Accurate cable sizing and voltage estimates are crucial to minimize this problem.

✓ **Reliability:** The dependability of a daisy chain wiring system is contingent upon the caliber of connections and the overall soundness of the wiring. Unstable connections or cables in a state of disrepair can cause a complete interruption in the sequence of events.

✓ **Control and Monitoring:** Daisy chain wiring may not be the optimal choice in systems requiring individual control or device monitoring, as it does not allow for individual addressing of units.

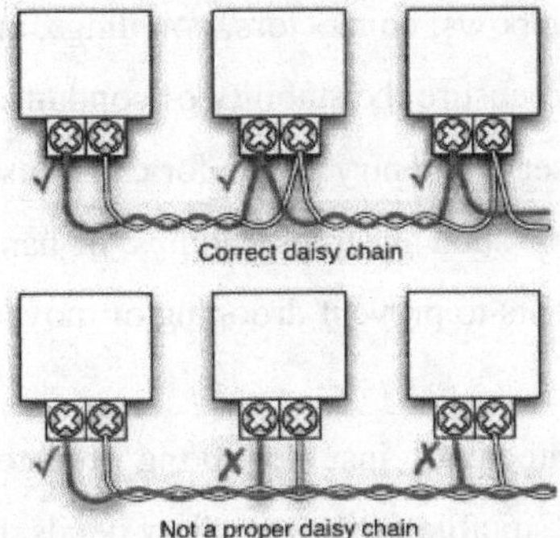

Figure 3.5 Daisy chain wiring

Daisy chain wiring (Fig.3.5) provides a direct and practical approach for sequentially linking electrical devices or components. It is frequently employed in diverse applications where the benefits of sequential connectivity and reduced wiring are favourable.

3.4.6 Conduit Wiring:

Conduit wiring (Fig.3.6) entails the process of routing cables via conduits made of either metal or plastic. This technique offers safeguarding to the wires and is frequently employed in industrial and commercial settings where wiring may be susceptible to physical harm. Conduit wiring is an electrical wiring technique that involves enclosing electrical lines within metal or plastic conduits to provide protection and facilitate organization. Conduit wiring has numerous benefits, such as improved safety, longevity, and simplified maintenance.

- **Types of Conduits:**

✓ **Metal Conduits:** Metal conduits, such as rigid metal conduit (RMC) and intermediate metal conduit (IMC), are constructed from steel or aluminium and offer a strong defence against physical harm, fire, and electromagnetic interference (EMI).

✓ **Plastic Conduits:** Plastic conduits, such as PVC and HDPE conduits, are lightweight, resistant to corrosion, and appropriate for less challenging settings.

- **Components of Conduit Wiring:**

✓ **Conduits:** Conduits are cylindrical structures that house electrical lines and safeguard against physical harm, dampness, and potential dangers in the surroundings.

✓ **Cables:** Electrical cables, such as insulated conductors or wires, transport electric current through conduits.

✓ **Fittings:** Fittings, such as elbows, connectors, couplings, and junction boxes, are utilized to establish connections and ensure the stability of conduits. They also enable the bending or turning of conduits and serve as entry points for cable installation and maintenance.

✓ **Supports:** Conduit supports, such as straps, clamps, or hangers, fasten conduits to walls, ceilings, or structural elements to prevent drooping or movement.

Installation Process:

✓ **Routing:** Conduits are strategically installed along predetermined routes, such as walls, ceilings, or floors, utilizing suitable fittings to allow bends, turns, or changes in direction.

✓ **Cable Installation:** Electrical cables are threaded into conduits using specialized instruments, such as fish tapes or cable pullers, and then joined to terminals or junction boxes at the desired places.

✓ **Connection:** Conduits are attached to electrical boxes, panels, or equipment using fittings and secured with suitable fasteners.

✓ **Grounding:** Metal conduits serve as a means to ground electrical systems and must be correctly bonded and grounded by electrical codes and standards.

Advantages of Conduit Wiring:

✓ **Enhanced Protection:** Conduit wiring offers durable safeguarding against physical harm, moisture, dust, and environmental dangers, hence minimizing the likelihood of electrical malfunctions, fires, and mishaps.

✓ **Durability:** Conduits provide durable shielding for electrical wires, especially in challenging or corrosive conditions, guaranteeing the dependability and durability of electrical systems.

✓ **Flexibility:** Conduit systems offer greater flexibility for making changes or additions to electrical circuits than open wiring approaches, enabling alterations and upgrades without causing significant interruption.

✓ **Safety:** Conduit wiring improves electrical safety by decreasing the chance of unintentional contact with live wires and offering a secure container for cables, reducing the probability of electrical dangers.

Applications:

✓ Conduit wiring is a prevalent electrical installation method in residential, commercial, industrial, and institutional buildings. It is utilized for the distribution of electricity, installation of lighting systems, the establishment of data communications, and implementation of control circuits.

✓ It is appropriate for various settings, such as indoor and outdoor installations, hazardous places, and areas that demand high levels of protection and dependability.

Conduit wiring is a highly adaptable and extensively employed technique for electrical wiring, providing exceptional safeguarding, longevity, and security for various uses.

Figure 3.6 Conduit wiring

3.4.7 Casing and Capping Wiring:

Casing and capping refer to routing cables through metal or plastic channels (casing) and enclosing them with a cap. This is a prevalent technique for wiring in both residential and commercial structures. Casing and capping wire, sometimes referred to as casing capping wiring or C&C wiring, is an electrical wiring technique that involves enclosing cables within plastic conduits (casings) and protecting them with a cap (capping) made of the same material. This wiring technique offers a straightforward and economical approach to safeguard electrical lines and streamline their installation on the surfaces of walls, ceilings, or floors.

- **Components of Casing and Capping Wiring:**

✓ **Casing:** Plastic conduits, commonly composed of PVC (polyvinyl chloride), are empty tubes that surround electrical cables and offer safeguarding from physical harm and environmental dangers.

✓ **Capping:** Capping refers to a flat strip or cover constructed from the same material as the casing (PVC). Its purpose is to enclose and securely fasten the conduits and wires. It offers a supplementary safeguard and a polished appearance to the wire installation.

✓ **Fittings:** Various fittings, including elbows, connectors, tees, couplings, and end caps, are utilized to establish connections and ensure the stability of conduits. These fittings also enable the bending or turning of conduits and serve as access points for the installation and maintenance of cables.

- ✓ **Cables:** Electrical cables, such as insulated conductors or wires, are guided through conduits and linked to terminals or junction boxes at the intended positions.

Installation Process:

- ✓ **Surface Preparation:** The surface on which the conduits will be installed undergo cleaning and preparation to provide optimal adhesion and stability.
- ✓ **Routing:** Conduits are directed along predetermined channels on the surface, utilizing suitable fittings to accommodate curves, angles, or alterations in direction.
- ✓ **Attachment:** Conduits are affixed to the surface using fastening methods like screws, anchors, or adhesive tapes. Conduits must be uniformly spaced and aligned by electrical laws and standards.
- ✓ **Cable Installation:** Electrical Specialised tools pull or push electrical cables into conduits, which are then linked to terminals or junction boxes at the desired places.
- ✓ **Capping:** After the wires are installed, the conduits are covered with a capping, which is fastened in place using screws or adhesive. The capping serves as an extra layer of safeguarding and imparts a tidy and polished look to the wiring installation.

Advantages of Casing and Capping Wiring:

- ✓ **Simplicity:** Casing and capping wiring installation is straightforward and entails minimal disturbance to pre-existing buildings and finishes. This is appropriate for retrofit installations or where covert wiring is not feasible.
- ✓ **Protection:** Conduits safeguard against physical harm, dampness, debris, and potential dangers in the surroundings, diminishing the likelihood of electrical malfunctions and mishaps.
- ✓ **Accessibility:** Conduits offer convenient electrical line access for maintenance, changes, or troubleshooting without requiring demolition or expensive repairs.
- ✓ **Flexibility:** Casing and capping wire systems offer greater adaptability for incorporating changes or additions to electrical circuits compared to conceal wiring methods. This allows for alterations and upgrades to be made without causing considerable disturbance.

Applications:

- ✓ Casing and capping wiring is widespread in residential, commercial, industrial, and institutional structures for power distribution, lighting systems, data connections, and control circuits.
- ✓ It is appropriate for installations that focus not on appearance, such as utility rooms, workshops, garages, or industrial facilities.
- ✓ Temporary installations, construction sites, or rented locations where permanent structural changes are not allowed also utilize casing and capping wiring.

Casing and capping wiring (Fig.3.7) provides a pragmatic and economical method for enclosing and safeguarding electrical lines in surface-mounted installations.

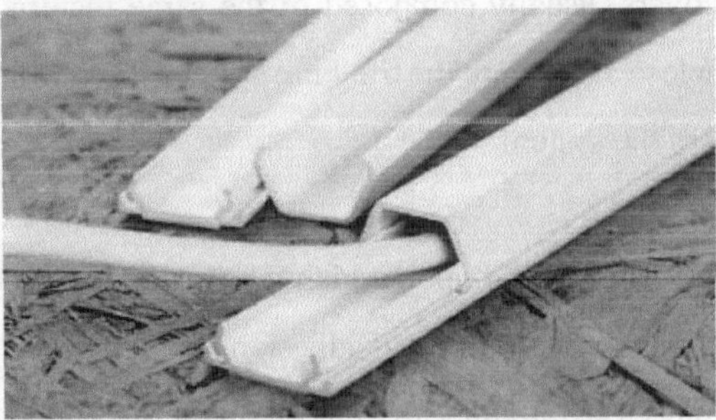

Figure 3.7 Casing and capping wiring

3.4.8 Trunking Wiring:

Trunking is a method in which insulated wires are contained within a framework, typically metal or plastic. It serves as a safeguarded conduit for the cables and is frequently employed in industrial environments. Trunking wiring, commonly called cable trunking or electrical trunking, is an electrical wiring technique that involves enclosing cables within a rigid or flexible trunking system constructed from metal or plastic. Trunking safeguards electrical cables from physical harm, dust, moisture, and other environmental risks while aiding in their arrangement and control.

Types of Trunking:

- ✓ **Metal Trunking:** Metal trunking, commonly constructed from steel or aluminium, provides sturdy shielding and is well-suited for challenging settings that necessitate enhanced resilience and longevity.
- ✓ **Plastic Trunking:** Plastic trunking, such as PVC (polyvinyl chloride) or ABS (acrylonitrile butadiene styrene), possesses the qualities of being lightweight, resistant to corrosion, and

simple to install. It is appropriate for situations and applications that require less demand.

Components of Trunking Wiring:

- ✓ **Trunking Sections:** Trunking sections are empty passages or chambers that surround electrical lines and offer safeguarding against physical harm and environmental dangers.
- ✓ **Fittings:** Various fittings, including elbows, connectors, tees, couplings, and end caps, are utilized to establish connections and ensure the stability of trunking sections. These fittings also enable the bending or turning of the trunking and serve as access locations for cable installation and maintenance.
- ✓ **Covers:** Trunking covers, usually composed of the same material as the trunking itself (metal or plastic), enclose and secure the trunking system. They offer further protection and give a polished look to the system.

Installation Process:

- ✓ **Surface Preparation:** The surface upon which the trunking will be affixed is thoroughly cleansed and prepped to guarantee optimal adhesion and stability.
- ✓ **Routing:** Trunking pieces are strategically placed on the surface, following predetermined paths and utilizing suitable fittings to allow bends, turns, or changes in direction.
- ✓ **Attachment:** Trunking portions are affixed to the surface using fasteners such as screws, anchors, or adhesive. Trunking sections must have consistent spacing and alignment, adhering to electrical norms and standards.
- ✓ **Cable Installation:** Electrical installation Specialised tools are utilized to pull or push electrical wires through trunking sections, and then they are linked to terminals or junction boxes at the desired locations.
- ✓ **Cover Installation:** Once the cables are installed, fitted, the trunking system is concealed by attaching trunking covers, which are firmly secured using screws or other fasteners. The covers offer further safeguarding and a tidy, polished look to the wiring installation.

Advantages of Trunking Wiring:

- ✓ **Protection:** Trunking offers robust safeguarding for electrical cables, shielding them from potential harm caused by physical impact, moisture, dust, and other environmental threats. This effectively minimizes the likelihood of electrical malfunctions and accidents.

- ✓ **Organization:** Trunking systems facilitate the organization and administration of electrical lines, resulting in a reduced mess and enhanced visual appeal of the wiring installation.
- ✓ **Accessibility:** Trunking facilitates convenient access to electrical lines for maintenance, changes, or troubleshooting, eliminating the need for destructive measures or expensive repairs.
- ✓ **Flexibility:** Trunking systems offer greater flexibility for making changes or additions to electrical circuits compared to concealed wiring methods. This allows for alterations and upgrades to be made without causing considerable interruption.

Applications:

- ✓ Trunking wiring is a prevalent method employed in residential, commercial, industrial, and institutional buildings to facilitate the distribution of power, lighting systems, data communications, and control circuits.
- ✓ It is appropriate for both indoor and outdoor installations, as well as for surface-mounted, flush-mounted, or suspended applications.
- ✓ Trunking wiring is utilized in temporary installations, such as construction sites or rented areas when permanent structural modifications are prohibited.

Trunking wiring (Fig.3.8) provides a pragmatic and adaptable method for enclosing and safeguarding electrical lines in diverse applications and surroundings. The selection of the wiring method is contingent upon criteria such as the building's kind, visual appeal, safety prerequisites, and the characteristics of electrical loads. Every approach possesses distinct benefits and is chosen according to the precise requirements of the installation.

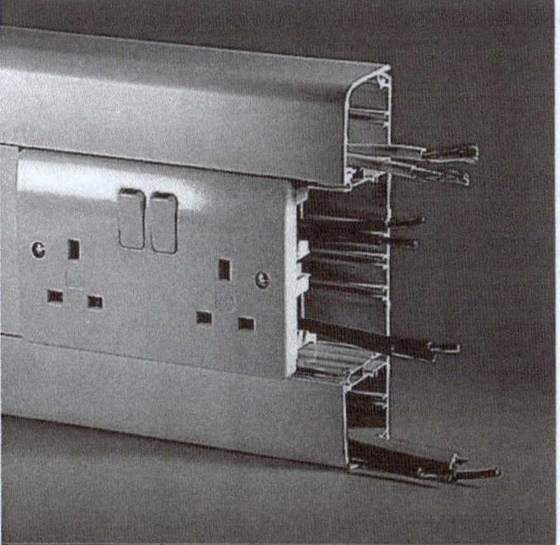

Figure 3.8 Trunking wiring

3.5 PVC Casing-Capping Wiring - IS: 732-1983:

PVC casing-capping wiring refers to enclosing electrical cables within PVC conduits, known as casings, and protecting them with PVC caps, known as capping, to ensure their safety and organization. This wiring technique is frequently employed in surface-mounted systems, offering a straightforward and economical approach to safeguard electrical lines while providing easy access for maintenance and alterations. The Indian Standard IS 732:1983 outlines the specifications for PVC casing-capping wire systems. Below is a summary of PVC casing-capping wiring according to the IS 732:1983 standard:

Components of PVC Casing-Capping Wiring:

- ✓ **PVC Conduits (Casings):** PVC conduits are cylindrical structures composed of polyvinyl chloride that serve as protective enclosures for electrical wires, safeguarding them from potential harm caused by mechanical forces, moisture, and various environmental risks.
- ✓ **PVC Caps (Cappings):** PVC caps are flat strips or coverings of polyvinyl chloride that envelop and fasten the PVC conduits and cables, offering supplementary safeguarding and a polished look.

Specifications and Requirements:

- ✓ **Material:** The PVC conduits and covers must be constructed from high-grade polyvinyl chloride material that satisfies the stipulated criteria for strength, flexibility, and durability.
- ✓ **Dimensions:** The PVC conduits and caps must adhere to the stated criteria in terms of their measurements, which include the outer diameter, wall thickness, and length. This is necessary to guarantee that they fit well and are compatible.
- ✓ **Colour:** The PVC conduits and caps can be supplied in various colours to meet specific needs, facilitating effortless recognition and alignment with the surrounding surroundings.
- ✓ **Installation:** Install PVC casing-capping wiring must adhere to the manufacturer's instructions and the applicable electrical rules and standards. Achieving a successful installation requires meticulous attention to surface preparation, routing, attachment, and cover installation.
- ✓ **Cable Installation:** Electrical cables must be put within PVC conduits using suitable techniques and tools, assuring correct routing, support, and termination at the intended positions.

- ✓ **Covering:** The PVC conduits must be enclosed with PVC caps, which should be firmly attached using screws or other suitable fasteners. The caps offer supplementary safeguarding and a tidy, polished appearance to the wiring installation.

Advantages of PVC Casing-Capping Wiring:

- ✓ **Protection:** PVC casing-capping wiring provides sturdy protection for electrical cables against mechanical damage, moisture, dust, and other environmental threats, minimizing the risk of electrical malfunctions and accidents.
- ✓ **Organization:** PVC conduits organize and oversee electrical cables, resulting in a reduction of mess and an enhancement of the visual appeal of the wiring installation.
- ✓ **Accessibility:** PVC casing-capping wiring, which allows for convenient access to electrical cables for tasks like maintenance, alterations, or troubleshooting. This access can be achieved without the need for destructive actions or lengthy repairs.
- ✓ **Flexibility:** PVC conduits offer a high degree of flexibility, enabling easy incorporation of changes or additions to electrical circuits. This allows for adjustments and upgrades to be made without causing considerable interruption.

Applications:

- ✓ PVC casing-capping wiring is extensively utilized in residential, commercial, industrial, and institutional buildings for power distribution, lighting systems, data communications, and control circuits.
- ✓ Surface-mounted installations on walls, ceilings, or floors are ideal for varied applications and conditions, offering a flexible and cost-efficient wiring option.

PVC casing-capping wiring systems, by IS 732:1983, provide a pragmatic and dependable approach for enclosing and safeguarding electrical cables in surface-mounted installations.

3.6 Conduit Wiring:

Conduit wiring is an electrical wiring technique that involves enclosing electrical lines within metal or plastic conduits to ensure their protection and organization. The conduits function as protective channels, safeguarding the cables against physical harm, moisture, dust, and other potential environmental dangers. Conduit wiring is frequently employed in residential and commercial constructions to guarantee electrical safety and adherence to building regulations.

3.6.1 Types of Conduit:

- ✓ **Surface Conduit Wiring:** Surface conduit wiring refers to the installation of conduits on the exterior of walls, ceilings, or floors through which electrical cables are routed. This approach is appropriate when hiding wire within the framework is neither practical nor preferred.
- ✓ **Concealed Conduit Wiring:** Concealed conduit wiring includes routing electrical cables within covert conduits embedded within the walls, ceilings, or floors. The concealed conduits ensure a tidy and visually appealing appearance in the area.
- ✓ **Casing and Capping Wiring:** Casing and capping wiring involves using PVC conduits (casings) and caps (cappings) to contain and safeguard electrical lines. The conduits are mounted on the surface of walls or ceilings, and the cables are placed within them. The caps provide additional protection and a complete appearance.
- ✓ **Batten Wiring:** Batten wiring is a method that utilizes wooden or bands of metal called battens to provide support and safeguard electrical cables. These cables are affixed to the surface of walls or ceilings. This approach is currently less prevalent but was extensively employed in previous deployments.
- ✓ **Conduit Wiring with Metal Conduits:** Conduit wiring can employ different types of metal conduits, including rigid metal conduit (RMC), intermediate metal conduit (IMC), or electrical metallic tube (EMT). Metal conduits provide durable safeguarding against physical harm and are frequently employed in industrial and commercial settings.
- ✓ **Conduit Wiring with PVC Conduits:** PVC conduit wiring uses PVC conduits that are lightweight, long-lasting, and impervious to corrosion. PVC conduits are frequently employed in residential and commercial installations where the priority lies in convenient installation and cost efficiency.
- ✓ **Flexible Conduit Wiring:** Flexible conduit wiring utilizes flexible conduits, such as flexible metal conduit (FMC) or liquid-tight flexible metal conduit (LFMC), to guide electrical cables through narrow or hard-to-reach areas. Flexible conduits offer adaptability and convenient installation in demanding settings.
- ✓ **Underfloor Conduit Wiring:** Underfloor conduit wiring refers to running electrical cables through conduits built beneath the surface of the floor. This technique is frequently

employed in commercial and industrial structures to disperse power and data cables while preserving a tidy and clear floor area.

3.6.2 Comparison between Metal and PVC conduit:

Metal and PVC conduit wiring are techniques to safeguard and direct electrical lines in diverse installations. Various conduits have distinct merits and drawbacks, rendering them appropriate for varying purposes.

- **Metal Conduit Wiring:**

✓ **Durability and Strength:** Metal conduits, such as rigid metal conduit (RMC), intermediate metal conduit (IMC), and electrical metallic tubing (EMT), possess exceptional endurance and strength. These devices offer a strong defence against physical harm, impact, and compression, rendering them appropriate for rugged settings and locations with substantial machinery or equipment.

✓ **Fire Resistance:** Metal conduits have intrinsic fire-resistant properties, aiding in the containment of electrical fires and impeding their propagation. These are commonly employed in structures where fire safety is of utmost importance, such as industrial complexes, tall buildings, and areas with potential dangers.

✓ **Mechanical Protection:** Metal conduits provide enhanced mechanical safeguarding for electrical wires, effectively isolating them from external forces, vibrations, and impact. This aids in reducing the likelihood of cable harm and guarantees the soundness of the electrical system.

✓ **Corrosion Resistance:** Commonly, metal conduits are constructed from corrosion-resistant materials like steel or aluminium. These materials can endure exposure to damp chemicals and caustic substances, making them well-suited for outdoor and hostile situations.

✓ **Rigidity and Stability:** Metal conduits exhibit rigidity and stability, ensuring their shape and structural integrity endure over time. These cables can bear substantial weight and withstand distortion, ensuring dependable safeguarding in challenging environments.

- **PVC Conduit Wiring:**

✓ **Cost-Effectiveness:** PVC conduits are typically less expensive than metal conduits, making them a cost-efficient choice for electrical installations, particularly in residential and light business environments.

- ✓ **Ease of Installation:** PVC conduits are low-weight and convenient to manipulate, sever, and set up. They may be rapidly built and affixed to walls, ceilings, or floors without requiring specialized tools or equipment.
- ✓ **Corrosion Resistance:** PVC conduits have exceptional corrosion, rust, and chemical degradation resistance. They are appropriate for indoor and outdoor installations, including regions subject to moisture, humidity, or caustic substances.
- ✓ **Flexibility:** PVC conduits provide flexibility and may be manipulated or directed around obstructions, rendering them appropriate for installations in confined or irregular areas. Flexible conduits provide enhanced flexibility compared to inflexible metal conduits and are more convenient to adjust according to evolving installation needs.
- ✓ **Insulation Properties:** PVC conduits offer electrical insulation for cables, minimizing the likelihood of electrical shock or short circuits. These materials are insulators, meaning they do not allow the flow of electricity, which improves safety in electrical installations.

Comparison Summary:

- ✓ **Strength and Durability:** Metal conduits possess superior strength and endurance compared to PVC conduits.
- ✓ **Fire Resistance:** Metal conduits provide intrinsic fire resistance, but PVC conduits have the potential to melt or combust when subjected to elevated temperatures.
- ✓ **Cost:** PVC conduits are typically more economical than metal conduits.
- ✓ **Ease of Installation:** PVC conduits are easier to install and handle than metal conduits.
- ✓ **Flexibility:** PVC conduits are easier to install and handle than metal conduits.
- ✓ **Corrosion Resistance:** PVC conduits provide enhanced flexibility and adaptability compared to stiff metal conduits.

3.6.3 Conduit Wiring Accessories:

Conduit wire accessories refer to various components and fittings utilized alongside conduits to simplify installing, organizing, and safeguarding electrical cables. These accessories augment conduit wiring systems' utility, safety, and efficacy.

a) **Conduit Connectors:** Conduit connectors link different conduit sections or connecting conduits to electrical boxes, fittings, or devices. They establish secure and dependable connections while preserving the integrity of the conduit system.

b) **Conduit Couplings:** Conduit couplings serve the purpose of connecting two lengths of conduit linearly. Conduits facilitate the seamless and uninterrupted passage of electrical cables while ensuring the proper alignment and stability of different portions.

c) **Conduit Elbows:** Conduit elbows are angular connectors utilized to alter the course of conduit installations, enabling the routing of wires around corners, obstacles, or structural components. They are available in several angles, such as 90 and 45 degrees, to suit diverse installation needs.

d) **Conduit Tees:** Conduit tees are fittings shaped like the letter T that are utilized to create branches in conduit runs, enabling the distribution of wires to numerous directions or devices. They offer a practical method for dividing conduit paths while ensuring continuity and organization.

e) **Conduit Bushings:** Conduit bushings safeguard cables against abrasion or harm when they enter or depart conduits. They offer a sleek, curved surface for cables to traverse, minimizing the likelihood of insulation erosion or wire fracture.

f) **Conduit Straps and Hangers:** Conduit straps and hangers fasten conduits to walls, ceilings, or structural supports. They aid with the prevention of drooping, shifting, or displacement of conduits and provide accurate support and alignment during installation.

g) **Conduit End Caps:** Conduit end caps serve the purpose of sealing the ends of conduits to prevent the entry of dust, moisture, or debris. They offer safeguarding for cables and uphold the structural soundness of the conduit system, particularly in outdoor or exposed installations.

h) **Conduit Adapters and Reducers:** Conduit adapters and reducers facilitate the transition between various sizes or types of conduits and connect conduits to different fittings or devices. They provide versatility and adaptability in conduit installations with diverse specifications.

i) **Conduit Pull Boxes and Junction Boxes:** Conduit pulls, and junction boxes are access locations for cable pulling, splicing, or termination in conduit systems. They serve as enclosures for electrical connections, junctions, or terminations, ensuring cables' orderly arrangement and safeguarding.

j) **Conduit Locknuts and Washers:** Conduit locknuts and washers fasten conduit fittings to electrical boxes or devices. They offer a firm and reliable connection, ensuring that fittings do not become loose or detached during installation or operation.

The following are typical conduit wiring accessories utilized in electrical installations. The choice of accessories is contingent upon the particular demands of the installation, encompassing factors such as conduit type, dimensions, arrangement, and prevailing climatic conditions.

3.6.4 BIS Rules for Metal and PVC Conduit Wiring:

The Bureau of Indian Standards (BIS) creates and publishes standards for diverse products and materials to assure quality, safety, and reliability. Particular norms and rules guide manufacturers and consumers of metal and PVC (Polyvinyl Chloride). Here is a comprehensive summary:

- **BIS Rules for Metal:**

a) **IS 1879: Part 1 (2003) - Non-ferrous Metal Conduits for Electrical Installations:**

This standard outlines the specific criteria that must be met by non-ferrous metal conduits used in electrical installations. This document encompasses the measurements, physical characteristics, and examination methods necessary to guarantee the excellence and security of these conduits.

b) **IS 9537 (Part 2) (1981) - Metal Conduit Fittings:**

This standard establishes standards for metal conduit fittings, guaranteeing their compatibility and quality for utilization in electrical installations. This resource encompasses a wide range of fittings used with metal conduits.

c) **IS 694: (1990) - PVC Insulated Cables for Working Voltages Up to and Including 1100 V:**

Although not explicitly designed for metal, this standard provides detailed guidelines for PVC-insulated cables, frequently employed with metal conduits. This aims to assure the safety and dependability of electrical wires by addressing the properties of conductors, insulation, and sheathing materials.

BIS Rules & Regulations

BIS Rules for PVC:

a) **IS 4985 (2000) - Unplasticized PVC Pipes for Potable Water Supplies:**

This standard outlines the specific criteria for unplasticized PVC pipes that distribute safe drinking water. This document encompasses the various aspects of PVC pipes used in water distribution systems, including their dimensions, physical features, and testing procedures. Its purpose is to guarantee the quality and safety of these pipes.

b) **IS 13592 (1992) - High-impact Strength PVC Sewerage Pipes:**

This standard primarily targets PVC pipes with high-impact strength and is explicitly designed for use in sewerage applications. The criteria encompass dimensions, physical characteristics, and testing protocols to guarantee the longevity and functionality of PVC sewerage pipes.

c) **IS 694: (1990) - PVC Insulated Cables for Working Voltages Up to and Including 1100 V:**

As previously stated, this standard is essential for PVC-insulated cables, which are extensively used in many applications. It guarantees that PVC-insulated wires adhere to predetermined standards for safety and performance.

d) **IS 10151 (2002) - Thermoplastic PVC Conduit for Electrical Installations:**

This standard pertains to thermoplastic PVC conduits utilized in electrical installations. It sets specifications for dimensions, qualities, and testing to ensure the quality and safety of PVC conduits in electrical applications.

The BIS standards ensure the quality and safety of metal and PVC materials used in various applications, including electrical installations, water supply, and sewage systems. Both manufacturers and customers can consult these standards to ensure that items adhere to specified requirements for performance and reliability.

BIS IS-4985

PVC Casing-Capping Wiring - Wiring Rules (IS: 732-1983):

PVC Casing-Capping Wiring:

- ✓ **IS 732:1983** is the Indian Standard that specifies the requirements for PVC-insulated cables used in casing-capping wiring systems.
- ✓ The standard offers instructions on using PVC-insulated wires for indoor electrical installations.
- ✓ The specifications encompass conductor dimensions, insulation depth, voltage capacity, and testing protocols to guarantee the safety and dependability of PVC casing-capping wire. Wiring.

Wiring Rules According to IS 732-1983:

- ✓ **Conductor Size:** Defines the acceptable range of conductor sizes for PVC casing-capping wiring.
- ✓ **Insulation Thickness:** Specifies the minimum thickness of PVC insulation necessary to comply with safety regulations.
- ✓ **Voltage Rating:** The voltage rating indicates the highest voltage the wires can safely handle during operation.
- ✓ **Testing Procedures:** Provides instructions for conducting tests on PVC-insulated wires to verify their compliance with predetermined standards.
- ✓ **Installation Guidelines:** Installation Guidelines suggest installing PVC casing-capping wire systems correctly.

IS: 732-1983

3.7 Comparison of Various Wiring Systems:

Diverse wire schemes are employed in electrical installations, each possessing distinct qualities, benefits, and drawbacks. Below is a comparison of various wiring schemes frequently employed in residential and commercial structures:

a) **Conduit Wiring:**

➢ **Advantages:**

✓ Offers exceptional safeguarding against physical harm, dampness, and environmental perils.

✓ It improves safety by confining electrical malfunctions and avoiding inadvertent contact with energized cables.

✓ Provides versatility in arranging wires and adapting to modifications or expansions in the electrical system.

✓ Applicable for diverse uses, encompassing industrial, commercial, and residential settings.

➢ **Disadvantages:**

✓ Installation necessitates extra materials (such as conduits and fittings) and labor, resulting in higher costs than alternative wiring methods.

✓ The installation process can be intricate and time-consuming, particularly for concealed conduit wiring.

✓ Periodic inspection and maintenance may be necessary to ensure integrity and regulation compliance.

b) **Casing and Capping Wiring:**

➢ **Advantages:**

✓ Casing and capping wiring is a straightforward and economical technique appropriate for surface-mounted installations.

✓ Offers fundamental safeguarding for wires from physical harm and adverse environmental conditions.

✓ It is effortless to install and alter, making it ideal for temporary or non-permanent installations.

➢ **Disadvantages:**

✓ Conduit wiring offers more protection than this method, making it more appropriate for challenging settings or regions with great danger.

✓ Proper organization and space allocation are necessary to prevent overcrowding and ensure tidiness.

✓ It is possible that this product may not satisfy the necessary regulations for specific uses, such as in industrial or commercial environments.

c) **Batten Wiring:**

➢ **Advantages:**

✓ Batten wire is a straightforward and cost-effective type of wiring, often employed in older systems or temporary settings.

✓ It offers fundamental cable safeguarding while facilitating straightforward identification and accessibility.

✓ Installation can be expedited and simplified without requiring specialized tools or equipment.

➢ **Disadvantages:**

✓ Provides limited safeguard against physical harm, dampness, or environmental threats.

✓ Limited flexibility and adaptability may render it unsuitable for intricate installations or future expansions.

✓ May not adhere to contemporary safety norms and regulations, presenting possible hazards in specific applications.

d) **Surface Conduit Wiring:**

➢ **Advantages:**

✓ Offers adequate cable protection while maintaining visibility and accessibility for maintenance purposes.

✓ Provides the ability to route wires along walls, ceilings, or floors without the requirement for concealment, allowing for increased flexibility.

✓ It may be rapidly and effortlessly installed, making it appropriate for retrofitting or temporary installations.

➢ **Disadvantages:**

✓ May offer insufficient safeguarding compared to concealed conduit wiring, particularly in high-risk settings.

✓ Proper support and alignment of conduits necessitate meticulous planning and spacing.

✓ Visible conduits can diminish the visual appeal of a location, particularly in residential or commercial environments.

e) **PVC Conduit Wiring:**

➢ **Advantages:**

✓ Lightweight, resilient, and cost-effective wiring solution that is ideal for a diverse array of applications.

- ✓ It is highly resistant to corrosion, moisture, and chemical damage, making it excellent for indoor and outdoor installations.
- ✓ It is simple to install and change, providing flexibility and agility to accommodate changing requirements.

➢ **Disadvantages:**

- ✓ It may offer less mechanical safeguarding than metal conduits, particularly in high-risk settings.
- ✓ Proper installation practices are necessary to prevent problems such as kinking, crushing, or degeneration over time.
- ✓ May not be appropriate for situations that demand significant fire resistance or mechanical durability.

f) **Overall Comparison:**

- ✓ **Cost:** Conduit wiring is typically the costliest option due to the supplementary materials and labour required. Casing, capping, and batten wiring are cost-effective options, although surface conduit and PVC conduit wiring are moderately priced.
- ✓ **Protection:** Conduit wiring provides optimal protection against physical damage, moisture, and environmental threats. PVC conduit wiring offers robust protection, whereas casing capping and batten wiring give little protection.
- ✓ **Installation:** Surface conduit and PVC conduit wire are typically more efficient and expeditious to install than concealed conduit wiring, casing, capping, or batten wiring.
- ✓ **Flexibility:** Conduit wiring systems provide optimal flexibility for routing wires and adapting to modifications or additions in the electrical system. PVC conduit wire exhibits greater flexibility compared to metal conduit wiring.
- ✓ **Aesthetics:** Concealed conduit and surface conduit wiring systems offer a neater and more visually appealing appearance than casing and capping or batten wiring, which can be visible and diminish the area's overall look.

3.8 General BIS Rules for Domestic Installations:

The Bureau of Indian Standards (BIS) establishes regulations and criteria for a wide range of products and procedures in India, encompassing household electrical systems. These standards guarantee the safety, quality, and efficiency of electrical systems utilized in residential structures. Below are many overarching regulations established by the Bureau of Indian Standards (BIS) for home installations:

a) **IS 732 (Part 1): Code of Practice for Electrical Wiring Installations (Part 1: General):**

✓ This standard offers comprehensive guidance and suggestions for the planning, implementing, and upkeep of electrical wiring systems in residential structures.

✓ It encompasses various electrical installation topics, such as wiring techniques, choosing appropriate equipment, ensuring safety, and adhering to applicable regulations and standards.

b) **IS 732 (Part 2): Code of Practice for Electrical Wiring Installations (Part 2: Wiring Rules):**

✓ Part 2 of IS 732 outlines precise regulations and criteria for electrical wire installations in residential structures.

✓ The document provides comprehensive instructions for wiring techniques, cable dimensioning, grounding configurations, protective mechanisms, and installation procedures to guarantee safety and adherence to regulations.

c) **IS 732 (Part 3): Code of Practice for Electrical Wiring Installations (Part 3: Electrical Equipment and Accessories):**

✓ Part 3 of IS 732 pertains to choosing, installing, and utilizing electrical equipment and accessories in residential settings.

✓ The document outlines the necessary criteria for switches, sockets, distribution boards, circuit breakers, and other electrical components to guarantee correct operation and safety in residential environments.

d) **IS 1646: Code of Practice for Fire Safety of Buildings (General):**

✓ This standard offers directives for implementing fire safety precautions in many types of buildings, encompassing residential edifices.

✓ This course encompasses a wide range of topics related to fire prevention, detection, suppression, and evacuation to reduce fire hazards and improve the safety of occupants.

e) **IS 8130: Code of Practice for Earthing:**

✓ IS 8130 outlines the specific criteria for earthing systems in electrical installations encompassing residential structures.

✓ This course encompasses designing, installing, and testing earthing systems to guarantee optimal protection against electric shock and fault currents.

f) **IS 3043: Code of Practice for Earthing:**

✓ This standard offers instructions for the design, implementation, and upkeep of earthing systems in electrical installations.

✓ The document guides conducting soil resistivity tests, selecting appropriate electrodes, determining conductor sizes, and meeting bonding standards to establish earthing systems with minimal impedance.

Electrical Safety Standards in India

g) **National Building Code (NBC):**

✓ The National Building Code (NBC) is a set of comprehensive principles that govern building design, construction, and safety in India.

✓ The document encompasses regulations about electrical installations, fire safety, structural design, plumbing, and other facets of building construction.

The BIS guidelines and standards offer crucial direction for the planning, setup, and upkeep of electrical systems in residential structures, guaranteeing safety, dependability, and adherence to requirements.

National Building Code (NBC)

3.9 Design, Working, and Drawing of Electrical Circuits

Designing, developing, and sketching electrical circuits entails crucial processes and considerations to guarantee electrical systems' secure and effective functioning.

3.9.1 Simple Light and Fan Circuit:

- **Design:**

✓ **Power Source:** Commence with a power source, usually an electrical panel or distribution board.

✓ **Switches:** Install individual switches for the light and fan. These switches can function as single-pole switches.

✓ **Wiring for Light:**

- Connect the live wire (often black in color) from the power source to one light switch terminal.
- Connect the remaining terminal of the light switch to the active terminal of the light fixture.
- Connect the neutral wire, typically identified by its white color, from the power supply straight to the neutral terminal of the light fixture.

✓ **Wiring for Fan:**

- Connect the live wire from the power source to one terminal of the fan switch.
- Connect the fan switch's remaining connection to the fan's active terminal.
- Connect the neutral wire from the power source straight to the neutral terminal of the fan.

✓ **Common Neutral:** Connect the standard neutral wire to the light and fan if the same source powers the light and fan.

✓ **Grounding:** Ensure adequate grounding by connecting the ground wire to the grounding terminals of the light fixture and fan.

- **Working:**

✓ Activating the light switch establishes a closed circuit, enabling the passage of electricity to the light fixture and causing the bulb to emit light.

✓ Activating the fan switch establishes a closed circuit, enabling the flow of electricity to power the fan and initiate its motor.

3.9.2 Staircase Wiring:

- **Design:**

✓ Install a two-way switch at the bottom and top of the staircase.

✓ Connect the shared terminals of the switches to the phase or live wire.

✓ Connect one of the remaining terminals of each switch to the bulb or lighting point.

✓ Connect the remaining terminal of the bottom switch to the phase or live wire.

✓ Connect the remaining terminal of the top switch to the lamp or lighting point on the higher floor.

Working:

✓ The enables the regulation of the light from both the lower and upper ends of the staircase.

3.9.3 Go-Down Wiring Circuit:

Design:

✓ Install a two-way switch at the go-down's entrance and exit locations.

✓ Connect shared terminals of the switches to the phase or live wire.

✓ Connect one of the remaining terminals of each switch to the bulb or lighting point.

✓ Connect the remaining terminal of the entry switch to the phase or live wire.

✓ Connect the remaining terminal of the exit switch to the lamp or lighting point.

Working:

✓ This feature allows the lighting to be controlled from both the entrance and departure points of the warehouse.

3.9.4 Bedroom Lighting Circuit:

Design:

✓ Utilize one or many switches to regulate the illumination in the bedroom.

✓ Connect the switches to the phase or live wire.

✓ Connect the shared terminals of the switches to the lamp or lighting fixtures in the bedroom.

✓ Connect the neutral wire to the lighting.

Working:

✓ Enables the regulation of bedroom illumination using switches.

3.9.5 Corridor Lighting Circuit:

Design:

✓ Utilize one or several switches to regulate the illumination of hallway lights.

✓ Connect the switches to the phase or live wire.

✓ Connect the shared terminals of the switches to the lamp or lighting locations in the hallway.

✓ Connect the neutral wire to the lighting.

Working:

✓ This system allows for the management of corridor illumination using switches.

3.9.6 Series-Parallel Circuit:

Design:

✓ Integrate both series and parallel connections inside a circuit.

✓ Connect components in a series configuration to provide the desired functionality.

✓ Utilize parallel connections to ensure the autonomous functioning of individual components.

Working:

✓ Enables the integration of both series and parallel circuits, offering versatility in managing various components.

3.9.7 Master Switch Control Circuit:

Design:

✓ Implement a master switch to regulate the flow of electricity to many circuits.

✓ Connect the main switch to the active wire of each circuit.

✓ Utilize separate switches to regulate specialized circuitry.

Working:

✓ The master switch is a central control, allowing or preventing power flow to different circuits.

3.9.8 Different Lighting Circuit using Intermediate Switch:

Design:

✓ Utilize an intermediate switch in combination with two-way switches.

✓ Connect the common terminal of the intermediate switch to the phase or live wire.

✓ Connect the remaining terminals of the intermediate switch to the lights or lighting locations.

✓ Install two-way switches at multiple locations to enhance control.

Working:

✓ Intermediate switches can control lights from multiple locations, hence increasing versatility.

3.9.9 Call Bell Circuit using Bell Indicator:

- **Design:**

✓ Implement a circuit that activates a bell indicator when a call is made.

✓ Connect the remaining terminal of the call bell switch to the bell indicator.

✓ Connect the bell indication to the neutral wire.

- **Working:**

✓ Depressing the call bell switch initiates the bell indicator, indicating a call.

3.9.10 Design of Wiring Circuits According to User's Requirement:

- **Design:**

✓ Comprehend the user's specifications for electrical circuits.

✓ Determine and identify the quantity and categories of switches, lights, or other necessary devices.

✓ Design the circuit arrangement according to the user's preferences and for ease of usage.

- **Working:**

✓ Customized wiring circuits are created to fulfill the user's requirements and preferences, guaranteeing the best possible functioning and ease.

Solved Numerical Problem

Example 3.1:

A residential building has a radial electrical wiring system. The main distribution board (MDB) is located in the basement, and it supplies power to various floors through radial circuits. The total length of the wiring from the MDB to each floor is as follows:

Ground Floor: 25 meters

First Floor: 30 meters

Second Floor: 35 meters

Third Floor: 40 meters

The total load on each floor is as follows:

Ground Floor: 15 kW

First Floor: 20 kW

Second Floor: 18 kW

Third Floor: 25 kW

The voltage drop permissible for each circuit is 5% of the rated voltage.

Assume the rated voltage is 230 volts.

Calculate:

1. The maximum permissible voltage drop.
2. The minimum allowable cross-sectional area of the conductors for each floor's radial circuit.

Solution:

1. Maximum Permissible Voltage Drop: Permissible voltage drop = 5% of rated voltage = 0.05×230 volts = 11.5 volts
2. Minimum Allowable Cross-sectional Area of Conductors: We will use the voltage drop formula:

Voltage drop (V_d) = $(2 \times R \times L \times I) / A$

Where:

R = Resistance of the conductor material (ohms/km)

L = Length of the conductor (meters)

I = current (amperes)

A = cross-sectional area of the conductor (mm^2)

We rearrange the formula to solve for the cross-sectional area (A):

$A = (2 \times R \times L \times I) / V_d$

Given:

R = Resistance per meter of the conductor material (ohms/m)

L = Length of the conductor (m)

I = Current (A)

V_d = Permissible voltage drop (V)

Let's assume the resistance of a copper wire is 0.0173 ohms per meter.

Using the formula, we calculate the minimum allowable cross-sectional area for each floor's radial circuit:

For Ground Floor: $A = (2 \times 0.0173 \times 25 \times 15{,}000) / 11.5$ A $\approx 210\ mm^2$

For First Floor: $A = (2 \times 0.0173 \times 30 \times 20{,}000) / 11.5$ A $\approx 300\ mm^2$

For Second Floor: $A = (2 \times 0.0173 \times 35 \times 18{,}000) / 11.5$ A $\approx 287\ mm^2$

For Third Floor: $A = (2 \times 0.0173 \times 40 \times 25{,}000) / 11.5$ A $\approx 434\ mm^2$

Therefore, the minimum allowable cross-sectional areas of conductors for each floor's radial circuit are approximately 210 mm^2 for the Ground Floor, 300 mm^2 for the First Floor, 287 mm^2 for the Second Floor, and 434 mm^2 for the Third Floor.

Example 3.2:

A commercial building requires conduit wiring for its electrical system. The building has four sections, each with different electrical load requirements and conduit lengths required. The following table summarizes the details:

Section	Length of Conduit (meters)	Load (kW)
Section 1	50	20
Section 2	60	15
Section 3	70	25
Section 4	80	18

The conduit manufacturer provides the following conduit size options along with their respective maximum allowable fill capacities:

1. 20mm conduit: Maximum allowable fill capacity = 25%
2. 25mm conduit: Maximum allowable fill capacity = 35%
3. 32mm conduit: Maximum allowable fill capacity = 40%

Determine the minimum conduit size required for each section to ensure the conduit fill does not exceed the maximum allowable capacity.

Solution: To determine the minimum conduit size required for each section, we need to calculate the fill percentage for each conduit size option and ensure that it is within the maximum allowable fill capacity.

Let's start by calculating the fill percentage for each section and each conduit size option:

1. For 20mm conduit:

   Section 1: Fill percentage = (Load / Maximum allowable fill capacity) × 100 = (20 / 25) × 100 = 80%

   Section 2: Fill percentage = (15 / 25) × 100 = 60%

   Section 3: Fill percentage = (25 / 25) × 100 = 100%

   Section 4: Fill percentage = (18 / 25) × 100 = 72%

2. For 25mm conduit:

   Section 1: Fill percentage = (20 / 35) × 100 = 57.14% (approximately)

Section 2: Fill percentage = (15 / 35) × 100 = 42.86% (approximately)

Section 3: Fill percentage = (25 / 35) × 100 = 71.43% (approximately)

Section 4: Fill percentage = (18 / 35) × 100 = 51.43% (approximately)

3. For 32mm conduit:

Section 1: Fill percentage = (20 / 40) × 100 = 50%

Section 2: Fill percentage = (15 / 40) × 100 = 37.5%

Section 3: Fill percentage = (25 / 40) × 100 = 62.5%

Section 4: Fill percentage = (18 / 40) × 100 = 45%

Based on the calculations, the minimum conduit size required for each section should be chosen so that the fill percentage does not exceed the maximum allowable capacity.

Considering the calculations for each section:

Section 1: 20mm conduit

Section 2: 20mm conduit

Section 3: 25mm conduit

Section 4: 20mm conduit

This ensures that the conduit fill is at most the maximum allowable capacity for each section.

Example 3.3:

A three-story residential building requires wiring for lighting, appliances, and power outlets. The building owner prefers surface-mounted wiring for its simplicity and cost-effectiveness. Each floor consists of four apartments, and each apartment has three rooms. Design the wiring layout for one apartment on the first floor and calculate the total length of surface-mounted wiring required.

Solution: For one apartment on the first floor:

Number of rooms = 3

Distance between rooms = 5 meters

Distance between main distribution panel and apartment = 20 meters

Total length of surface-mounted wiring required = Number of rooms × Distance between rooms + Distance between main distribution panel and apartment

= (3 rooms × 5 meters) + 20 meters = 15 meters + 20 meters = 35 meters

Example 3.4:

A commercial office space requires wiring for computers, printers, and lighting fixtures. The office manager prefers concealed conduit wiring for its sleek appearance and safety. There are four workstations in the office space, and each workstation requires power and network connectivity. Design the wiring layout for the office space and calculate the total length of concealed conduit wiring required.

Solution: For the office space:

Number of workstations = 4

Distance between workstations = 3 meters

Distance between main distribution panel and office space = 15 meters

Total length of concealed conduit wiring required = Number of workstations × Distance between workstations + Distance between main distribution panel and office space

= (4 workstations × 3 meters) + 15 meters = 12 meters + 15 meters = 27 meters

Example 3.5:

A warehouse facility requires wiring for overhead lighting and power outlets for machinery. The warehouse manager prefers overhead conduit wiring for its durability and ease of installation. The warehouse is 50 meters long and 30 meters wide. Calculate the total length of overhead conduit wiring required for the warehouse facility.

Solution: For the warehouse facility:

Length of the warehouse = 50 meters

Width of the warehouse = 30 meters

Total perimeter of the warehouse = 2 × (Length + Width)

= 2 × (50 meters + 30 meters) = 2 × 80 meters = 160 meters

Total length of overhead conduit wiring required = Total perimeter of the warehouse = 160 meters

Example 3.6:

A laboratory requires wiring for specialized equipment and lighting fixtures. The laboratory manager prefers flexible conduit wiring for its adaptability and ease of maintenance. The laboratory has five workstations, and each workstation requires power and data connections. Design the wiring layout for the laboratory and calculate the total length of flexible conduit wiring required.

Solution:

For the laboratory:

Number of workstations = 5

Distance between workstations = 4 meters

Distance between main distribution panel and laboratory = 25 meters

Total length of flexible conduit wiring required = Number of workstations × Distance between workstations + Distance between main distribution panel and laboratory

= (5 workstations × 4 meters) + 25 meters = 20 meters + 25 meters = 45 meters

Exercise

Multiple Choice Questions:

1. Which of the following wiring methods is commonly used for residential wiring due to its ease of installation and cost-effectiveness?

   a) Conduit Wiring b) Surface Wiring c) Armored Cable Wiring d) Busbar Trunking.

2. In conduit wiring, which type of conduit provides the highest mechanical protection?

   a) PVC Conduit b) Rigid Metal Conduit (RMC) c) Flexible Metal Conduit (FMC) d) Electrical Metallic Tubing (EMT).

3. Which of the following is a characteristic of busbar trunking?

   a) It provides mechanical protection to cables. b) It is flexible and easy to install. c) It is commonly used in residential wiring. d) It allows for easy access and modification of wiring.

4. Which wiring layout is commonly used for industrial and commercial installations due to its flexibility and ease of maintenance?

   a) Loop-in Wiring b) Radial Wiring c) Ring Main Wiring d) Grid Wiring.

5. Which of the following is a benefit of using surface wiring?

   a) It provides excellent mechanical protection. b) It is hidden within the walls, making it aesthetically pleasing. c) It is easy to install and modify. d) It is the most cost-effective wiring method.

6. Which of the following wiring methods is best suited for areas where the environment is corrosive or hazardous?

   a) Surface Wiring, b) Conduit Wiring, c) Armored Cable Wiring, d) Non-Metallic Sheathed Cable Wiring.

7. Which type of conduit is commonly used for underground installations due to its resistance to moisture and corrosion?

   a) PVC Conduit b) EMT Conduit c) Rigid Metal Conduit (RMC) d) Flexible Metal Conduit (FMC).

8. what is the purpose of using cable trays in electrical installations?

a) To provide mechanical protection to cables b) To organize and support cables in an open-air environment c) To conceal cables within walls d) To facilitate easy access for cable termination.

9. Which wiring method provides the highest degree of protection against physical damage and environmental factors?

a) Surface Wiring, b) Conduit Wiring, c) Armored Cable Wiring, d) Non-Metallic Sheathed Cable Wiring.

10. In ring main wiring, what is the characteristic configuration of the wiring layout?

a) Each outlet is connected directly to the main distribution board (MDB)

b) Outlets are connected in a loop configuration, allowing multiple paths for the current flow

c) Outlets are connected in a radial configuration, with each outlet branching out from the MDB

d) Outlets are connected in a series configuration, where the current flows sequentially through each outlet

11. What is an advantage of using armored cable wiring?

a) It is lightweight and flexible. b) It provides excellent mechanical protection. c) It is the most cost-effective wiring method. d) It is easy to install in tight spaces.

12. What is the primary purpose of conduit fill requirements in electrical installations?

a) To ensure proper grounding of the electrical system b) To prevent electromagnetic interference c) To limit the number of conductors placed within a conduit d) To increase the current-carrying capacity of the conductors.

13. Which wiring method is commonly used in exposed areas where the wiring needs to be concealed for aesthetic purposes?

a) Surface Wiring, b) Conduit Wiring, c) Armored Cable Wiring, d) Non-Metallic Sheathed Cable Wiring.

14. How is each outlet connected to the main distribution board (MDB) in radial wiring?

a) Outlets are connected in a loop configuration. b) Outlets are connected in a series configuration. c) Each outlet is connected directly to the MDB. d) Outlets are connected in a parallel configuration.

15. Which wiring method is often preferred in areas with a risk of moisture or dampness exposure?

a) Surface Wiring, b) Conduit Wiring, c) Armored Cable Wiring, d) Non-Metallic Sheathed Cable Wiring.

Answers:

1. b) Surface Wiring
2. b) Rigid Metal Conduit (RMC).
3. d) It allows for easy access and modification of wiring.
4. c) Ring Main Wiring.
5. c) It is easy to install and modify.
6. b) Conduit Wiring.
7. c) Rigid Metal Conduit (RMC).
8. b) To organize and support cables in an open-air environment.
9. c) Armored Cable Wiring.
10. b) Outlets are connected in a loop configuration, allowing multiple paths for current flow.
11. b) It provides excellent mechanical protection.
12. c) To limit the number of conductors placed within a conduit.
13. a) Surface Wiring.
14. b) Outlets are connected in a series configuration.
15. b) Conduit Wiring.

Short Answer Type Questions:

1) What is conduit wiring, and why is it commonly used in electrical installations?
2) Explain the difference between surface wiring and concealed wiring.
3) What are the main components of a conduit wiring system?
4) Describe the purpose of armoured cable wiring and its typical applications.
5) What factors should be considered when selecting the appropriate wiring method for an electrical installation?
6) What is the significance of conduit fill-in conduit wiring installations?
7) Explain the concept of radial wiring and provide an example of its application.
8) What is the purpose of using cable trays in electrical installations, and how do they differ from conduit systems?
9) Describe the critical characteristics of ring main wiring and its advantages over radial wiring.

10) How does the choice of wiring method impact the safety and reliability of an electrical system?

Long Answer Type Questions:

1) Compare and contrast conduit wiring and surface wiring. Discuss their respective advantages, disadvantages, and typical applications in electrical installations.
2) Describe a conduit wiring system's components and installation process in detail. Explain the different types of conduits used and their specific applications in various environments.
3) Discuss the importance of proper wiring layout design in residential, commercial, and industrial buildings. Highlight the key factors influencing wiring layout decisions and the implications of poor design design.
4) Explain the concept of radial wiring and ring main wiring. Compare and contrast these two wiring layouts regarding their structure, installation requirements, and suitability for different types of buildings.
5) Investigate the role of wiring methods in ensuring electrical safety and compliance with building codes and regulations. Discuss the importance of selecting the proper wiring method based on environmental factors, load requirements, and maintenance considerations.

Numerical Problems:

1) Conduit Wiring: Calculate the voltage drop in a 30-meter-long conduit wiring system supplying a load of 15 kW at 220 volts. The conduit has a resistance of 0.018 ohms per meter. (0.0162A)
2) Surface Wiring: Determine the voltage drop in a surface wiring system with a total length of 25 meters, supplying a load of 12 kW at 240 volts. The resistance of the wiring is 0.025 ohms per meter. (7.5V)
3) Armored Cable Wiring: Calculate the voltage drop in an armored cable wiring installation with a length of 40 meters, supplying a load of 18 kW at 230 volts. The resistance of the armored cable is 0.022 ohms per meter. (15.84V)
4) Ring Main Wiring: Determine the maximum allowable resistance for a ring primary wiring circuit supplying a load of 20 kW at 240 volts, with a permissible voltage drop of 4%. (9.6ohm)

5) Radial Wiring: Calculate the voltage drop in a radial wiring system supplying a load of 30 kW at 220 volts. The total length of the radial wiring is 50 meters, and the resistance per meter is 0.015 ohms. (0.045V)
6) Cable Tray Design: Design a cable tray system for a three-floor commercial building. Each floor has a total load of 40 kW. Determine the minimum cable tray size required if the permissible voltage drop is 6%.
7) Conduit Fill Calculation: Calculate the percentage fill of a conduit containing 10 conductors with a diameter of 10 mm each, given that the conduit diameter is 40 mm.
8) Conduit Sizing: Determine the minimum conduit size required to accommodate 15 conductors, each with a diameter of 8 mm.
9) Cable Routing Length: A cable must be routed from the main distribution board to a machine 60 meters away. Calculate the voltage drop if the cable resistance is 0.02 ohms per meter and the current passing through it is 25A. (30V)
10) Conduit Resistance: Determine the total resistance of a conduit made of PVC with a length of 35 meters and a resistance of 0.015 ohms per meter. (0.525)

Unit Summary

The unit summary for wiring methods and wiring layout is written below:

- Introduction to Wiring Methods
- Types of Wiring Methods
- Wiring Materials and Components
- Installation Techniques and Best Practices
- Wiring Layout and Design
- Code Compliance and Regulations

This unit provides a comprehensive overview of wiring methods, installation techniques, wiring layout design principles, code compliance, troubleshooting, and emerging trends in electrical wiring systems.

Practical

1. Demonstrate the installation of surface-mounted raceways or wire channels on walls or ceilings.
2. Describe the steps involved in installing a PVC conduit for an electrical circuit. Include the tools and materials required.

3. Design a cable tray layout for a commercial building, considering factors such as cable types, support spacing, and load distribution.
4. Demonstrate installing a wire channel on a wall and route wires inside it while maintaining proper separation and support.
5. Identify common mistakes or errors during wire termination and how to avoid them for reliable electrical connections.
6. Conduct a safety inspection checklist for a wiring installation, highlighting key areas such as grounding, insulation, and proper labelling of circuits.

Know More

Sl No	Topic	QR Code
1	Indian Standards Building Service	
2	National Electrical Code 2011	
3	Electrical Safety and National Electrical Code 2011	

4	Indian Standard Details	
5	Electrical Measurement and Instrumentation	
6	Utilization of Electrical Energy	

References and Suggested Readings

1. W.E. Steward, and T.A. Stubbs, *Modern wiring practice*, Routledge, 2005.
2. D. Locke, *Guide to the wiring regulations: IEE wiring regulations (BS 7671: 2008)*, John Wiley & Sons, 2008.
3. A.J. Pansini, *Electrical distribution engineering*, River Publishers, 2020.

UNIT 4 RESIDENTIAL BUILDING ELECTRIFICATION

UNIT SPECIFICS

This unit elaborately discusses the following topics:

- ✓ Introduction to building drawing, electrical installation plan, and electrical diagrams of residential building electrification
- ✓ Exploring the difference between residential and industrial load
- ✓ Classification of electrical load assessment
- ✓ Residential installation design, drawing, estimating, and price.
- ✓ Various types of testing of wiring installation as per IS:732-1982
- ✓ Exploring various types of residential building service connection

This section provides a comprehensive examination of the significance of domestic dwellings and residential buildings, the analysis of civil engineering building designs, the interpretation of electrical installation plans and diagrams, and the utilization of electrical symbols by IS: 732. This subject covers the distinction between residential and industrial loads, the regulations and criteria for lighting load in electrical systems, the placement of equipment, and the lighting and power circuits. This report presents the wiring and circuit schematic diagram by IS: 2042(Part-I)-1962. In addition to covering load assessment, protective switch gear, central switch rating, and conductor size selection, this unit also covers these topics. This report provides an overview of the design, drawing, estimates, and costing of a residential installation with a maximum load capacity of 5 kW. It outlines the sequential steps in creating an estimate, calculates the length of wire and other materials, and includes information on labour costs. This paper examines the various service connections in residential buildings, including underground and overhead. Additionally, it explores the computation of materials necessary for establishing service connections.

RATIONALE

This fundamental unit on creating detailed building drawings, electrical installation plans, and electrical diagrams for residential building electrification is multifaceted and crucial for ensuring electrical systems' safety, compliance, efficiency, and effective management. The electrical installation plan details the design and layout of the residential building's electrical components, circuits, distribution boards, grounding systems, and protective devices. It includes load calculations, wire sizing, voltage requirements, and equipment specifications based on user needs and electrical codes. Based on the load assessment, the design phase involves creating a layout for electrical circuits, distribution panels, wiring paths, outlets, switches, and grounding systems within the residential premises. Electrical designs and drawings adhere to applicable building codes, electrical standards (e.g., BIS, NEC, IEC), and local regulations. Estimation involves calculating the electrical installation's materials (wires, cables, conduits, devices, etc.) based on the design and load requirements. Costing includes pricing these materials and estimating labour costs for installation.

PRE-REQUISITES

Basics of Electrical Engineering

UNIT OUTCOMES

The list of outcomes of this unit is as follows:

U4-O1: Describe various building drawings, electrical installation plans, and electrical diagrams.

U4-O2: Explain the classification of electrical load and electrical load assessment.

U4-O3: Explain a residential installation's design, drawing, estimation, and cost.

U4-O4: Describe various types of wiring installation and residential building service connection.

Unit-4 Outcomes	Expected Mapping with Programme Outcomes (*1- Weak Correlation; 2- Medium correlation; 3- Strong Correlation*)					
	CO-1	**CO-2**	**CO-3**	**CO-4**	**CO-5**	**CO-6**
U4-O1	3	3	1	3	1	2
U4-O2	3	3	1	3	1	2
U4-O3	3	3	1	3	1	2
U4-O4	3	3	1	3	1	2

4.1 Introduction: Domestic Dwellings/Residential Buildings

Every electrical setup commences with a detailed plan and design, whether in a residential, commercial, or industrial environment. Factors and elements such as structure sorts, purposes, and physical properties are considered when planning and installing a structure. Electrical design is creating, constructing, testing, and installing electrical equipment according to set regulations. This study focuses on the role of an electrical contractor in installing residential buildings by following safety measures, complying with regulations, and executing plans as instructed. Various factors were considered during the design process. Some factors to consider include installation cost, flexibility, durability, and safety. This job focuses on estimating and calculating loads, installing energy meters based on loads, and calculating and implementing earthing. Building often involves various instances, such as mechanical and electrical installations necessitating specialized tools and techniques. Electrification measures for buildings are considered dependent on structural measures.

4.1.1 The Steps of Electrification of the Building:

Determining the building's construction type is necessary for developing the structural and electrical systems. Subsequently, the electrical engineer will develop a comprehensive design outlining all specifics, such as component dimensions and types, tailored to meet the building's needs and energy usage. Electrical engineers and electricians are encompassed in the following concept.

- ✓ Developing the electrical engineer's blueprint for the residential building, installing the electrical panel, and arranging the apparatus according to the map's plan.
- ✓ Constructing the framework required for installing hoses, channels, fittings, and other types of cold bending and polyamide electrical pipes.
- ✓ Set up an earth well system and connect the required hardware to the facility map.
- ✓ Choose the wire and cable type according to the map's layout and path through the electrical conduit and channel, create panels, and divide boxes to disperse them across the building.
- ✓ Install a compact fuse box with internal safety switches to link cables and wires per the installation blueprint.
- ✓ The little box outputs to the lights, switches, and socket connection points.
- ✓ Installing a parking door jack, lighting system, switches, and connections.

✓ The lift seller is responsible for installing the equipment and implementing the lift.

✓ Establishing connections between the building's engine room, mechanical systems, and electrical wiring.

4.1.2 Reading of Civil Engineering Building Drawing:

Engineering drawings are interpreted using a standardized set of symbols and conventions. Technical drawings may need clarification first due to specialized symbols and codes. Engineering drawings are produced following technical standards and professional conventions to ensure they are easily understood and interpreted by all viewers. To comprehend an engineering drawing, one needs to understand the following parts. The elements and codes function as the fundamental components of an engineering drawing. Interpreting civil engineering building drawings for electrification entails analyzing different drawings and papers about the electrical architecture and installation inside the building.

- **Electrical Floor Plans**:

✓ **Layout**: Examine the electrical floor plans to comprehend the arrangement of electrical fixtures, outlets, switches, and other elements across the building.

✓ **Symbol Legend**: Consult the caption or key on the illustrations to comprehend the symbols representing electrical components.

✓ **Room Designations**: Determine the positions of electrical fixtures in every room, such as lighting fixtures, outlets, and switches.

✓ **Circuit Designations**: Identify the circuits by their numbers or markings that match specific electrical loads.

- **Electrical Schematics**:

✓ **Circuit Diagrams**: The control systems, circuits, and wiring connections for electrical loads, such as lights and outlets, can be better understood by reviewing electrical diagrams.

✓ **Component Identification**: Note the connections and ratings of electrical components such as transformers, circuit breakers, switches, and relays.

✓ **Control Logic**: Learn how various electrical parts, such as switches, sensors, and control panels, work together and how the control logic works.

- **Panel Schedules**:

✓ **Distribution Panels**: Examine panel schedules to comprehend the arrangement of electrical circuits in distribution panels or breaker boxes.

- ✓ **Circuit Details**: Record the specifics of each circuit, such as circuit numbers, wire gauges, breaker sizes, and loads connected.
- ✓ **Load Balancing**: Ensure electrical loads are evenly distributed among circuits to avoid overloading and maintain safety.

Electrical Details:

- ✓ **Construction Details**: Before building or installing any electrical fixtures, conduits, wiring, or equipment, it is essential to review the electrical specifics.
- ✓ **Mounting Requirements**: Remember how to mount things like electrical outlets, switches, and panel boards, such as the necessary heights, clearances, and ways of support.
- ✓ **Conduit and Wiring Details**: Gain knowledge of the necessary steps for installing and routing electrical conduits, raceways, and wire systems, which encompass bends, connections, and supports.

Specifications and Notes:

- ✓ **General Requirements**: Examine specifications and notes concerning electrical work to grasp overall needs, standards, and compliance criteria.
- ✓ **Material Specifications**: Specify electrical materials, fixtures, equipment, and finishes with details such as product types, sizes, ratings, and manufacturers.
- ✓ **Installation Requirements**: Understand installation requirements, procedures, and quality standards for electrical work, including testing, labelling, and documentation.

Coordination with Other Disciplines:

- ✓ **Architectural**: Align electrical schematics with architectural blueprints to guarantee seamless integration of electrical elements with architectural structures and designs.
- ✓ **Structural and MEP**: Align electrical schematics with structural and MEP blueprints to guarantee accurate positioning and reinforcement of electrical machinery, conduits, and fixtures.

Understanding civil engineering building plans for electrification necessitates meticulous attention to detail, knowledge of electrical symbols and protocols, and collaboration with other fields.

4.1.3 Interpretation of Electrical Installation Plan and Electrical Diagrams:

Understanding electrical installation plans and diagrams is crucial for grasping the arrangement, elements, and interconnections of electrical systems in structures.

- **Electrical Installation Plan:**
- ✓ **Layout**: Inspect the building plan and pinpoint the positions of electrical elements such as outlets, switches, lighting fixtures, and panel boards.
- ✓ **Room Designations**: Identify the names of each room or section in the building and the electrical fixtures and outlets.
- ✓ **Circuit Designations**: Identify the circuits for lighting, power outlets, appliances, and other electrical demands. Take note of the circuit numbers or labels for each circuit.
- ✓ **Wiring Routes**: Comprehend the layout of electrical wiring in the building, including the routes of conduits, raceways, and cable trays.
- ✓ **Dimensional Information**: Refer to the plan's dimensions and measurements to position and correctly space electrical components.
- ✓ **Legend**: Consult the legend or key on the plan to comprehend symbols, acronyms, and notes denoting electrical components and features.
- ✓ **Elevation Views**: Examine elevation views to comprehend the vertical positioning of electrical components, including outlets, switches, and fixtures on walls and ceilings.
- **Electrical Diagrams:**
- ✓ **Circuit Diagrams**: Examine circuit diagrams to comprehend the electrical connections, components, and control logic for specific circuits in the building.
- ✓ **Component Identification**: Identify electrical components shown in the diagram, including switches, relays, circuit breakers, transformers, and wiring connections.
- ✓ **Symbols and Notations**: Comprehend the symbols, notations, and norms in the schematic for electrical components, connections, and control logic.
- ✓ **Control Systems**: Identify the control systems and components shown in the diagram, such as switches, sensors, timers, and programmable logic controllers (PLCs).
- ✓ **Power Distribution**: Identify how power is distributed from the main service panel or distribution boards to individual circuits, outlets, and fixtures throughout the building.
- ✓ **Load Balancing**: Observe the allocation of electrical loads among circuits to avoid overloading and maintain a balanced power distribution.
- ✓ **Safety Devices**: Fuses, circuit breakers, ground fault circuit interrupters (GFCIs), surge protectors, and other safety equipment are depicted in the diagram.
- ✓ **Voltage and Current Ratings**: Take note of the voltage and current ratings of electrical components and circuits shown in the diagram to guarantee compatibility and safety.

Coordination and Integration:

- ✓ **Architectural and Structural Coordination**: Align electrical plans and diagrams with architectural and structural designs to guarantee accurate positioning and incorporation of electrical components with building features and structural elements.
- ✓ **Mechanical, Electrical, and Plumbing (MEP) Coordination**: Coordinate MEP by aligning electrical plans and diagrams with mechanical and plumbing plans to ensure accurate placement and integration of electrical systems with mechanical and plumbing systems.
- ✓ **Code Compliance**: Ensure electrical plans and diagrams adhere to local building requirements, regulations, and industry standards for electrical installations and safety.

Understanding electrical installation plans and diagrams necessitates knowledge of electrical symbols, conventions, terminology, and collaboration with other disciplines in building design and construction. One can comprehend the following.

- The role of symbols in logic diagrams
- Identify electrical components
- Understand abbreviations, symbols, and terminology in electrical drawings
- Recognize various types of electrical drawings
- Understanding electrical, earthing, and cabling arrangements
- The process of creating a block map of mounted equipment
- The data should be transferred to the one-line diagram.
- Comprehending ladder diagrams, logic symbols and how to understand and use them
- Check electrical printouts and maintain regulatory-compliant electrical diagrams.
- Drawings are currently undergoing revision and updating.

Electrical drawings are a precise and formal way of conveying information regarding electrical systems' layout, size, characteristics, and accuracy. Engineering's universal language is drawing. Understanding the typical standards, regulations, and symbols used in different electrical designs is necessary to interpret them effectively. The goal of this course is to teach those who work in the electrical field how to read and understand schematics and drawings, as well as the rules and regulations that govern their interpretation. The participants will receive an introduction to electrical symbols, as well as one-line and three-line electrical schematics, along with their respective content, which encompasses crucial layouts and legends. The course will help participants comprehend the symbols and terminology used in electrical drawings following worldwide standards and norms. It aims to provide a comprehensive grasp of the many types of electrical drawings used in the industry and their purposes and applications.

4.1.4 Residential Wiring Diagram:

A wiring diagram is a visual depiction of an electrical circuit, showcasing the constituent elements of the circuit and the interconnections among devices and the power supply through simplified shapes adhering to established rules. A domestic wiring diagram is a house's electrical wiring system schematic. A domestic wiring diagram is a simplified schematic of an electric circuit in your home, designed for easy understanding and following by an electrician or homeowner. The diagram comprises interconnections between circuit elements and their connections to and from the power source. It considers the room's dimensions and the selected input and output locations, prioritizing easy access. House wiring diagrams are schematic representations of electrical circuits within a house.

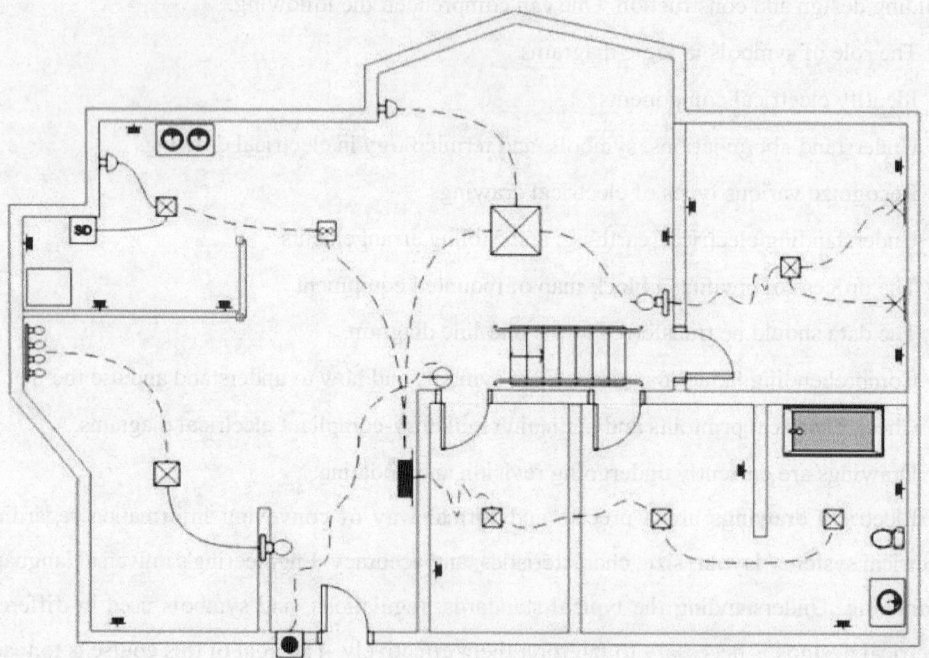

Figure 4.1 House wiring diagram

The electrical wiring diagram (Fig.4.1) is a graphic representation of a house or room's electrical wiring system or circuitry. It helps create the system to distribute energy and power appliances and equipment throughout the home. The diagram includes electrical outlets, meter bases, switches, and breakers.

The essential elements of a simple home wiring schematic will be covered in this essay. We will understand what it is, why we need it, what advantages it can offer, and the concepts and symbols involved in the procedure. Lastly, some examples demonstrating how to create a wiring diagram for a house will be examined.

Usage of House Wiring Diagram:

Undoubtedly, the significance of a house wiring diagram cannot be overstated. However, we must address the necessity for one. As previously stated, it is an essential component of wiring your residence. The following explanations are provided:

- ✓ **Financial:** An electrician's ability to create an electrical system is limited to approximating the required wiring. Conversely, the house wiring diagram might offer a precise value. An electrician who does not utilize a house wiring diagram may be ineffective. There are two likely results: Without precise measurements, your electrician may employ more wiring, resulting in unnecessary installation costs and financial waste.

Conversely, having less wire might lead to more time spent acquiring additional wiring, causing delays and higher costs.

- ✓ **Timescale:** Both the budget and the timeline are negatively impacted. In the event that additional wire is required, it is necessary to procure additional wire or await the arrival of the new installation. It has the potential to cause delays in completing large-scale projects such as an apartment complex or a condominium.
- ✓ **Prevent Injuries:** When dealing with electrical systems, we are all instructed to prioritize safety. Injuries can occur in various locations and at any time, resulting from numerous causes. An informative house wiring diagram can indicate to electricians the locations of problems throughout the entire network. It can assist them in steering clear of sharp objects, exposed live wires, and various other dangerous situations. An injury could result in a significant delay and a costly hospital bill. A house wiring diagram is the most secure remedy for such situations.

With a house wiring diagram, constructing a proper electrical system is nearly possible, even for the most skilled electrician.

4.1.5 Electrical Symbols as per IS:732:

IS 732 is an Indian Standard that specifies graphical symbols for electrical wiring and layout diagrams in electrical engineering. It offers a collection of standardized symbols depicting different electrical components and devices in diagrams and drawings. Here are some prevalent electrical symbols according to IS 732:

a) **Switches**:

- ✓ Single pole switch
- ✓ Double pole switch
- ✓ Push button switch
- ✓ Knife switch
- ✓ Selector switch

b) **Circuit Breakers and Fuses**:

- ✓ Circuit breaker
- ✓ Fuse
- ✓ Thermal overload relay
- ✓ Earth leakage circuit breaker (ELCB)
- ✓ Miniature circuit breaker (MCB)

c) **Contactors and Relays**:

- ✓ Contactor
- ✓ Relay
- ✓ Normally open contact
- ✓ Normally closed contact
- ✓ Changeover contact

d) **Indicators and Lights**:

- ✓ Pilot lamp
- ✓ Indicator lamp
- ✓ Fluorescent lamp
- ✓ Incandescent lamp
- ✓ LED (Light Emitting Diode)

e) **Transformers and Motors**:

- ✓ Transformer
- ✓ Single-phase motor

- ✓ Three-phase motor
- ✓ Synchronous motor
- ✓ DC motor

f) **Capacitors and Resistors**:

- ✓ Capacitor
- ✓ Variable capacitor
- ✓ Resistor
- ✓ Potentiometer
- ✓ Rheostat

g) **Power Supply and Distribution**:

- ✓ Power supply
- ✓ Battery
- ✓ Earth connection
- ✓ Ground symbol
- ✓ Busbar

h) **Wires and Connections**:

- ✓ Conductor
- ✓ Coaxial cable
- ✓ Shielded cable
- ✓ Optical fiber cable
- ✓ Connection point

i) **Miscellaneous Symbols**:

- ✓ Grounding electrode
- ✓ Lightning arrestor
- ✓ Junction box
- ✓ Terminal block
- ✓ Meter

The symbols depict electrical components and devices in several electrical diagrams, such as circuit diagrams, wiring diagrams, layout diagrams, and schematics. Standard wiring symbols are shown in Fig.4.2.

IS: 732 (1989)

Standardized Wiring Diagram Symbols

STANDARDIZED WIRING DIAGRAM SYMBOLS

ANTENNAS: GENERAL, DIPOLE, LOOP, FERRITE TYPE

BATTERIES: SINGLE-CELL, MULTI-CELL

BELL

BUZZER

CAPACITORS: FIXED MICA OR PAPER, ELECTROLYTIC, VARIABLE CAPACITORS GANGED, TRIMMER OR PADDER, SPLIT-STATOR, FEED-THRU

T-DESIGNATES TRIMMER OR PADDER

COILS: FIXED R.F. COIL, COIL WITH FIXED TAP, COIL WITH VARIABLE TAP, VARIABLE COIL OR CHOKE, IRON CORE OR CHOKE, BIFILAR, SLUG-TUNED COIL

CRYSTALS: CRYSTAL DETECTOR, PIEZOELECTRIC CRYSTAL

FUSE

GROUNDS: WIRING, CHASSIS

HEADPHONES: DOUBLE, SINGLE

JACKS: OPEN CIRCUIT, CLOSED CIRCUIT, 2 CIRCUIT, SHORTING TYPE 2 CIRCUIT, PHONE PLUG, PIN PLUG, PIN JACK

METER

MOTOR

MICROPHONE: SINGLE BUTTON, DOUBLE BUTTON, CAPACITOR, DYNAMIC, VELOCITY, CRYSTAL

NEON BULB

PILOT LIGHT

PHONO PICKUPS: CRYSTAL, MAGNETIC

PLUG OR SOCKET

RECEPTACLE 117V.

RECORDING HEAD: MAGNETIC

RECTIFIER: SELENIUM TYPE

RELAYS: S.P.S.T. NORMALLY OPEN, S.P.S.T. NORMALLY CLOSED, S.P.D.T.

RESISTORS: GENERAL, TAPPED OR ADJUSTABLE, POTENTIOMETER OR RHEOSTAT, CONTINUOUSLY VARIABLE

SHIELDING: DOTTED LINES INDICATE SHIELDING COULD BE AROUND ANY COMPONENT OR GROUPS

SPEAKERS: PM, ELECTRO-MAGNETIC

SWITCHES: S.P.S.T., S.P.D.T., PUSH BUTTON, D.P.S.T., D.P.D.T., BOTH TYPES GANGED, ROTARY TYPE S.P.6.P., WAFER

TELEGRAPH KEY

TRANSFORMERS: AIR CORE, IRON CORE, AIR CORE VARIABLE, IRON CORE VARIABLE, POWDERED IRON CORE, AUTO

PRIMARY, 117V., C.T., FILAMENTS, HIGH VOLTAGE, FILAMENTS, POWER TRANSFORMER

TUBES: PLATE, GRID, DIODE, HEATERS OR FILS., CATHODE, TRIODE, TETRODE, SURPRESSOR, SCREEN, CONTROL, PENTODE, PLATE CAP, GRID CAP, DIODE PLATE, DUO TRIODE, COLD CATHODE, DESIGNATES GAS FILLED, PHOTO-ELECTRIC

TRANSISTOR (JUNCTION): BASE, COLLECTOR, EMITTER, P-N-P TYPE, N-P-N TYPE SAME SYMBOL EXCEPT ARROW IS REVERSED

VIBRATOR

WIRES: CONNECTION, NO CONNECTION

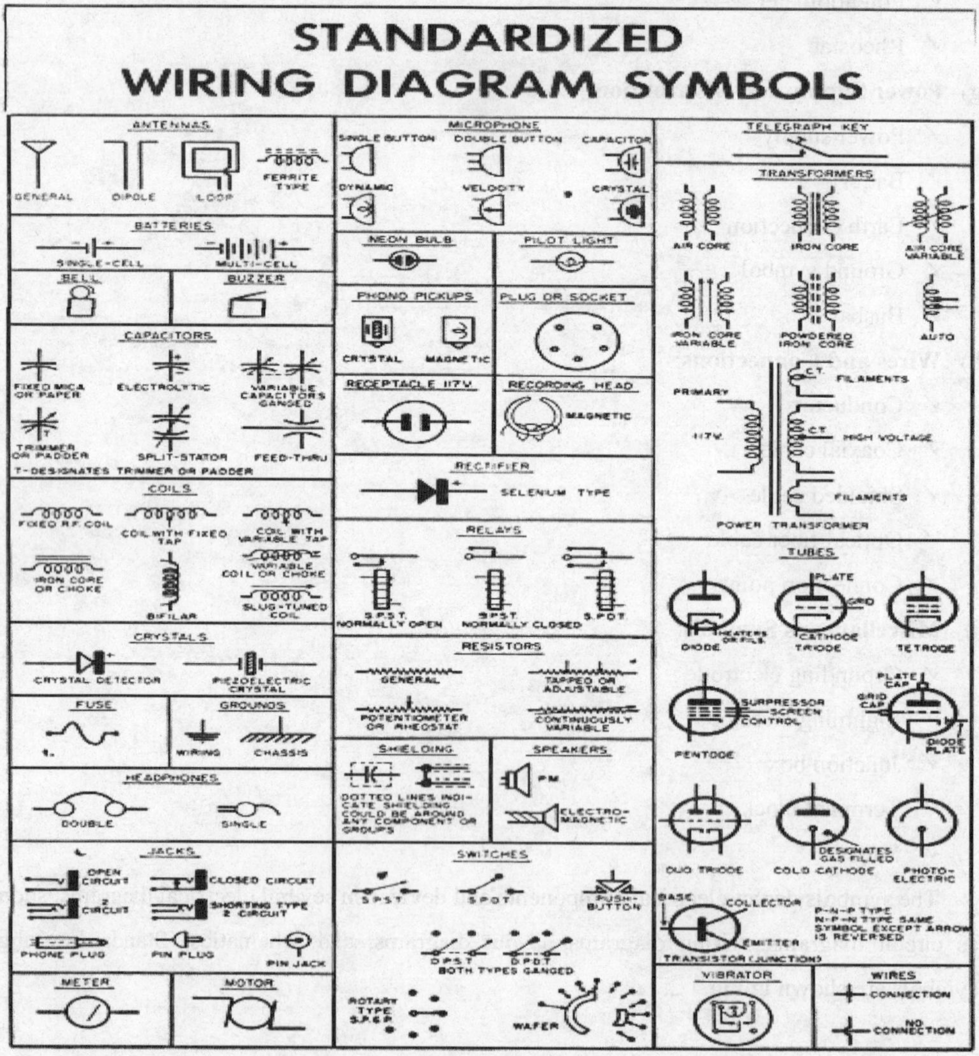

Figure 4.2 Standard wiring symbols

Sl. No	Elements	Symbols
1	Electrical switch box	
2	Single pole switch	S
3	Three-way switch	S_3
4	1P switch	
5	2P switch	
6	1DP switch	
7	2DP switch	
8	Water tap	
9	Wall light	
10	circuit breaker	
11	Light bar	
12	Junction box	J
13	Lamp holder	L
14	Water heater	WH
15	Electrical panel	
16	Battery	
17	Fire alarm	

18	Doorbell	●
19	Smoke detector	SD
20	Ground	⏚

Aside from these 20 symbols, a house wiring diagram may include lighting, electrical, telecom symbols, wall, shell, or structure symbols.

4.2 Electrical Installation for Residential Building as per Part I Section 9 of NEC-2011:

Section 9 of the 2011 National Electrical Code (NEC) edition discusses "Tables." It includes different tables concerning electrical installations, such as conductor sizes, conduit fill, motor circuit conductors, and other essential information. Nevertheless, it does not explicitly provide information on electrical installations in residential structures.

It is common practice to consult the NEC's provisions about residential wiring standards while doing electrical work in residential buildings, such as houses, apartments, and condominiums. Listed below are a few of the most critical sections of the NEC that pertain to homeowner electrical wiring.

- ✓ **General Wiring Methods (Article 300)**: This article outlines the specifications for wiring methods, encompassing wiring within buildings, raceways, and enclosures. The subjects covered include installation criteria, conductor insulation, and permissible wiring methods.
- ✓ **Branch Circuits (Article 210)**: Branch circuits, responsible for supplying electricity to appliances, lights, and outlets in residential buildings, are discussed in Article 210. It covers load calculations, outlets, overcurrent prevention, and conductor sizing.
- ✓ **General Requirements for Electrical Installations (Article 110)**: Article 110 outlines the fundamental criteria for electrical installations, covering the installation and use of equipment, space needed for work, and specifications for electrical enclosures.
- ✓ **Grounding and Bonding (Article 250)**: Essential for electrical safety in residential structures; this article contains regulations for grounding and bonding. This covers the necessary steps for grounding electrical equipment, metal water pipe systems, and electrodes.

- ✓ **Overcurrent Protection (Article 240)**: Article 240 discusses regulations for overcurrent protection, detailing guidelines for choosing and setting up circuit breakers, fuses, and other safety devices for household circuits.
- ✓ **Appliances (Article 422)**: Article 422 outlines the criteria for installing appliances such as stoves, dryers, dishwashers, and garbage disposals in residential structures.
- ✓ **Lighting Systems (Article 410)**: This article outlines the criteria for installing lighting fixtures and systems in residential structures, covering regulations for fixture support, wiring techniques, and clearance specifications.
- ✓ **Receptacles, Cord Connectors, and Attachment Plugs (Article 406)**: Article 406 outlines specifications for receptacles in residential buildings, such as spacing, position, GFCI protection, and tamper-resistant features.

Referring to the specific criteria in the NEC and local building rules and regulations is crucial when designing and carrying out electrical installations in residential structures.

NEC-2011

4.2.1 Examples of House Wiring Diagram:

This section will examine three house wiring diagrams for a bedroom, bathroom, and basement.

Bedroom Wiring Diagram:

Imagine a simple bedroom with two lights and a solitary ceiling fan. Under these circumstances, the house wiring diagram would appear as follows. The Live, Neutral, and Earth wires are color-coded as red, black, and green, respectively, for easy identification. Switch one is connected to bulb one, and switch three is connected to bulb two. The connections are direct, with the live and neutral wires correctly secured. Switch two is linked to the ceiling fan using a dimmer switch or a regulator to adjust the fan speed. The bedroom wiring plan is shown in Fig.4.3.

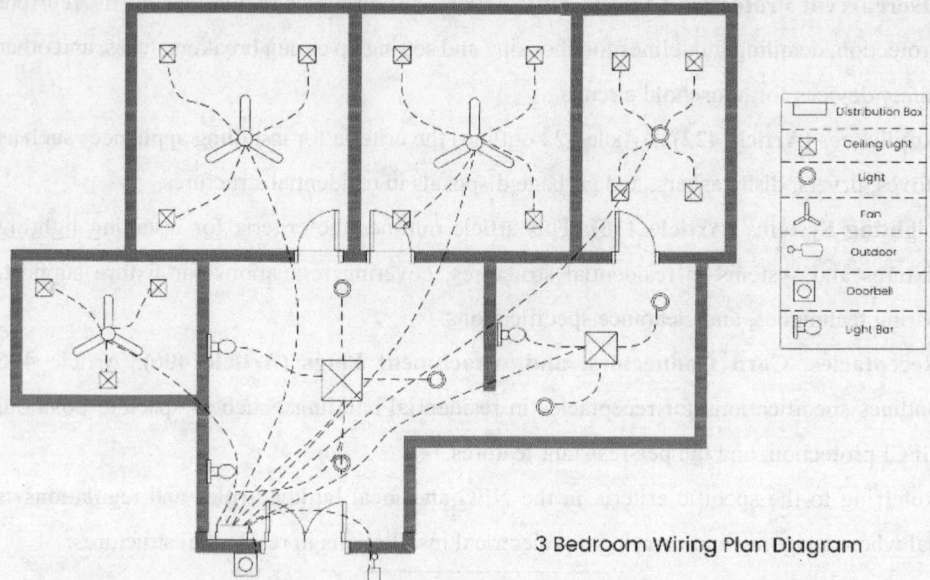

Figure 4.3 Bedroom wiring plan

- **Bathroom Wiring Diagram:**

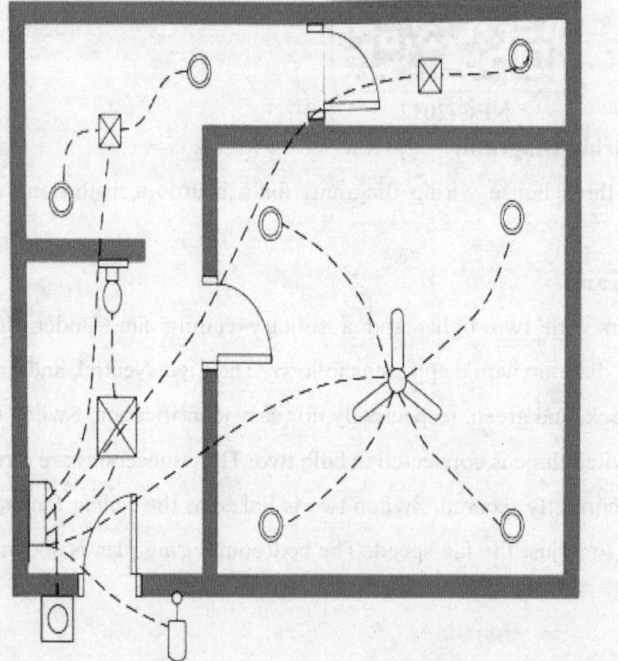

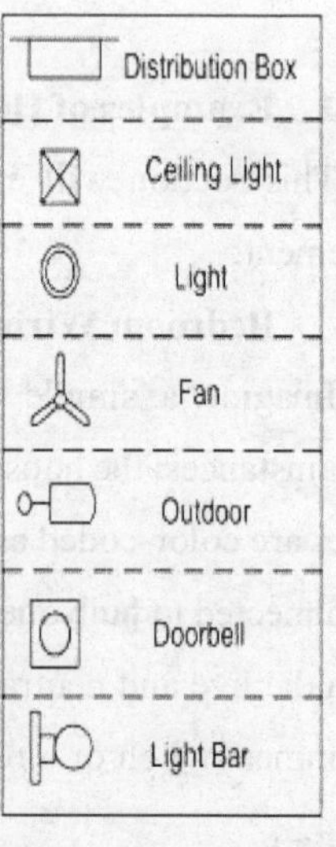

Figure 4.4 House wiring plan

Examine the bathroom wiring schematic provided below. Three switches are displayed on the central switchboard. Turn off one of the four 60-watt lamps on the left side above the washing machine. Switch two is connected to the central ceiling fan, while switch three is connected to two 50-watt lights. The underfloor heating device is linked to the 180-volt thermostat line voltage. The house wiring plan is shown in Fig.4.4.

- **Basement Wiring Diagram:**

Square patterns in the Basement Wiring Diagram represent switches, whereas circular or rectangular patterns represent light fixtures. A basic wiring diagram illustrates electrical components' series and parallel connections like lighting. The various colored sections do not correspond to wire color labeling. The basement wiring plan is shown in Fig.4.5.

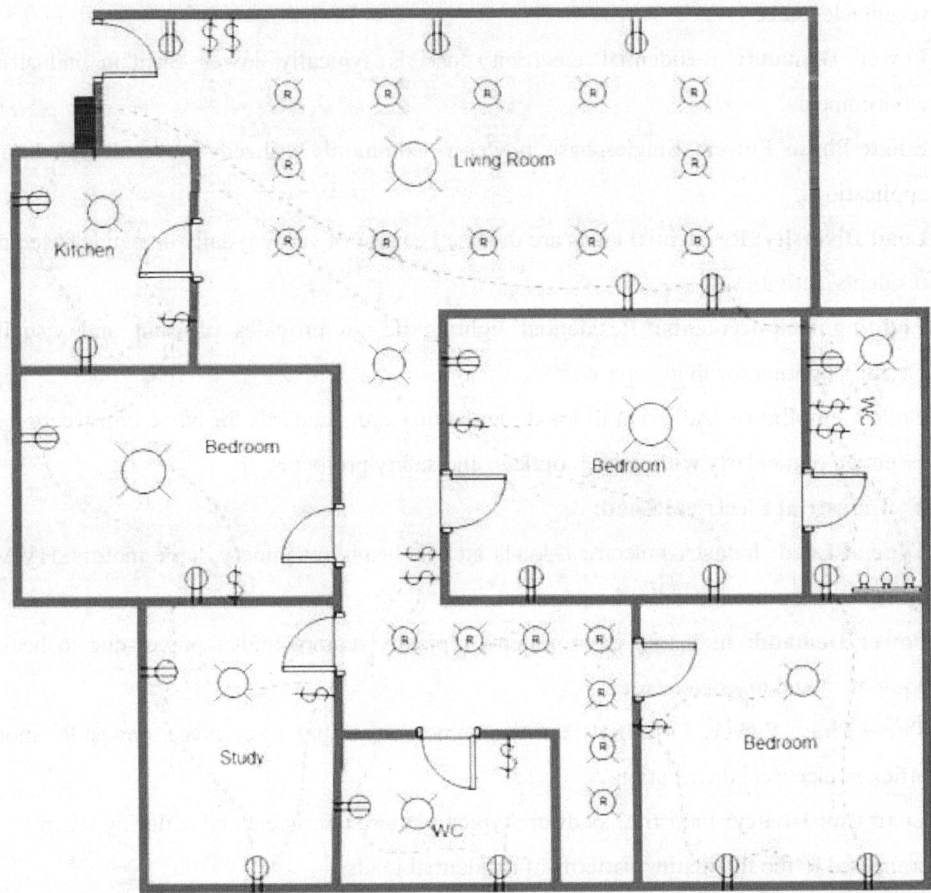

Figure 4.5 Basement wiring plan

4.2.2 How to Draw an Efficient House Wiring Diagram:

- ✓ Ensure the blueprints contain adequate bedroom, bathroom, and kitchen outlets. You must retain a minimum of three items in every scenario.
- ✓ Consider the placement of your furniture and the location of the outlets.
- ✓ Outdoor lighting is essential; hence, position the outlets near the doors.
- ✓ Decide thoughtfully on the arrangement of the wiring. To prevent leaks, cross-connections, or other safety concerns. There must be no exposed live wires present.

4.3 Difference Between Residential and Industrial Load:

4.3.1 Residential Electrical Load:

- ✓ **Type of Load:** Residential electrical loads often include lighting, small appliance, and receptacle loads.
- ✓ **Power Demand:** Residential electricity use is typically lower than in industrial environments.
- ✓ **Single-Phase Power:** Single-phase power is commonly utilized for residential home applications.
- ✓ **Load Diversity:** Residential loads are diverse because of varying daily demands based on residents' activities.
- ✓ **Lighting Requirements:** Residential lighting design provides pleasant and visually pleasing lighting for living spaces.
- ✓ **Code Compliance:** Adhering to local regulations and standards for home construction is essential, particularly with wiring, outlets, and safety protocols.

4.3.2 Industrial Electrical Load:

- ✓ **Type of Load:** Industrial electrical loads include heavy machinery, large motors, HVAC systems, and other specialized equipment.
- ✓ **Power Demand:** Industrial environments typically require higher power due to heavy equipment and machinery use.
- ✓ **Three-Phase Power:** Industrial facilities sometimes utilize three-phase power for more efficient electrical distribution.
- ✓ **Load Consistency:** Industrial loads are typically more consistent and enduring when compared to the fluctuating patterns of residential loads.
- ✓ **Lighting Requirements:** Industrial lighting is designed to be both safe and practical. This could include high-bay lighting and emergency lighting systems.

✓ **Code Compliance:**

✓ Specific industrial electrical regulations and standards must be adhered to, including guidelines for equipment grounding, overcurrent protection, and hazardous zones.

4.3.3 Differences:

Residential and industrial loads vary significantly in their characteristics, needs, and usage patterns. Below are some essential distinctions between residential and industrial loads:

✓ **Size and Scale**: Residential loads are often smaller and less extensive than industrial loads. Residential buildings, such as houses and flats, usually require less electricity than industrial facilities, such as factories, warehouses, and manufacturing plants that contain a lot of electrical equipment and machinery.

✓ **Usage Patterns**: Residential loads exhibit consistent usage patterns, with peak demand usually happening in the evenings when residents are present and using equipment like lights, televisions, and kitchen appliances. Industrial loads vary based on production schedules, shifts, and operating needs.

✓ **Types of Equipment**: Most residential loads are gadgets and appliances used throughout the house, including refrigerators, air conditioners, TVs, computers, and light fixtures. On the other hand, industrial loads encompass a diverse array of specialized apparatus and equipment, including motors, pumps, compressors, heaters, welding machines, and heavy-duty processing platforms.

✓ **Power Quality Requirements**: Compared to residential loads, industrial loads typically have more stringent demands for electricity stability, reliability, and quality. Voltage regulators, surge protectors, and uninterruptible power supplies (UPS) are examples of power conditioning equipment that may be necessary in industrial settings due to the sensitivity of specific processes and equipment to power interruptions, harmonic distortion, and voltage variations.

✓ **Three-Phase Power**: Residential buildings usually utilize single-phase power distribution; however, industrial facilities frequently have three-phase power systems to accommodate the increased power requirements of industrial equipment and machinery. Three-phase electricity enables a more efficient and balanced distribution of electrical power, which decreases voltage drop and enhances the operation of industrial motors and machines.

✓ **Load Diversity**: Residential loads show high load variety, indicating that appliances and equipment do not all run at maximum capacity simultaneously. Industrial loads often have increased levels of simultaneous operation, particularly in manufacturing and processing plants where numerous machines may run simultaneously to achieve output goals.

- ✓ **Regulatory Compliance**: Industrial facilities must adhere to regulatory mandates and criteria for electrical safety, energy efficiency, and environmental conformity. Adhering to laws like the National Electrical Code (NEC) and Occupational Safety and Health Administration (OSHA) requirements is crucial for industrial operations to guarantee worker safety and regulatory adherence.
- ✓ **Rules and Requirements Related to Lighting Load Followed in Electrical Installations:** Regulations concerning lighting load in electrical systems are usually determined by electrical codes and standards set by regulatory agencies or organizations in charge of electrical safety and adherence. Below are typical regulations and criteria for lighting load in electrical setups:
- ✓ **Maximum Circuit Load**: Electrical rules dictate the highest permissible load for lighting circuits to avoid overloading. The limit is usually specified in watts or amperes and can be influenced by factors like wiring type, circuit protection devices, and electrical panel capacity.
- ✓ **Circuit Protection**: Lighting circuits should have circuit protection devices like circuit breakers or fuses to reduce overheating and fire risks caused by overcurrent situations. Choose the rating of the circuit protection device according to the highest expected load of the lighting circuit.
- ✓ **Wire Size and Type**: Lighting circuits must adhere to the established standards in electrical codes regarding the size and kind of wiring utilized. Factors to consider include wire gauge, insulation type, and ampacity ratings to guarantee the safe and dependable functioning of the lighting circuit.
- ✓ **Voltage Drop**: Electrical regulations may set restrictions on voltage drop for lighting circuits to maintain voltage levels at lighting fixtures within acceptable parameters. High voltage drops can lead to decreased lighting efficiency and early malfunction of lighting devices.
- ✓ **Grounding and Bonding**: Grounding and bonding lighting circuits allow fault currents to flow safely and reduce the likelihood of electrical shocks. Electrical codes usually outline the necessary specifications for grounding, which typically involve grounding electrodes, equipment bonding, and grounding conductors.

- ✓ **Fixture Installation Requirements**: Specifications for mounting heights, clearances, and fixture spacing may be outlined in electrical codes for installing lighting fixtures. These standards ensure that indoor and outdoor lighting installations are adequately illuminated, accessible, and safe.
- ✓ **Energy Efficiency**: Modern electrical rules and standards now guide energy-efficient lighting design and installation methods. This may involve mandating energy-efficient lighting systems like LED fixtures, daylighting controls, and occupancy sensors to decrease energy usage and encourage sustainability.
- ✓ **Emergency Lighting**: Emergency lighting systems may be mandated by electrical codes in specific occupancies, including commercial buildings, healthcare facilities, and industrial plants, to offer illumination in case of power failures or emergencies. Emergency lighting systems must adhere to precise design, installation, testing, and upkeep guidelines to guarantee dependability and adherence to safety regulations.
- ✓ **Accessibility Requirements**: Lighting installations must adhere to accessibility regulations outlined in building codes and standards, which include guidelines for lighting controls, switches, and fixtures to guarantee usability for people with disabilities.
- ✓ **Documentation and Labelling**: Electrical codes frequently mandate the recording and labelling lighting circuits, fixtures, and controls to facilitate identification, troubleshooting, and maintenance. This may involve circuit schematics, panel schedules, lighting fixtures, and control labelling.

Residential Lighting Regulations:

- ✓ Compliance with home electrical codes.
- ✓ Outlets and switches are installed following construction codes.
- ✓ Encouraging the use of energy-efficient lighting is advisable.

Industrial Lighting Regulations:

- ✓ Compliance with industrial electrical regulations.
- ✓ Sturdy and efficient lighting fixtures.
- ✓ Ensure adherence to hazardous area regulations as needed.

In conclusion, lighting considerations are necessary for residential and industrial electrical loads, but the criteria and standards are tailored to the unique circumstances of each setting. Residential lighting focuses on comfort and aesthetics, while industrial lighting prioritizes safety and functionality in complex and challenging environments. Adherence to relevant electrical codes is mandatory in both situations.

4.3.4 Positioning of Equipment

Proper electrical equipment placement in a structure is essential for safety, accessibility, functionality, and adherence to electrical regulations and standards. General criteria exist for the placement of electrical equipment.

- **Accessibility**: Proper electrical equipment placement in a structure is essential to provide safety, accessibility, functionality, and adherence to electrical regulations and standards. General criteria exist for the placement of electrical equipment.
- **Clearance Requirements**: Adhere to the clearance distances required by electrical codes and standards for all electrical equipment. This involves ensuring adequate spacing from walls, ceilings, floors, and other obstacles to guarantee optimal ventilation, heat dissipation, and the safe functioning of the equipment.
- **Ventilation**: Ensure proper ventilation for electrical equipment to prevent overheating and maintain peak performance. Do not place equipment in small spaces or block ventilation apertures.
- **Protection from Environmental Factors**: To ensure the best performance and most extended lifespan for your electrical equipment, keep it free from any sources of moisture, humidity, corrosive chemicals, dust, or other environmental contaminants. When installing outside, use enclosures or covers resistant to the elements.
- **Accessibility for Service**: Position electrical equipment for convenient servicing, repair, and component replacement. Avoid installing equipment in areas that necessitate significant disassembly or disruption of other building systems for access.
- **Height and Reach**: To facilitate ease of use and accessibility for building occupants, place outlets, switches, and other electrical equipment that can be easily operated properly. When deciding on the reach and height of electrical devices, consider the needs of people with impairments.
- **Safety Considerations**: Place electrical equipment in positions that reduce the likelihood of unintended contact or harm. Avoid situating equipment near heat sources, flammable chemicals, or high-traffic areas vulnerable to physical damage from vehicles or other equipment.

- **Load Distribution**: Strategically place electrical equipment throughout the building to balance loads, reduce voltage drop, and optimize the efficiency of electrical distribution systems. Consider elements such as the distance between equipment and power sources, the placement of loads, and the arrangement of electrical wires.
- **Labelling and Identification**: Ensure that electrical equipment is well marked with suitable signs and identification to aid operation, upkeep, and problem-solving. Employ uniform labelling practices for circuits, panels, breakers, switches, and other components to guarantee precise and consistent identification.
- **Compliance with Codes and Standards**: Ensure electrical equipment placement meets relevant electrical codes, laws, and industry standards. Seek guidance from certified electrical engineers, designers, or inspectors to ensure adherence to regulations and handle any unique installation needs.

4.4 Lighting and Power Circuits:

Lighting and power circuits are two primary electrical circuits frequently present in buildings. The lighting circuit is utilized for low-power devices, whereas the power circuit is for high-power devices.

- **Lighting Circuits**:

✓ Lighting circuits are tailored to supply power to lighting fixtures such as lamps, bulbs, fluorescent tubes, and LEDs.

✓ These circuits generally function at lower currents and voltages than power circuits.

✓ Lighting circuits are commonly connected in parallel, enabling separate control of each fixture using switches.

✓ They might include dimmer switches or sensors for automated control and energy efficiency.

✓ Lighting circuits can be built with various wire layouts, like loop-in or junction box wiring, based on the building's layout and needs.

- **Power Circuits**:

✓ Lighting circuits are built for the express purpose of supplying power to lighting devices like lamps. Power circuits are created to provide electrical power to different outlets and appliances within a building, including wall sockets, kitchen appliances, HVAC systems, and other electrical devices.

- ✓ These circuits generally function at elevated current and voltage levels compared to lighting circuits.
- ✓ Power circuits can be wired in several forms, such as radial and ring circuits.
- ✓ Radial circuits consist of a single cable connecting each outlet or appliance to the distribution board. In contrast, ring circuits create a loop from the distribution board, linking numerous outlets or appliances in series.
- ✓ Ring circuits are prevalent in certain regions, providing redundancy and reduced voltage drop compared to radial circuits.
- ✓ Power circuits can have specialized outlets for specialized uses, such as dedicated circuits for large appliances like refrigerators, air conditioners, and washing machines.
- ✓ Ground fault circuit interrupters (GFCIs) or residual current devices (RCDs) are commonly used in power circuits to offer extra safety against electrical shock.

Lighting and power circuits are vital elements of a building's electrical system, enabling illumination, equipment functioning, and general electrical usage.

4.4.1 Ceiling Fan and Light Wiring Circuit Diagram:

A typical wiring circuit in homes and offices is a single switch controlling a fan and a light. In this circuit, the phase line is linked to one of the SPST switches' poles, and the neutral line is linked to the fan's and light's neutral terminals in parallel. There is a direct connection between the bulb's phase terminal and the switch's S2 terminal. From the switch S1 connector, a regulator is linked in series with the fan to control its speed. Grounding the fan in the circuit shown is optional; often, ceiling fan setups consist of two wires without a grounding connection. Human contact with the metal structure of a ceiling fan is highly improbable. A leak through the roof with modern wiring with sensitive RCCBs will cause the circuit to trip immediately. Extra protection is provided by earthing, which stops earth from leaking through ceilings and other conductive materials. Figure 4.6 shows a wiring plan depicting the circuit for the ceiling fan and lights.

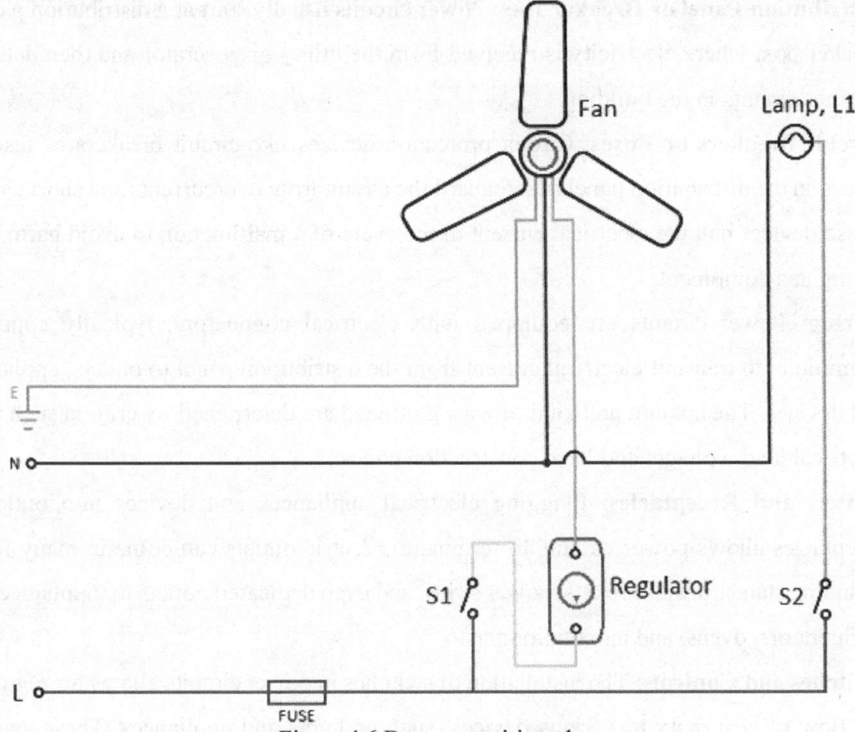

Figure 4.6 Basement wiring plan

➢ **Materials Required:**

Component	Specification	Quantity
MCB	230V,50Hz	1
Lamp	230V,50Hz	1
Fan	230V,50Hz	1
Switch	SPST,230V,5A	2
Regulator	230V,50Hz	1

4.4.2 Power Circuit:

Electrical outlets, appliances, and other equipment in a structure get their electricity from a specific kind of circuit called a power circuit. These circuits are in charge of delivering power to run machinery, appliances, and other electrical devices. Every building's electrical distribution system relies on power circuits, whether home, business, or factory.

- ✓ **Distribution Panel or Breaker Box**: Power circuits usually start at a distribution panel or breaker box, where electricity is received from the utility or generator and then delivered to other circuits in the building.
- ✓ **Circuit Breakers or Fuses**: Circuit protection devices like circuit breakers or fuses are placed in the distribution panel to safeguard the circuit from overcurrents and short circuits. These devices halt the electrical current in the event of a malfunction to avoid harm to the wiring and equipment.
- ✓ **Wiring**: Power circuits are equipped with electrical conductors, typically copper or aluminium, to transmit electrical current from the distribution panel to outlets, appliances, and devices. The amount and kind of wiring utilized are determined by criteria such as the electrical load, voltage, and local construction codes.
- ✓ **Outlets and Receptacles**: Plugging electrical appliances and devices into outlets or receptacles allows power circuits to terminate. Electric outlets can come in many forms, including standard wall sockets, power strips, and even dedicated outlets for appliances like refrigerators, ovens, and air conditioners.
- ✓ **Switches and Controls**: The installation of switches in power circuits allows for regulating the flow of electricity to various devices, such as lights and appliances. These switches could be built in or mounted on the wall depending on the equipment.
- ✓ **Grounding**: A grounding system is a standard component of power circuits that allows electrical currents to flow safely in the event of a malfunction. The use of grounding devices lessens the likelihood of electrical fires and shocks.
- ✓ **Voltage and Current Rating**: Power circuits are engineered to manage precise voltage and current ratings according to the electrical load they sustain. The dimensions and capability of the circuit elements, such as wiring, breakers, and outlets, are chosen appropriately.

Power circuits are critical for distributing electrical power safely and reliably within a building, allowing the functioning of various electrical appliances, equipment, and gadgets used for daily activities and business activities.

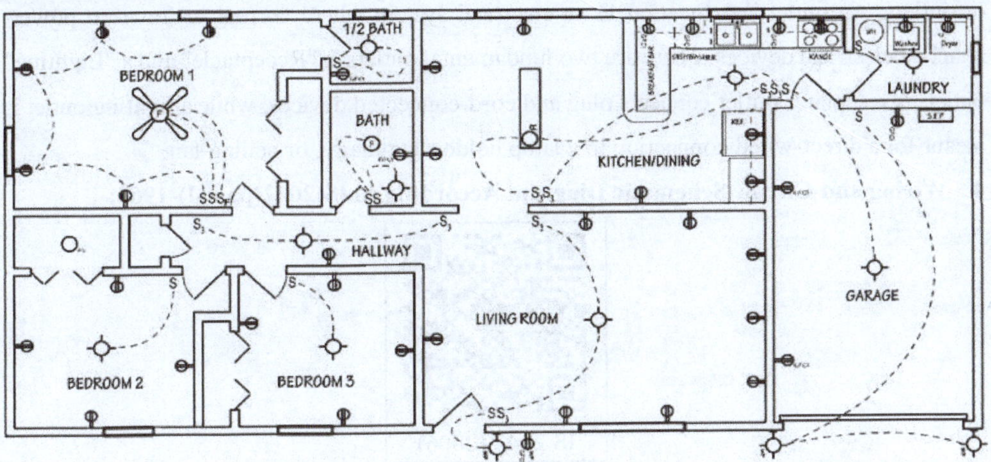

Figure 4.7 Home wiring plan

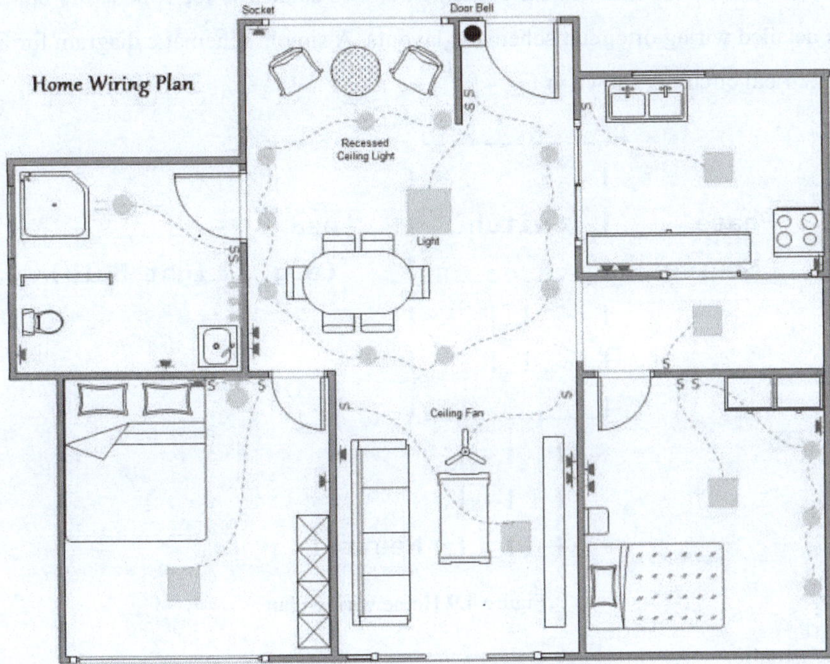

Figure 4.8 Home wiring plan

Electrical symbols on home electrical wiring diagrams indicate the position, control point(s), and type of electrical equipment needed at those locations. The icons depicted on the floor layout represent lighting outlets, receptacle outlets, special-purpose outlets, fan outlets, and switches. Dotted lines connect symbols to indicate which switches operate particular lights or outlets. An outlet is a location in an electrical system where current is drawn to power-attached electrical devices. There are two fundamental outlets: a "Receptacle" and a "Lighting" outlet. A receptacle outlet connects plug and cord-connected devices, while a lighting outlet is meant for a direct-wired connection to a lamp holder, luminaire, or ceiling fan.

4.5 Wiring and Circuit Schematic Diagram According to IS:2042(part-I)-1962:

IS 2042 (2006)

The Indian Standard Specification IS:2042(part-I)-1962 pertains to conductors used in insulated electric cables and flexible cords. It outlines standards for conductors but needs to provide detailed wiring or circuit schematic layouts. A simple schematic diagram for a single-phase electrical circuit is shown in Fig.4.9.

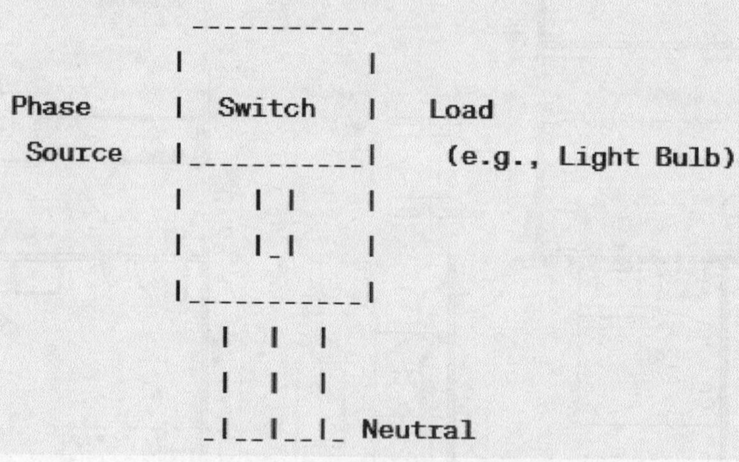

Figure 4.9 Home wiring plan

In this schematic:

- ✓ The Phase Source symbolizes the active wire, typically colored red or brown.
- ✓ The Neutral is the pathway for the current to return, typically colored black or blue.
- ✓ The switch regulates the electrical flow to the load.
- ✓ The load is the electrical equipment or power appliance, such as a light bulb.

Here's how the circuit operates:

- ✓ When the switch is turned to the "ON" position, the circuit is completed, enabling the current to pass from the phase source through the switch to the load.
- ✓ The load, such as a light bulb, transforms the current it receives into light or other types of energy.
- ✓ The current flows back to the power source through the neutral wire, thus closing the circuit.

4.5.1 Multiline & Single Line Representation:

Electrical wiring and circuit diagrams are commonly represented using single- and multiline styles. However, standards for these representations are provided by the American National Standards Institute (ANSI) and the International Electrotechnical Commission (IEC) (ANSI Y32.2/IEEE 315 and IEC 60617, respectively).

Single-Line Representation:

Electrical connections and components are shown in single-line diagrams with a single line. This is a single-line representation of the same circuit:

```
(Phase Source) --(Switch)----(Light Bulb)---- (Neutral)
```

In this representation:

- ✓ **(Phase Source)** and **(Neutral)** represent the live and neutral wires, respectively.
- ✓ **(Switch)** represents the switch.
- ✓ **(Light Bulb)** represents the light bulb.
- ✓ The single line symbolizes the electrical connections among the components.

Single-line diagrams are commonly utilized in technical illustrations and electrical schematics to enhance simplicity and clarity, particularly in intricate systems.

Multiline Representation:

Each electrical component in multiline diagrams is symbolically represented on individual lines, with connections indicated by lines or wires. Here is a basic illustration of a circuit with multiple lines showing a switch that controls a light bulb:

```
Phase Source ----[Switch]----[Light Bulb]---- Neutral
```

In this representation:

- ✓ **Phase Source** and **Neutral** represent the live and neutral wires, respectively.
- ✓ **[Switch]** represents the switch.

✓ **[Light Bulb]** represents the light bulb.

The dashed lines represent wires connecting the components.

Methods for representation of wiring diagrams:

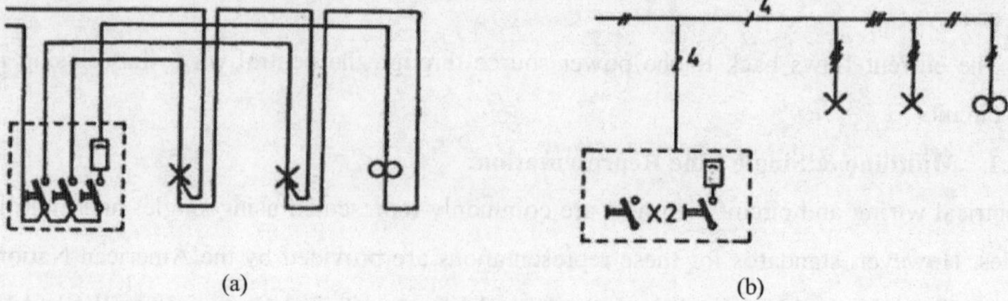

(a) (b)

Figure 4.10 Different wiring diagrams

- ✓ A wiring diagram can be displayed in a multiline or single-line format. Fig.4.10(a) illustrates a wiring diagram in a multiline layout. Fig.4.10(b) shows a simplified version of the diagram.
- ✓ Fig.4.10(b) The multiline form in Fig.4.10(a) displays lines that converge into a single line in the single-line version.
- ✓ An equal number of small, thin, slanting lines are placed perpendicular to the single line to represent the quantity of wires. When more than three wires run together, depict a single diagonal line across the line and indicate the number of cables with a numeral (Fig.4.10(b)).
- ✓ A single symbol denotes many similar accessories arranged next to each other in one place, with a numerical value indicating the quantity of these accessories used. Fig.4.10(b) depicts two switches using a symbol for one switch and a number.
- ✓ Multi-line wiring designs are rarely used in practice. Single-line diagrams offer wiremen the necessary information to identify required materials and execute the installation.

4.6 Load assessment:

Assessing the load for building electrification entails calculating the electrical demand needed to operate different systems and appliances in a building. Here is an essential procedure for performing a load assessment for building electrification:

- ✓ **Identify Electrical Systems and Appliances**: Compile a detailed inventory of all electrical systems and appliances present in the building. This encompasses lighting, HVAC systems, water heaters, kitchen appliances, computers, servers, lifts, and other electrical devices.

- ✓ **Determine Power Requirements**: Determine the power requirements in watts (W) or kilowatts (kW) for each item on the list. This information can be found in the manufacturer's specs or obtained by measuring it with power meters.
- ✓ **Estimate Usage Patterns**: Determine the average power consumption of all electrical systems and appliances. You might find some that run all the time, while others have more sporadic or seasonal use. Think about things like building function, weather, and occupancy schedules.
- ✓ **Calculate Energy Consumption**: To find the energy consumption per unit of time (e.g., kWh/day, kWh/month), multiply the power needs of each system or appliance by its projected usage time. The overall energy demand can be calculated by combining all the products' energy consumption.
- ✓ **Consider the Diversity Factor**: Not every system and appliance in a structure runs at maximum efficiency all the time. To compensate for this, a diversity factor should be applied. By dividing the total connected load by the diversity factor, we can get a better idea of the demand simultaneously.
- ✓ **Account for Future Growth**: Consider any future building expansions or additions that could raise the electrical load. Provide flexibility in the electrical infrastructure to support future expansion.
- ✓ **Include Safety Margins**: Include safety margins in your load calculations to guarantee the electrical system can manage peak loads and sudden increases in demand without overloading or causing circuit breakers to trip.
- ✓ **Consult Relevant Codes and Standards**: Ensure your load assessment meets your area's electrical laws and standards. These may consist of national electrical codes, municipal building codes, and industry standards.
- ✓ **Document the Assessment**: Record all computations, presumptions, and factors considered throughout the load evaluation. This paperwork is crucial for planning, decision-making, regulatory compliance, or certification.
- ✓ **Review and Validation**: Discuss the load assessment with key stakeholders, including building owners, engineers, architects, and electrical contractors, to confirm assumptions and verify accuracy.

4.6.1 Selection of Size of Conductors:

Choosing the correct size of conductors for building electrification requires considering the electrical load, voltage drop, ambient temperature, and relevant electrical norms and standards.

- ✓ **Determine Electrical Load**: Utilise the prior load assessment to calculate the overall electrical load that the cables must support. The load is usually quantified in kilowatts (kW) or amperes (A).
- ✓ **Choose Conductor Material**: Copper and aluminium are the predominant conductor materials utilized in constructing wiring systems. Copper has higher conductivity and cost, but aluminium is cheaper but needs more extensive cross-sectional areas to transmit the same current.
- ✓ **Refer to the National Electrical Code (NEC) or Relevant Standards**: Refer to the NEC or any pertinent electrical rules and standards relevant to your area. The rules include instructions and standards for determining the appropriate size of conductors, considering *aspects including ampacity, insulation type, and installation conditions.*
- ✓ **Determine Allowable Ampacity**: Calculate the maximum current-carrying capacity of the chosen conductor size according to the insulation temperature rating and the mode of installation (such as conduit, cable tray, and direct burial). The NEC offers tables and methods for determining the permissible ampacity based on these variables.
- ✓ **Consider Voltage Drop**: Determine the voltage decrease across the conductor to ensure it stays within permissible boundaries. Voltage drop can impact the functionality of electrical devices and machines, mainly when transmitted over extended distances. The NEC generally suggests a maximum voltage loss of 3% for branch circuits and 5% for feeder circuits.
- ✓ **Account for Ambient Temperature**: Consider the ambient temperature of the installation area since elevated temperatures can decrease the ampacity of the conductors. Adjust the conductor size as needed to accommodate ambient temperature conditions.
- ✓ **Size Up for Future Growth**: Choosing conductors slightly larger than the minimum requirement is advisable for future expansions or additions to the electrical load. This offers adaptability and prevents the necessity for expensive future updates.
- ✓ **Consult Manufacturer's Data**: Consult the manufacturer's data sheets for certain conductors to confirm compatibility with your application and get more details on ampacity, voltage drop, and other pertinent parameters.
- ✓ **Document and Verify**: Record your conductor sizing calculations and validate them with pertinent stakeholders, such as engineers and electricians, to guarantee precision and adherence to relevant rules and standards.

4.6.2 Selection of Rating of Main Switch:

Choosing the appropriate rating for the main switch in building electrification requires careful study to guarantee the safety and efficiency of the electrical system.

- ✓ **Determine Total Connected Load**: Determine the aggregate connected load of all electrical devices and appliances within the structure. This encompasses lighting, HVAC systems, appliances, and other electrical loads.
- ✓ **Consider Future Expansion**: Consider any future increases or additions to the building's electrical demand. Choosing a primary switch that is slightly larger than needed is recommended to allow for future expansion without requiring quick improvements.
- ✓ **Voltage and Phase**: Locate the building's electrical panel and check the voltage and phase. Both three-phase (480V/277V) and single-phase (120V/240V) setups are common.
- ✓ **Refer to Applicable Codes and Standards**: Refer to your area's appropriate electrical codes and standards. This would involve adhering to the National Electrical Code (NEC) in India.
- ✓ **Determine Load Diversity Factor**: Consider the variety factor, which considers the probability that not all electrical loads will simultaneously run at their maximum capacity. This factor assists in determining the optimum size for the main switch.
- ✓ **Select a Rating**: Choose a primary switch with a rating higher than the calculated load, considering the total connected load, future expansion plans, voltage, phase, and diversity factor. Main switches are commonly found in conventional ratings like 100A, 200A, and 400A.
- ✓ **Consider Safety and Reliability**: Verify that the chosen main switch meets safety and reliability criteria. Seek switches from reliable manufacturers that adhere to applicable safety requirements.
- ✓ **Coordination with Overcurrent Protection Devices**: Align the rating of the primary switch with the ratings of downstream overcurrent safety devices, such as circuit breakers or fuses. The primary switch must be able to interrupt the highest expected fault current in the system safely.
- ✓ **Consult with Electrical Engineers or Professionals**: If you are still determining the choosing process or if your building has specific requirements, it is advisable to seek advice from electrical experts or professionals specializing in building electrification.

- ✓ **Document and Verify**: Record the selecting process, detailing the calculations and factors considered. Confirm the chosen primary switch with appropriate stakeholders to ensure it meets project specifications and safety regulations.

4.6.3 Selection of Rating of Protective Switchgear:

Choosing the rating of protective switchgear for building electrification requires guaranteeing the safety and dependability of the electrical system. Protective switchgear, such as circuit breakers and fuses, are essential components that safeguard electrical circuits from overloads, short circuits, and malfunctions.

- ✓ **Determine Load Characteristics**: Determine the attributes of the electrical loads in the building, such as their voltage, phase, and current specifications. This information will assist in determining the suitable type and rating of protective switchgear required.
- ✓ **Calculate Maximum Fault Current**: Calculate the peak fault current that the protective switchgear must be able to interrupt during a fault. This estimate considers variables including the electrical service's size, the impedance of the electrical distribution system, and the attributes of upstream power sources.
- ✓ **Consider Load Diversity**: Consider the variety of electrical loads present in the facility. The protective switchgear should be sized to handle the peak demand with some leeway, considering that not all loads would operate at maximum capacity simultaneously.
- ✓ **Select the Type of Protective Switchgear**: Consider the electrical system's unique needs while selecting the protective switchgear. Electric safety devices such as thermal-magnetic, electronic, or digital circuit breakers, fuses, ground fault circuit interrupters (GFCIs), or residual current devices (RCDs) are standard.
- ✓ **Refer to Applicable Codes and Standards**: Please refer to the appropriate electrical regulations and standards. These standards aim to establish criteria for choosing and setting up protective switchgear.
- ✓ **Determine Rating**: When choosing protective switchgear, ensure the rating suits the electrical system's voltage and phase and meets or surpasses the maximum fault current calculated. It is common practice to provide the voltage, current, and breaking capacity when describing ratings for protective switchgear.

✓ **Consider Coordination**: Ensure synchronization among protective devices across the electrical distribution system. This coordination eliminates excessive tripping and removes problems promptly and selectively to minimize downtime.

✓ **Evaluate Selectivity and Discrimination**: Assess the protective switchgear's ability to selectively and accurately discriminate between electrical faults. Selectivity guarantees that the closest protection device to the fault will activate, preventing other system areas from being impacted.

✓ **Consider Remote Monitoring and Control**: Consider using protective switchgear with remote monitoring and control capabilities based on the complexity of the building's electrical system. This enables the immediate tracking of electrical metrics and the remote control of safety mechanisms.

✓ **Document and Verify**: Record the selecting process, detailing calculations, specs, and factors considered. Confirm the chosen protective switchgear with appropriate stakeholders to verify it meets project specifications and safety regulations.

Following these processes and considering elements like load characteristics, fault current, diversity, coordination, selectivity, and monitoring capabilities allows for selecting protective switchgear ratings. The maximum voltage that Low Voltage (LV) switchgear can handle is 1KV. This category includes Earth Leakage Circuit Breakers, HRC Fuses, Offload Isolators, Miniature Circuit Breakers, and Moulded Case Circuit Breakers. A rating capacity of 3KV to 36KV is typical for medium voltage (MV) switchgear.

4.7 Design And Drawing:

It is necessary to create comprehensive plans and drawings to install electrical systems in buildings or other structures, which is what electrical wiring design and drawing are all about.

Initial Planning:

✓ Collect data regarding the project's specifications, such as the building's dimensions and plan, electrical loads, and the expected function of the area.

✓ Locate electrical components such as outlets, switches, and lighting fixtures by building codes, safety standards, and user requirements.

Electrical Load Calculation:

✓ Determine the building's overall electrical load by considering the quantity and kind of electrical appliances, devices, and equipment that will be put in place.

- ✓ To accommodate the expected load, find out how big the electrical service panel or distribution board has to be.

Circuit Design:

- ✓ Create the electrical circuits for the building's lights, power outlets, appliances, and other electrical loads.
- ✓ Calculate the required number of circuits, design the circuit layout, and allocate loads appropriately to achieve optimal electrical distribution and balance.
- ✓ Utilise dedicated circuits for high-power appliances and equipment to avoid overloading.

Electrical Wiring Layout:

- ✓ Create a layout plan for electrical wiring that illustrates the path of wires, cables, and conduits within the building.
- ✓ Ascertain the locations of electrical enclosures such as junction boxes, distribution panels, and others where wires will be terminated and connected.
- ✓ Verify that all wire pathways adhere to all applicable safety regulations, architectural limitations, and building requirements.

Drawing Creation:

- ✓ Create precise electrical wiring diagrams, blueprints, and schematics using computer-aided design (CAD) software or drafting tools.
- ✓ Create floor layouts, elevation views, and section views to demonstrate the positioning of electrical components and wiring in the building.
- ✓ Ensure that dimensions, labels, symbols, and annotations effectively communicate information precisely and clearly.

Symbol Standardization:

- ✓ Utilise standardized electrical symbols and conventions to depict electrical components, devices, and connections in the drawings.
- ✓ Consult industry standards and norms like ANSI/IEEE or IEC for advice on choosing and depicting symbols.

Compliance and Safety:

- ✓ Ensure the electrical wiring design meets relevant building requirements, electrical laws, and safety standards.
- ✓ To guarantee electrical safety, consider wire sizing, insulation ratings, grounding requirements, and protection devices like circuit breakers, fuses, and GFCIs.

Review and Validation:

- ✓ Review the electrical wiring designs with stakeholders, such as architects, engineers, electricians, and building owners, to verify accuracy and completeness.
- ✓ Validate the design against project needs, specifications, and regulatory requirements before commencing construction or installation.

Documentation and Distribution:

- ✓ Create detailed paperwork, such as electrical drawings, specifications, timetables, and item lists, for construction and installation purposes.
- ✓ Distribute the finished drawings and documents to all stakeholders engaged in the project, such as contractors, electricians, inspectors, and building authorities.

Revision and Updates:

- ✓ Revise electrical wiring plans to incorporate any alterations, adjustments, or amendments during the building or installation phase.
- ✓ Keep precise documentation and record all changes for future reference and maintenance.

4.7.1 Estimation and Costing of a Residential Installation Having Maximum 5 KW Load:

To calculate the cost of installing a home system with a 5 kW load, you must consider various aspects like supplies, labor, licenses, and any other expenses. Here is a basic overview of the cost estimation:

- ✓ **Solar Panels:** For a 5 kW system, you may require around 15-20 solar panels, which will vary according to the efficiency and wattage of the panels. Solar panel prices can range from ₹40 to ₹80 per watt. For a 5 kW system, the cost of solar panels might vary between 2 lakh and 4 lakh.
- ✓ **Inverter:** The inverter transforms the direct current (DC) electricity generated by the solar panels into alternating current (AC) electricity utilized in your household. An inverter of high quality for a 5 kW system could range in price from ₹64k to ₹1.2k.
- ✓ **Mounting and Racking:** This encompasses the supplies and labor needed to securely install the solar panels on the roof or the ground. The cost could range from ₹4k to ₹1.2lakh, depending on the complexity of the installation.
- ✓ **Balance of System (BOS) Components:** The components required to link the solar panels to your home's electrical system are wires, switches, circuit breakers, and other electrical components. Allocate around ₹80k to ₹1.6lakh for these components.

- ✓ **Labor Costs:** Installation labor costs can vary widely based on your location, the complexity of the installation, and the contractor's rates. Expect to pay anywhere from ₹80k to ₹2.4lakh for labor.
- ✓ **Permits and Inspection Fees:** You will probably require licenses from your local government and might have to pay inspection fees. The rates vary considerably based on your region, ranging from ₹40k to ₹1.2lakh.
- ✓ **Additional Expenses:** Additional incidental expenses, such as transportation costs for equipment, equipment rental fees, or unforeseen expenses, could increase the overall cost by an additional 40k to 80k.

The total projected cost for a domestic solar installation with a maximum 5 kW load ranges from ₹6.24 lakh to ₹13.20 lakh. These figures are approximate, and expenses may differ for several reasons. Obtaining quotations from several solar installers in your vicinity is advisable to understand the costs involved better.

4.7.2 Sequence to be followed for preparing estimate:

Preparing an estimate for a residential installation with a maximum 5 kW load involves several vital steps to ensure accuracy and completeness. Here's a sequence to follow:

- ✓ **Initial Consultation:** Arrange a meeting with the client to review their needs, desires, and financial constraints for the residential installation project. Comprehend their requirements for electrical appliances, lighting, HVAC systems, and other electrical loads.
- ✓ **Site Visit and Assessment**: Perform a thorough on-site inspection to evaluate the current electrical infrastructure, wiring, and panel capacity and identify any possible obstacles or restrictions affecting the installation process. Make a record of the positions of electrical sockets, distribution panels, and any specific needs for the project.
- ✓ **Load Calculation**: Use the client's specifications and the wattage ratings of all electrical devices—including appliances, lighting, HVAC systems, and more—to determine the overall electrical load for the home installation. Verify that the sum of all loads is not above the electrical service panel's maximum capacity.
- ✓ **Material List**: Compile a detailed inventory of all necessary supplies for the installation, such as wiring, circuit breakers, outlets, switches, lighting fixtures, conduits, junction boxes, and any other essential electrical components needed to finish the project.
- ✓ **Labor Estimate**: Determine the labor hours needed for installation by considering the project's complexity, the residential property's size, and the scope of work. Consider

elements like wiring routes, fixture installations, panel enhancements, and any specific requests or modifications from the client.

- ✓ **Material Pricing**: Get quotes for all necessary materials for the installation from reliable suppliers or wholesalers. Consider variables like volume discounts, shipping fees, and taxes when determining material expenses.
- ✓ **Labor Costs**: Calculate the labor expenses by multiplying the estimated labor hours needed for installation by the current labor rates for electricians or electrical contractors in the region.
- ✓ **Overhead and Contingencies**: Consider indirect charges like insurance, permits, administrative costs, and other overhead costs related to the project. Incorporate contingencies to address unexpected events, alterations in project size, or other difficulties that could occur while installing.
- ✓ **Markup for Profit**: Add a markup or profit margin to the total expected cost to account for the contractor's overhead charges and earn a profit. The markup percentage might fluctuate based on market conditions, the competitive environment, and the contractor's business style.
- ✓ **Prepare the Estimate Document**: Consolidate all cost data, computations, assumptions, and documentation into a detailed and thorough estimate document. Provide a detailed description of the work, a clear breakdown of expenses, payment terms, and other pertinent details for the client to evaluate and approve.
- ✓ **Review and Finalize**: Thoroughly examine the estimate to verify its precision, comprehensiveness, and alignment with the client's specifications and budget. Consult with the client to resolve any inquiries, issues, or modifications required before completing the estimate.
- ✓ **Submit the Estimate**: Submit the completed estimate to the client for their assessment, consent, and potential additional discussion. Be ready to respond to inquiries, offer explanations, and make modifications as necessary to fulfill the client's expectations and needs.

By systematically following these procedures, you may create a thorough and precise estimate for a residential installation project with a maximum 5 kW load, guaranteeing its successful execution within budget and schedule limitations.

4.7.3 Calculation of length of wire and other materials:

To determine the amount of wire and other materials required for a residential installation with a maximum 5 kW load, many criteria must be considered, such as wire type, distance from power supply to load, voltage, and current rating of wires.

Here's a general outline of how you can approach this calculation:

- ✓ **Determine the Load**: You specified a 5 kW maximum load. Convert this load to amperes by considering the electrical system's voltage for greater accuracy. To calculate power in watts for a standard home voltage of 120V or 240V in the US, use Ohm's Law formula: P=VI, where P is power in watts, V is voltage in volts, and I is current in amperes.
- ✓ **Select Wire Type and Size**: The wire type and size are determined by criteria including current carrying capacity, insulation, and installation conditions. Refer to local electrical codes for the correct wire type and size.
- ✓ **Calculate Voltage Drop**: To fulfill the relevant standards, calculate the voltage drop to guarantee the load receives the necessary voltage. The voltage drop is determined by the wire's length, the current passing through it, and the wire gauge.
- ✓ **Determine the Length of Wire**: You can determine the maximum permissible length of wire by considering the voltage drop constraints and the wire's current carrying capabilities.
- ✓ **Other Materials**: In addition to wire, you will want items like conduit (if necessary), connectors, junction boxes, circuit breakers, etc. The amount of these materials required will vary based on your project's particular layout and style.

Here's a simplified example calculation:

- Let's assume a 240V residential system.
- Using $P=VI$, for a 5 kW load, $I=P/V=5000/240\approx 20.83$ amperes.
- Selecting appropriate wire gauge based on current carrying capacity and voltage drop considerations.
- Calculating the voltage drop and determining the maximum allowable length of wire.
- Estimating the quantity of other materials based on the layout and design of the installation.

To ensure accurate calculations and adherence to local electrical regulations and safety protocols, it is advisable to get guidance from a certified electrician or electrical engineer. Local rules may also have exceptional standards about wire types, diameters, and installation techniques.

4.7.4 labour cost:

Labor expenses for a domestic electrical setup can fluctuate significantly based on variables including geographical location, the intricacy of the project, prevailing local labor charges, and the expertise of the electrician or electrical contractor. Nevertheless, I can offer a rough estimate of how labor expenses could be calculated for a 5 kW household setup.

Labor costs typically include:

- ✓ **Installation Labor**: This involves labor hours to install wiring, outlets, switches, circuit breakers, and other electrical components.
- ✓ **Testing and Inspection**: The system must be inspected after installation to ensure it meets safety standards and local electrical codes.
- ✓ **Cleanup and Miscellaneous**: This includes cleanup of the work area and any miscellaneous tasks associated with the installation.

Hourly or fixed rates for the entire project often determine labor costs. Prices might fluctuate significantly based on the geographical area and the individual contractor. Electricians in certain regions may bill between 2k to 10k per hour or higher, determined by their expertise and the intricacy of the task. For a residential installation with a 5 kW load, labor costs could range anywhere from a few hundred to a few thousand rupees, depending on the factors mentioned earlier.

For precise labor cost estimates, it is advisable to contact local electricians or electrical contractors in your vicinity and solicit quotes tailored to your project. They can offer a comprehensive cost breakdown depending on the necessary scope of work. Furthermore, ensure to ask about any required permissions or inspection fees for the installation, as these can impact the total cost.

4.8 Testing of Wiring Installation As Per IS:732-1982:

The inside electrical installation of a building includes wiring and associated components such as switches, sockets, distribution boards, and light fixtures. IS 732:1982 is the Indian Standard code that pertains to the "Code of Practice for Electrical Wiring Installations." It offers instructions for placing electrical wiring in structures to guarantee safety and efficient operation. Testing wire installations according to IS 732:1982 requires multiple stages to confirm adherence to the standard and ensure the electrical system's safety. Below are a few typical tests conducted as part of the testing procedure:

- ✓ **Insulation Resistance Test**: This test is performed to quantify the resistance of the insulation material that encases the conductors. It aids in detecting insulation defects or degradation. The test is usually conducted using a megger or insulation resistance tester.
- ✓ **Earth Continuity Test**: This test ensures the protective earthing conductor is continuous throughout the installation. It guarantees that all metal components of the electrical system are properly grounded to avoid electric shock in the event of a malfunction. The test is conducted with a continuity tester or a multimeter.
- ✓ **Polarity Test**: This test verifies the proper connection of line, neutral, and earth conductors at different locations inside the installation. It guarantees that the electrical system is connected correctly and minimizes the chances of electrical dangers. Polarity testing is often conducted with a polarity tester or a multimeter.
- ✓ **Earth Resistance Test**: This test assesses the earth electrode system's resistance to verify its compliance with earthing standards. It is beneficial to confirm that the earth electrode offers a low-resistance route to the ground for fault currents. The test is performed via a specialized earth resistance tester.
- ✓ **Voltage Drop Test**: This test assesses the voltage decrease across conductors under load. To preserve the proper operation of electrical equipment and appliances, it is essential to ensure that the voltage drop remains within acceptable levels. Voltage drop testing is conducted with a voltmeter or a dedicated voltage drop tester.
- ✓ **Continuity Test**: This test ensures the uninterrupted flow of electricity via the conductors and connections in the installation. It guarantees the absence of open circuits or loose connections that may cause malfunctions or safety risks. Continuity testing is commonly performed with a continuity tester or a multimeter.
- ✓ **Functional Test**: This test assesses the whole operation of the electrical system, encompassing switches, outlets, circuit breakers, and other elements. It guarantees that the system functions appropriately and securely within standard operating conditions.

These are some of the primary tests carried out while inspecting wiring installations according to IS 732:1982. Adhering to specified protocols and safety precautions is crucial when conducting these tests to guarantee the dependability and safety of the electrical infrastructure.

IS 732

4.8.1 Insulation Resistance Test - Between Earth and Conductors:

Testing insulation resistance between earth and conductors is essential to verify the quality of the insulation around electrical conductors and to avoid leakage currents or short circuits to the ground.

- ✓ **Preparation**: Disconnect the electrical circuit from the power source before testing to prevent the risk of electric shock. All devices linked to the circuit must be powered down.
- ✓ **Isolation**: Isolate the circuit segment tested to avoid interference from parallel paths.
- ✓ **Measurement Setup**: Connect one terminal of the insulation resistance tester, such as a merger, to the earth or ground connector of the circuit. Connect the other terminal to each conductor to be tested individually. Each conductor in multi-core cables should be individually checked against the earth conductor.
- ✓ **Testing**: Use the insulation resistance tester to apply the test voltage. Usually, the manufacturer's recommendations or applicable standards will provide the test voltage, which should suit the tested system. 500V or 1000V DC are the most common test voltages.
- ✓ **Measurement**: Obtain the insulation resistance value shown on the insulation resistance tester. This value represents the impedance between the conductor and the ground. Compare the measured resistance with the tested system's minimum insulation resistance value.
- ✓ **Interpretation**: A high insulation resistance value signifies good insulation condition, whereas a low value indicates potential insulation degradation or flaws. If the measured insulation resistance is notably lower than anticipated or falls below the prescribed minimum threshold, additional research is required to pinpoint and correct any insulation defects.
- ✓ **Recording**: Document the test results, including the insulating resistance values measured and any observations or irregularities observed during the testing.
- ✓ **Repeat**: Conduct the insulation resistance test individually on each conductor in the circuit, providing comprehensive testing of all conductors and connections.

✓ **Reconnection**: Once insulation resistance testing is finished and defects are ruled out, reconnect the circuit to the power supply and confirm its correct functioning.

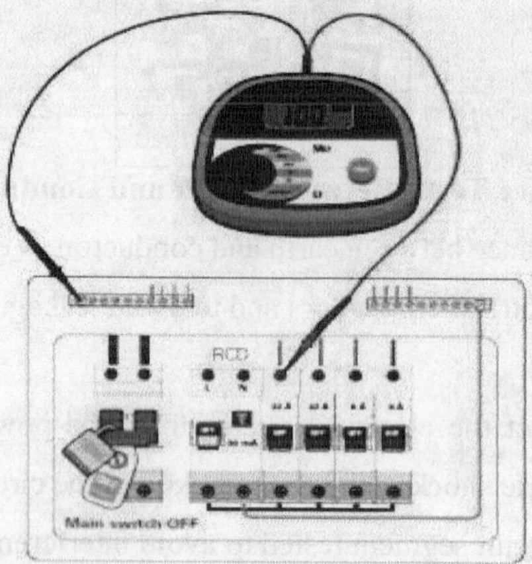

Figure 4.11 Testing Insulation Resistance

Testing insulation resistance (Fig.4.12) between earth and conductors is crucial for detecting any insulation problems, guaranteeing electrical safety, and upholding the electrical system's reliability.

4.8.2 Insulation Resistance Test - Between Conductors:

The Insulation Resistance Test evaluates the integrity of the insulation between conductors in an electrical system. This test detects insulation flaws such as fractures, moisture penetration, or degradation that may cause electrical risks or system malfunctions. Here are the steps to perform the Insulation Resistance Test on conductors:

- **Preparation**:

✓ Make sure the electrical circuit is disconnected and separated from any power sources.

✓ Disconnect all devices or appliances linked to the circuit.

✓ Ensure that all safety measures are implemented, including using suitable personal protective equipment (PPE).

- **Identification**: Determine the conductors for testing. These may consist of phase-to-phase conductors in three-phase systems or any other conductors that are anticipated to be electrically separated from one another.
- **Testing Equipment**: Utilise a megger or insulation resistance tester to deliver a high voltage (usually 500V or 1000V) to the conductors and measure the insulation resistance.

✓ Make sure that the testing equipment is accurately calibrated and functioning well.

- **Connections:**
  - ✓ Connect The insulation resistance tester has two terminals; connect one to one conductor and the other.
  - ✓ To avoid parallel paths, ensure no other conductors are connected throughout the test.
- **Testing:**
  - ✓ Use the insulation resistance tester to apply the test voltage. To prevent insulation damage, the voltage should be delivered gradually.
  - ✓ Find out how much insulating resistance the tester is showing you. We see the insulation resistance between the two tested conductors as a resistance value.
  - ✓ Let the insulating resistance reading stabilize for an adequate amount of time.
- **Interpretation:**
  - ✓ Evaluate the recorded insulation resistance value about the permissible boundaries laid out by applicable standards or local ordinances.
  - ✓ Insulation resistance between conductors should generally be high, preferably several megaohms or greater, to indicate that the insulation is in good condition.
  - ✓ It may be necessary to conduct more investigations and take remedial measures if the insulation resistance reading is low, as this could indicate insulation defects or moisture intrusion.
- **Recording:**
  - ✓ Record: Document all test findings, such as the insulation resistance value, the tested conductors, and any unusual observations.
  - ✓ Keep records for reference and to ensure compliance in the future.
- **Verification**: To ensure you've thoroughly tested and verified the insulation integrity, repeat the Insulation Resistance Test between each pair of conductors in the electrical system.
- **Corrective Action**: Investigate and rectify any insulation resistance readings below permissible levels or indicating potential defects by repairing or replacing damaged insulation or improving environmental conditions to avoid moisture penetration.
- **Reconnection and Testing**: After finishing the Insulation Resistance Test and resolving any detected concerns, reconnect the electrical circuit to the power source and conduct functional testing to verify appropriate operation.

Regularly conducting Insulation Resistance Tests between conductors is crucial for upholding the safety and dependability of electrical systems, particularly in settings where insulation integrity is vital.

4.8.3 Polarity Test of Single Pole Switches:

The polarity test of a switch checks if the switch is connected correctly in the electrical circuit, verifying that the line and load connections are aligned as per the wiring schematic. This test aids in averting electrical risks and guarantees the safe functioning of the switch.

- ✓ **Preparation**: Ensure that the electrical circuit is not energized and that all power sources to the switch are switched off. Utilize a voltage tester or multimeter to verify the absence of voltage at the switch terminals.
- ✓ **Identify Terminals**: Determine the terminals on the switch. Typical switch terminals consist of line (hot), load (switched hot), and ground.
- ✓ **Reference Wiring Diagram**: Consult the wiring diagram accompanying the switch or refer to the manufacturer's instructions to identify the accurate polarity and terminal connections.
- ✓ **Test Continuity**: Verify that the line terminal of the switch is in continuity with the line conductor (hot wire) in the electrical box using a continuity tester or a multimeter set to continuity mode. These two points ought to be continuous with one another.
- ✓ **Test Load Connection**: Verify that the electrical box's load conductor (the switched hot wire) is connected to the switch's load terminal. These two points ought to be continuous with one another.
- ✓ **Verify Grounding**: You need to check that the grounding terminal of the switch is connected to the grounding conductor in the electrical box if it has one.
- ✓ **Check Switch Position**: As you conduct the continuity tests, ensure the connections stay consistent by toggling the switch to the ON and OFF positions.
- ✓ **Interpretation**: If the switch works as it should and there is continuity between the terminals that match the wiring diagram, then the switch's polarity is correct. Additional inquiry is needed to find and fix any wiring mistakes or issues if continuity is irregular or nonexistent.
- ✓ **Labelling**: After making sure the polarity is correct, mark the conductors and terminals of the switch for easy reference and repair.

✓ **Reconnection and Testing**: Reconnect the switch to the power source and confirm its functionality when the polarity check is finished and any problems have been fixed.

Conducting a polarity test (Fig.4.13) on a switch is crucial during electrical installation or maintenance to verify the switch's safety and correct operation within the electrical system.

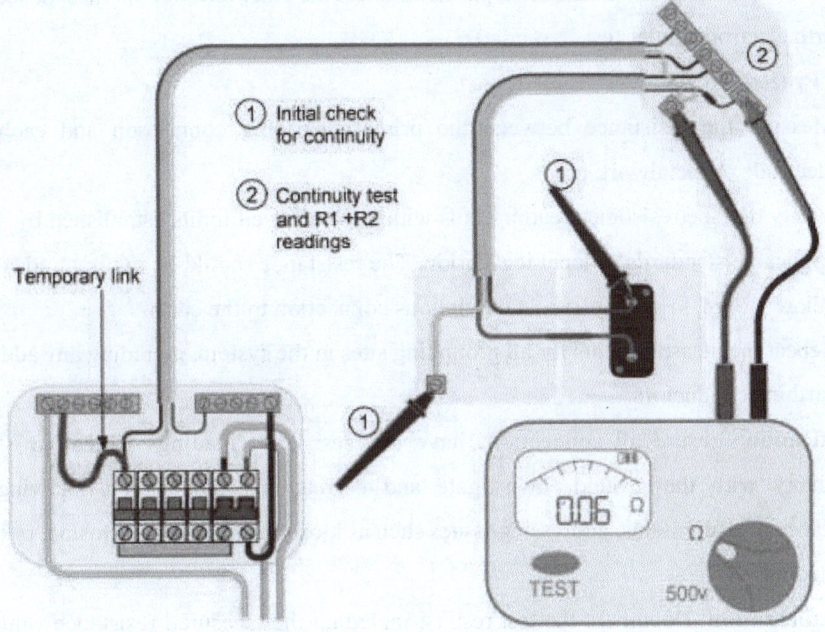

Figure 4.12: Polarity test of switch

4.8.4 Testing of Earth Continuity Path:

The Earth Continuity Test is an essential process carried out during electrical installation or maintenance to verify the effectiveness of the protective earthing system. This test ensures that the earthing components have low resistance channels to effectively transfer fault currents to the ground, preventing electric shocks and enabling the correct functioning of safety devices like circuit breakers or fuses. Below are the steps for doing an Earth Continuity Test.

✓ **Preparation**: Ensure that the electrical circuit is not energized and that all power sources in the system are switched off. Disconnect all devices or appliances linked to the circuit.

✓ **Identify Earthing Points**: Determine all grounding locations, such as earth electrodes, grounding conductors, metallic enclosures, and exposed metal structures in the system.

✓ **Testing Equipment**: Utilise a low-resistance ohmmeter or a dedicated earth continuity tester to conduct the test. Make sure that the testing equipment is accurately calibrated and functioning well.

✓ **Connections**: Connect one terminal of the test instrument to the primary grounding terminal or busbar in the distribution panel. Connect the other terminal to the exposed metal or earth electrode under test.

✓ **Test Procedure**:

- Measure the resistance between the primary earthing connection and each earth electrode or metalwork point.
- Verify that the resistance reading falls within the allowed limits established by applicable standards or local legislation. The resistance should be minimal, ideally below 1 ohm, to show a robust continuous connection to the earth.
- Repeat the measurements for all grounding sites in the system, including any additional earthing conductors.

✓ **Verification**: Ensure all connections have low resistance readings indicative of good continuity with the ground. Investigate and correct any connections showing high resistance measurements, addressing issues such as loose connections, corrosion, or broken conductors.

✓ **Documentation**: Document the test results, including the measured resistance values and the locations of the tested earthing sites. This documentation is crucial for keeping records of the electrical installation and guaranteeing adherence to safety rules.

✓ **Reconnection and Testing**: After conducting the Earth Continuity Test and addressing any problems, reconnect the electrical circuit to the power source and perform functional testing to verify the correct operation.

Regularly conducting Earth Continuity Tests is crucial for ensuring the safety and dependability of electrical installations. It aids in detecting possible flaws or vulnerabilities in the earthing system, enabling prompt repairs or enhancements to be implemented.

4.8.5 Earth Electrode Resistance Test:

A basic test to determine how well the grounding system performs in offering a low-resistance path to the Earth is the Earth Electrode Resistance Test. This test verifies that the electrical installation is grounded correctly, which is crucial for system safety and fault protection. The following is a written procedure for doing an Earth Electrode Resistance Test:

- **Preparation**:

✓ Make Ensure every electrical circuit linked to the grounding system has been disconnected and de-energized.

✓ Unplug any appliances or devices that are attached to the system.

✓ Determine which earth electrodes or electrodes require testing. Earth electrodes are frequently made of rods, pipes, plates, and grids.

- **Testing Equipment**:

✓ For precise measurements, use an earth resistance tester that is specialized and also referred to as a ground tester or earth tester.

✓ Make sure the testing apparatus is in good operating order and calibrated correctly.

- **Connections**:

✓ Attach the earth resistance tester's test leads to the earth electrode that is being tested. The current (C) and potential (P) connectors are usually connected to different terminals on the tester.

✓ Assemble the potential (P) terminal to a distant earth electrode or a stake in the ground sufficiently far away from the tested electrode, then connect the current (C) terminal to the earth electrode.

- **Measurement Procedure**:

✓ The earth resistance tester is typically programmed to run tests using the "Earth Electrode Resistance Test" or the "Two-Point Method."

✓ The earth electrode being tested should be subjected to a test current through the tester.

✓ The voltage drop between the tester's current (C) and potential (P) terminals needs to be measured.

✓ Calculate the earth electrode resistance using Ohm's Law: Resistance (R) = Voltage (V) / Current (I).

- **Repeat and Average**:

✓ Repeat the measurement several times to guarantee precision and uniformity.

✓ Calculate the mean earth electrode resistance based on the various data taken.

- **Interpretation**:

✓ Compare the measured earth electrode resistance with the permitted limits of relevant standards or local legislation.

✓ For most installations, the ground electrode resistance should be low, ideally less than 10 ohms. However, particular criteria could change based on the kind of installation and local laws.

- **Documentation**:

✓ Record test results, including earth electrode resistance measurements and installation data.

✓ Maintain paperwork for future reference and compliance.

- **Corrective Action**:

✓ Investigate and address potential concerns such as inadequate electrode depth, low soil conductivity, or corroded connections if the observed earth electrode resistance exceeds permissible limits.

- **Verification**:

✓ After making adjustments or improvements, conduct the Earth Electrode Resistance Test again to confirm compliance with standards and assure the grounding system's functionality.

Regularly conducting Earth Electrode Resistance Tests (Fig.4.14) is crucial for ensuring the safety and dependability of electrical systems, particularly in regions susceptible to electrical malfunctions or lightning strikes.

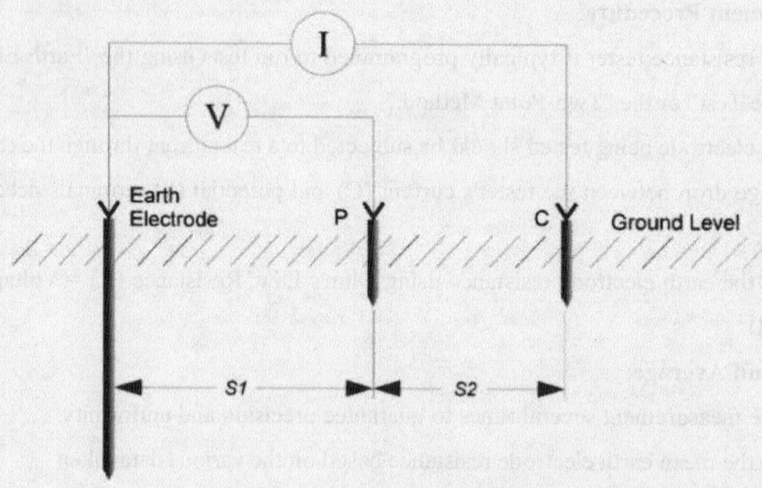

Figure 4.13: Earth electrode resistance test

- The figure shows that two additional earth electrodes are placed at an appropriate distance from the test electrode.

- A current is transmitted from the tested electrode 'A' to an auxiliary electrode 'C.' The difference in potential between electrode 'A' and the auxiliary potential 'B' is calculated.
- The resistance of test electrode 'A' is then calculated using the formula: R = V/I
- Where R is the resistance of the test electrode in ohms, V is the voltmeter reading in volts, and I is the ammeter reading in amps.
- A manual generator eliminates stray currents in the soil that lead to substantial mistakes in measuring earth resistance.
- The instrument pointer fluctuates when the supply frequency of the hand-driven generator matches the frequency of the stray current.
- This will disappear if the generation pace is altered.
- While the test is being conducted, the test electrode must not be connected to the grounding system.
- An auxiliary electrode, a mild steel rod with a diameter of 13 mm, is inserted 1 m into the ground for testing.
- Ensure that the resistance zones of all three electrodes are positioned in a manner separate from each other.
- The auxiliary current electrode 'C' and auxiliary potential electrode 'B' should be placed 30 metres apart and interspersed with each other if the test electrode is a pipe, plate, or rod.

Assume that there needs to be more consistency in three consecutive resistance measurements taken from test electrodes. After that, we'll implant electrode B and extend the distance between A and C to 50 metres to start the test.

4.9 Residential Building Service Connection:

A residential building electrical service connection delivers electricity from the utility provider to a residential property. Here is a summary of the stages for connecting electrical service to a residential building:

Application and Documentation:

- ✓ The property owner or developer applies for an electrical service connection with the local utility supplier.
- ✓ The application usually contains facts such as property address, owner information, building blueprints, load specifications, and any other necessary papers the utility supplier requires.

Utility Assessment:

- The utility supplier evaluates the application and performs a site inspection to examine the possibility of supplying electrical service to the residential building.
- Factors including distance to current electrical infrastructure, capacity on the distribution network, and voltage needs are considered in the evaluation.

Service Design:

- The utility supplier designs the residential building's electrical service connection based on the assessment.
- This entails figuring out the service entry point, installing distribution infrastructure such as poles, transformers, and switchgear, sizing conductors, and positioning metering equipment.

Permitting and Approvals:

- The utility supplier acquires regulatory authorities' required permissions and approvals to construct electrical service connections.
- Building permits and clearances from local authorities are necessary to comply with zoning restrictions, building codes, and safety standards.

Installation and Construction:

- Once permits are acquired, the utility supplier installs electrical service connections to the residential building.
- Installation may include overhead or underground power line placement, utility pole installation, transformers, service entrance panel setup, meter box installation, and grounding system implementation.

Metering and Activation:

- Upon installation, the utility company places electrical meters at the specified metering locations in the residential building.
- The utility supplier checks the integrity of the installations, conducts meter tests, and activates the electrical service connection to begin invoicing for power usage.

Final Inspection and Compliance:

- Regulatory bodies may perform final inspections to verify adherence to safety, quality, and environmental regulations.
- The property owner or developer has the right to check the installations to ensure all standards are completed before the residential building may be occupied.

Ongoing Maintenance and Support:

- ✓ The utility supplier is in charge of the continuous maintenance, repair, and support of the electrical service connections to the residential building.
- ✓ Property owners or tenants can contact the utility supplier for help with meter readings, billing questions, service outages, or emergencies concerning electrical services.

Residential structures can provide safe and dependable electrical service connections by following these measures, meeting inhabitants' needs, and adhering to regulatory standards.

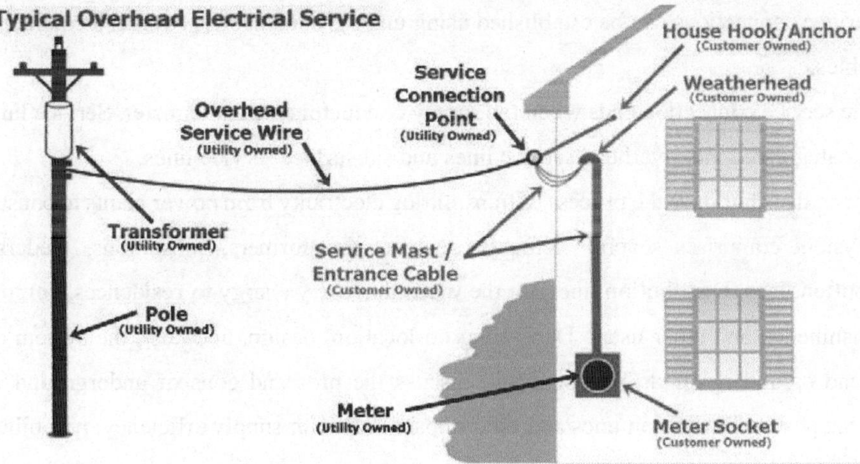

Figure 4.14 Typical overhead electrical service

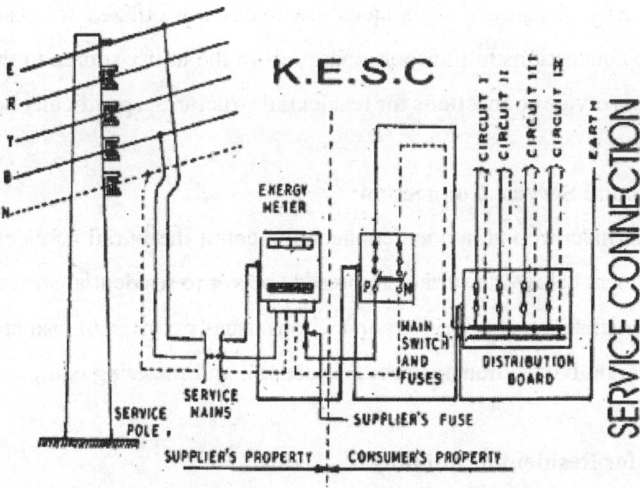

Figure 4.15 Typical service connection

Inspect the service connection for its type (voltage, amperage), general condition, and whether the total power is sufficient for the facility's requirements. The equipment must be clean and free of overgrown vegetation and debris.

- ✓ Either a bare conductor or a weatherproof cable made of aluminum or copper is used to provide the service connection from the nearest pole to the consumer's dwelling.
- ✓ The service connection is the line that delivers electricity from the supplier's low-voltage distributor to the energy meter at the consumer's location.
- ✓ Service connections can be established using underground cables, overhead conductors, or cables.
- ✓ The service connection ends when the supply conductors reach the meter. Service lines can be categorized into overhead service lines and subsurface service lines.

Power distribution is the process of transmitting electricity from power plants to consumers. The system comprises several elements, such as transformers, substations, feeders, and distribution lines. Distribution lines are the wires that carry energy to residences, commercial establishments, and other users. Depending on location, design, and cost, the system can be overhead or underground. This post will discuss the pros and cons of underground versus overhead power distribution lines and their impact on power supply efficiency, reliability, and safety.

4.9.1 Types of Underground and Overhead:

Both overhead and underground techniques are frequently utilized for residential building electrical service connections to transport energy from the utility source to the property. This is a comparison of service connections for residential structures, specifically between overhead and underground:

- **Underground Service Connection:**

a) **Description:** Underground service connections entail the burial of electrical cables and related equipment below the surface to provide power to residential structures.

b) **Components:** Underground service connections usually consist of insulated power cables, conduits, junction boxes, transformers (if needed), and metering equipment buried below the surface.

c) **Advantages for Residential Buildings:**

- ✓ **Aesthetics:** Subterranean service connections are hidden from view above ground, maintaining the visual appeal of residential areas.

✓ **Safety:** Underground cables are less prone to inadvertent touch or interference, diminishing the likelihood of electrical risks for residents.

✓ **Reliability:** Underground cables are more resistant to weather-induced harm, such as wind, lightning, and tree fall, reducing power interruptions.

✓ **Quiet Operation:** Underground service connections are quieter than overhead lines, which might benefit residential areas due to reduced noise levels.

d) **Challenges:**

✓ **Cost:** Underground service connections are usually costlier to set up and upkeep because of the need for excavation, trenching, and cable installation.

✓ **Accessibility:** Identifying and fixing issues in underground cables is typically more complex and time-intensive than overhead lines.

✓ **Coordination:** Underground service connections may need to be coordinated with other subterranean utilities such as water, gas, and telecommunications to prevent conflicts and reduce disruptions.

Overhead Service Connection:

a) **Description:** Overhead service connections refer to installing electrical cables on utility poles or towers above ground level to provide power to residential structures.

b) **Components:** Overhead service connections include conductors, insulators, utility poles or towers, transformers (if needed), and metering equipment erected above ground.

c) **Advantages for Residential Buildings:**

✓ **Cost-Effectiveness:** Overhead service connections are typically more cost-effective than underground systems for residential areas due to lower installation and maintenance costs.

✓ **Accessibility:** Overhead wires are more accessible for maintenance and repairs, making locating and fixing faults simpler.

✓ **Flexibility:** Overhead wires can be readily adjusted or extended to meet shifts in demand or infrastructural needs.

d) **Challenges:**

✓ **Aesthetics:** Overhead lines can diminish the aesthetic charm of residential neighbourhoods, especially in locations with beautiful views or architectural importance.

✓ **Vulnerability:** Overhead lines are vulnerable to weather conditions and external sources, heightening the likelihood of outages and damage.

- ✓ **Safety Concerns:** Overhead lines present a danger of electrocution and harm from unintended contact, particularly in regions with dense vegetation or when lines are situated close to the ground.

Deciding between subsurface and overhead service connections for residential structures is influenced by cost, aesthetics, safety, reliability, and regulatory mandates.

4.9.2 Calculation of Material Required For Service Connection:

The materials needed for an electrical service connection can differ based on criteria such as the service type (residential, commercial, industrial), load requirements, local construction codes, and the building or facility layout. Nevertheless, I can offer a standard list of materials commonly required for a fundamental household electrical service connection:

- ✓ **Service Entrance Cable**: This cable links the electrical service panel to the service drop or lateral. The distance between the panel, the utility connection point, and the service size, usually specified in amperage, determine the cable size. American Wire Gauge (AWG) sizes 2/0, 3/0, and 4/0 are common for home services.
- ✓ **Service Panel**: This is the primary hub for distributing electrical circuits throughout the facility. The system incorporates circuit breakers or fuses to safeguard the circuits. The expected load requirement of the building determines the panel size. Typical home panels have amperage ratings between 100 and 400 amps.
- ✓ **The Grounding System** consists of grounding rods, clamps, and conductors. It guarantees the electrical system's safety by establishing a route to the ground for fault currents.
- ✓ **Meter Socket**: This is the housing where the electric utility's meter is installed to quantify the electricity consumed by the building.
- ✓ **Conduit and Fittings**: Electrical wiring is protected and routed using conduit. Local building codes may require a rigid metal conduit, PVC, or EMT (Electrical Metallic Tubing). Connectors, couplings, and elbows are fittings used to unite conduit sections.
- ✓ **Wire and Cable**: This refers to the electrical cabling that connects the service panel to the building's outlets, switches, and light fixtures. The gauge and material of the wire are determined by the electrical demand of the circuit and the regulations outlined in the local electrical code.
- ✓ **Connectors and Fasteners**: The items consist of wire nuts, terminal blocks, screws, and other necessary hardware for establishing connections and fastening components.

✓ **Disconnect Switch**: This device interrupts the electricity supply to the building for maintenance or emergencies. It is usually positioned between the meter socket and the service panel.

✓ **Weatherhead and Service Mast**: The components bolster and safeguard the service entrance cable when it links to the structure and shifts from overhead to underground.

✓ **Miscellaneous Supplies**: These products may include electrical tape, wire markers, junction boxes, and conduit straps.

Note that this is a generic list, and the materials needed for your project may differ. Before starting any electrical work, seeking advice from a certified electrician and familiarising yourself with local building standards and laws is crucial.

➢ **Estimation Costing of Overhead Service**

Initially, I will elucidate the concept of overhead service mains.

The term "overhead" denotes the arrangement of conductors in a suspended manner above the level of the ground. The overhead transmission line comprises several components such as service cables, reel insulators, broken insulators, messenger wire, GI pipe, service pipe and additional elements.

Overhead service mains provide advantages such as quick installation, easy fault detection, and lower cost. However, they also have downsides, including safety concerns, susceptibility to weather conditions, and unattractive looks.

➢ **Single Phase Overhead Service Mains Calculation**

Q: Prepare a list of materials and estimate the cost for providing service connection to a building at 240 V single phase 50 Hz having a light and fan load of 2 KW and heating load of 2 KW. The supply will be provided by an overhead line 20 m away from the building.

Solution: Materials used for service connection:

Calculate the wire size by first determining the total connected load in the house and then calculating the load current of the house using the appropriate calculation.

Connected load = lighting load + heating load = 2 KW + 2 KW = 4000 W

See, the total load of the house is 4000 W, and multiply it by a factor of safety; the factor of safety means we should consider future load. Now, I will consider the future load double the present so that I will multiply by 2.

Considering the future additional demand (say 100%), let the future load be twice the present load (i.e., a factor of safety of 2). i.e. 2 × 4000 W = 8000 W.

Load current in amperes = Load in watts/Rated voltage = 8000/240 = 33 Amps.

The load current of the house is 33 Amps; now, by seeing the wire table, you can find the wire size. The size of the aluminium cable is ten sq mm from the wire table.

(a) The length of cable required = (20 + 6 + 3 + 4) × 2 = 66 m

We must determine the cable length from low-tension lines to the consumer meter board. Below, I will present formulas for calculating the cable length.

Note: Length of the cable = Horizontal length (20 m) + Looseness and meter board connection (6 m) + Service pipe length (3 m) + Pole side connection (4 m)

Horizontal length – Length of cable from a pothead up to service pipe.

Looseness and meter board connection – cable length from the end of the service pipe to the meter board.

Service Pipe – The wires are pulled into the service pipe; therefore, we also need to consider that length.

Pole side connection – We must evaluate the cables' length to link low-tension lines to the pothead.

(b) No of Break Insulators = 2 Nos

2 insulators are necessary. Overhead service mains use only two break insulators. One insulator is located at the pole side, while another is at the meter board side. Break insulators separate the messenger wire from the pole and support the messenger wires.

(c) No of Reel Insulators = 2 × (20 + 6) = 52 Nos

The number of reel insulators required is 52 No. Reel insulators are designed to create a separation between two wires to prevent short circuits in case the insulation of both conductors is compromised.

(d) 10 SWG G.I wire for messaging = 22m = 2 kg.

The purpose of messenger wire is to support the reel insulators and service cables.

(e) 30 mm PVC pipe of 2 m and bends of 3 Nos.

(f) 40 mm G.I pipe of 3 mm thick 1.8 m for supporting.

(g) 500 mm × 500 mm × 45 mm WVTWB – 1 No.

(h) 40 mm G.I pipe 2.5m and 1 kg 8 SWG G.I. wire for earthing.

(i) 30 A fuse of 2 Nos and Miscellaneous materials.

Miscellaneous materials mean nuts, bolts, etc.

These are the Materials required.

Costing of Single Phase Overhead Service Mains

SI. No	Particulars	Unit	Quantity	Rate In Rupees	Cost In Rupees
01	10 sq mm aluminium 11000 V grade double sheathed PVC cable	100 M (coil)	66M	1800	1188.00
02	1.1 KV porcelain break insulator	No's	2	10.00	20.00
03	Reel insulators	No's	52	2.00	104.00
04	10 SWG GI wire for messaging	Kg	2	100.00	200.00
05	40 mm 3 mm thick GI pipe for supporting	M	1.8	375.00	675.00
06	30 mm PVC pipe 2 mm thick for servicing	3 M	2 M	65.00	43.33
07	PVC bends	No's	3	10.00	30.00
08	500 × 500 × 45 mm well-seasoned varnished teak wood (WVTB)	No	1	600.00	600.00
09	40 mm GI pipe 3 mm thick for earthing	M	2.5	375.00	937.50
10	8 SWG GI wire for earthing	Kg	1	150.00	150.00
11	30 A 500 voltage grade porcelain fuse	No's	2	100.00	200.00
12	12.5 mm dia 300 eye bolt nuts of M.S	No's	2	50.00	100.00
13	Miscellaneous materials include binding wire clamps for pole and service pipe bolts, nuts, coal, salt, etc.	Lump sum			525.67
14	Labor charges				
	(a) for serving and earthwork				700.00
15	Contingencies and unforeseen things @ 5% on the above amount.				260.10
				Total=	5460.00

➢ **Three-Phase Overhead Service Mains Calculation**

Q: Three phases of overhead service connection: Select the size and prepare an estimate for the overhead service connection to feed the power supply to a factory with a 10 HP load for a distance of 10 meters.

Solution: Materials used for service connection.

Assuming a supply voltage of 400 V, p.f = 0.8, and the efficiency of the load is 80%.

Current = (10×735.5)÷(0.8×0.8×400×√3) = 16.6 Amps

Keeping a view of the future, additional demand says 100%. Let the future load be twice the present load or factor of safety as 2; the current rating is 33A.

The size of the aluminium cable required is 10 sq mm.

(a) Length of the cable = 10+3+3+4 = 20 × 4 = 80 m.

Note: Length of the cable = Horizontal length (10 m) + Looseness and meter board connection (3 m) + Service pipe length (3 m) + Pole side connection (4 m).

(b) G.I Messaging wire = 10 + 1 = 11m = 1kg.

(c) No of Break Insulators = 2 Nos.

(d) No of Reel Insulators = (10+3)4 = 13 × 4 = 52 Nos.

(e) 30 mm PVC pipe of 2 m and 3 bends.

(f) 50 mm G.I pipe of 3 mm thick 1.8 m for supporting.

(g) 500 × 500 × 65 m WVTWB – 1 No.

(h) 40 mm G.I pipe 2.5 m and 1 kg 8 SWG GI wire for earthing.

(i) 30 A fuse unit – 3 Nos.

(j) M.S. clamps with nuts and bolts – 2 sets and miscellaneous materials.

Costing of Three Phase Overhead Service Mains

SI. No	Particulars	Unit	Quantity	Rate in rupees	Amount in rupees
01	10 sq mm aluminium cable of 1.1 KV voltage grade, double-sheathed	100M (coil)	80 M	1800.00	1440.00
02	1.1 KV porcelain break insulators	No's	2	10.00	20.00
03	Porcelain reel insulators	No's	52	2.00	104.00
04	10 SWG GI messaging wire	Kg	2	100.00	200.00
05	30 mm PVC pipe 2 mm thick, for servicing	3 M	2	65.00	130.00
06	PVC bends for 30 mm PVC pipe	No's	3	10.00	30.00
07	50 mm GI pipe 3 mm thick 1.8M for supporting	M	1.8	495.00	891.00
08	500 × 500 × 45 mm WVTWB	No	1	600.00	600.00

09	30 A porcelain fuse unit 500 Vg	No	3	100.00	300.00
10	40 mm GI pipe 3 mm thick	M	2.5	375.00	937.00
11	8 SWG GI wire for earthing	Kg	1	150.00	150.00
12	12.5 mm dia 300 mm eye bolt nuts of M.S	No	2	50.00	100.00
13	Miscellaneous materials include binding wire, nuts, bolts, cement, salt, coal, etc.	Lump sum			298.00
14	Labour charge				
	(a) For earthing				300.00
	(b) For servicing				700.00
15	Contingencies and unforeseen things @ 5% on the above amount				260.00
				Total=	6460.00

Solved Numerical Problem

Example 4.1:

You are designing the electrical system for a residential building. The building has the following electrical appliances and loads:

Refrigerator: 500 watts, Television: 200 watts, Air conditioner: 1500 watts, Washing machine: 1000 watts, Lights: 10 bulbs, each 60 watts, Microwave oven: 800 watts

Calculate the total connected load and the minimum capacity of the main circuit breaker required for this residential building.

Solution:

To find the total connected load, we add up the power consumption of all the appliances and lights:

Total connected load = Refrigerator + Television + Air conditioner + Washing machine + Lights + Microwave oven

Total connected load = 500 + 200 + 1500 + 1000 + (10 × 60) + 800

Total connected load = 500 + 200 + 1500 + 1000 + 600 + 800

Total connected load = 4600 watts

We need to consider a safety margin and account for future expansions.

Let's assume a safety factor of 25%.

Safety factor = 25% of total connected load

Safety factor = 0.25 × 4600 Safety factor = 1150 watts

Total load with safety factor = Total connected load + Safety factor Total load with safety factor = 4600 + 1150 Total load with safety factor = 5750 watts

Now, we need to select the minimum capacity of the main circuit breaker. The circuit breaker should be able to handle the total load with the safety factor.

A typical circuit breaker capacity in residential installations would be 20% greater than the total load with a safety factor to accommodate starting currents and other variations.

Circuit breaker capacity = 1.20 × Total load with safety factor Circuit breaker capacity = 1.20 × 5750 Circuit breaker capacity = 6900 watts

Therefore, the minimum capacity of the main circuit breaker required for this residential building would be 6900 watts or 6.9 kW.

Example 4.2:

Calculate the total power consumed by a lighting circuit comprising 15 lamps, each rated at 100 watts.

Solution:

The total power consumed by the lighting circuit = Number of lamps × Power rating of each lamp Total power consumed = 15 lamps × 100 watts/lamp = 1500 watts

Therefore, the total power consumed by the lighting circuit is 1500 watts.

Example 4.3:

Determine the total power consumed by a power circuit with the following appliances:

Refrigerator: 500 watts, Microwave oven: 800 watts, Washing machine: 1200 watts, Television: 200 watts

Solution:

Total power consumed by the power circuit = Sum of power ratings of all appliances Total power consumed = 500 watts + 800 watts + 1200 watts + 200 watts = 2700 watts

Therefore, the total power consumed by the power circuit is 2700 watts.

Example 4.4:

A residential building has the following lighting loads:

Living room: 6 bulbs, each 60 watts

Bedroom: 4 bulbs, each 40 watts

Kitchen: 8 bulbs, each 75 watts

Calculate the total power consumed by the lighting loads in the building.

Solution:

Total power consumed by lighting in the living room = 6 bulbs × 60 watts/bulb = 360 watts

Total power consumed by lighting in the bedroom = 4 bulbs × 40 watts/bulb = 160 watts Total power consumed by lighting in the kitchen = 8 bulbs × 75 watts/bulb = 600 watts

The total power consumed by lighting in the building = Sum of energy consumed in each area.

Total power consumed = 360 watts + 160 watts + 600 watts = 1120 watts.

Therefore, the total power consumed by the lighting loads in the building is 1120 watts.

Example 4.5:

Selection of Conductor Size Given a residential building with the following loads:

Lighting load: 2000 watts, Power load: 5000 watts, Total length of wiring: 50 meters

Voltage drop limit: 3%

Calculate the minimum size of the copper conductor required for the wiring.

Solution:

Total load = Lighting load + Power load = 2000 watts + 5000 watts = 7000 watts

Voltage drop = 3% of the voltage (assumed to be 230V)

Voltage drop = 0.03 × 230V = 6.9V

Using the voltage drop formula:

Voltage drop = (2 × Length × Resistance × Load) / (Cross-sectional area × Conductivity) 6.9V = (2 × 50m × Resistance × 7000W) / (Cross-sectional area × Conductivity)

Assuming standard conductivity for copper (58 S/m):

Cross-sectional area × Resistance = (2 × 50m × 7000W) / (6.9V × 58 S/m) Cross-sectional area × Resistance = 0.1984 $m^2 \times S^2$

We can select a conductor with a cross-sectional area that meets this requirement based on standard conductor tables.

Example 4.6:

Selection of Rating of Main Switch Given a residential building with a total load of 10,000 watts, select the appropriate rating for the main switch.

Solution:

The rating of the main switch should be slightly higher than the total load to accommodate the appliances' starting current. Therefore, a 16A main switch would be suitable for a 10,000-watt load.

Example 4.7:

Selection of Rating of Protective Switchgear Given a residential building with a total load of 15,000 watts, select the appropriate rating for the protective switchgear.

Solution:

Like the main switch, the protective switchgear should have a rating slightly higher than the total load. Considering safety and future expansion, a 20A protective switchgear would be appropriate for a 15,000-watt load.

Example 4.8:

Selection of Conductor Size for Industrial Use

Given an industrial facility with a total load of 30,000 watts, a total length of wiring of 100 meters, and a voltage drop limit of 5%, calculate the minimum size of the copper conductor required.

Solution:

Follow a similar process as in Solution 1.5, considering the specific requirements and standards for industrial wiring.

Example 4.9:

Selection of Rating of Main Switch for Commercial Building Given a commercial building with a total load of 50,000 watts, select the appropriate rating for the main switch.

Solution: Similar to Solution 1.6, select a central switch rating slightly higher than the total load to ensure reliability and safety. A 63A main switch would suit a 50,000-watt load in a commercial building.

Example 4.10:

Selection of Rating of Protective Switchgear for Large Industrial Plant Given a large industrial plant with a total load of 200,000 watts, select the appropriate rating for the protective switchgear.

Solution:

Again, choose a protective switchgear rating slightly higher than the total load. For a large industrial plant with a 200,000-watt load, a 250A protective switchgear would be appropriate.

Exercise

Multiple Choice Questions:

1. What is most countries' standard voltage level for residential buildings?

   a) 110V b) 220V c) 440V d) 660V

2. Which is not typically considered a residential electrical load?

   a) Air conditioner b) Refrigerator c) Industrial machinery d) Television

3. Which wire color is commonly used for the neutral wire in residential wiring?

   a) Red b) Blue c) Black d) White

4. What is the purpose of a circuit breaker in a residential electrical system?

   a) To regulate voltage b) To interrupt power flow in case of overload or short circuit c) To step up voltage d) To provide grounding

5. Which type of electrical outlet is commonly used in kitchens and bathrooms for safety?

   a) GFCI outlet b) USB outlet c) Standard outlet d) Surge-protected outlet

6. What is the typical frequency of alternating current (AC) power supply in residential buildings?

   a) 50 Hz b) 60 Hz c) 100 Hz d) 120 Hz

7. What is the purpose of an earth/ground wire in residential electrical wiring?

   a) To carry the load current b) To provide a return path for fault currents c) To regulate voltage d) To provide insulation

8. Which of the following is a safety device used in residential electrical systems to protect against electric shock?

   a) Capacitor b) Diode c) Resistor d) GFCI (Ground Fault Circuit Interrupter)

9. In residential electrical systems, what is the purpose of a junction box?

   a) To connect wires and cables b) To regulate voltage c) To interrupt power flow d) To store electrical appliances

10. According to standard codes, What is the minimum height for electrical outlets above the floor level in residential buildings?

    a) 12 inches b) 18 inches c) 24 inches d) 36 inches

11. What is a commonly used wiring method in residential buildings?

    a) Conduit wiring b) Grounding rod wiring c) Overhead wiring d) Solar panel wiring

12. What is the purpose of a junction box in residential electrical wiring?

a) To distribute power to different areas of the building b) To provide a safe enclosure for wire connections c) To regulate voltage d) To act as a switch

13. Which of the following is a commonly used material for residential electrical wiring?

a) Copper b) Aluminum c) Steel d) Iron

14. Which of the following is not a typical component of a residential electrical panel?

a) Circuit breaker b) Voltage regulator c) Main switch d) Neutral bus bar

15. Which type of lighting is most commonly used in residential buildings due to its energy efficiency?

a) Incandescent lighting b) Halogen lighting c) Fluorescent lighting d) LED lighting

16. In residential electrical wiring, what is the purpose of a switch?

a) To regulate voltage b) To interrupt power flow to a specific circuit c) To carry the load current d) To provide insulation

17. Which of the following is a safety feature of a GFCI outlet?

a) Surge protection b) Overload protection c) Ground fault protection d) Short circuit protection

18. What is not typically considered a residential electrical appliance?

a) Toaster b) Air conditioner c) Industrial compressor d) Microwave oven

19. What is the primary purpose of the grounding system in a residential electrical installation?

a) To protect against lightning strikes b) To provide a safe path for fault currents to earth c) To regulate voltage d) To increase electrical resistance

20. Which of the following is a standard safety practice when working with residential electrical systems?

a) Working with wet hands b) Overloading electrical outlets c) Using insulated tools d) Standing on wet surfaces

Answer:

1. b) 220V
2. c) Industrial machinery
3. d) White
4. b) To interrupt power flow in case of overload or short circuit
5. a) GFCI outlet

6. b) 60 Hz
7. b) To provide a return path for fault currents
8. d) GFCI (Ground Fault Circuit Interrupter)
9. a) To connect wires and cables
10. b) 18 inches
11. a) Conduit wiring
12. b) To provide a safe enclosure for wire connections
13. a) Copper
14. b) Voltage regulator
15. d) LED lighting
16. b) To interrupt power flow to a specific circuit
17. c) Ground fault protection
18. c) Industrial compressor
19. b) To provide a safe path for fault currents to earth
20. c) Using insulated tools

Short Answer Type Questions:

1. Define GFCI and explain its importance in residential electrical systems.
2. What is the purpose of grounding in residential electrical installations?
3. Differentiate between a circuit breaker and a fuse.
4. Explain the significance of voltage drop in residential wiring.
5. What standard wire sizes are used in residential electrical wiring, and how are they denoted?
6. Describe the function of a neutral wire in residential electrical circuits.
7. What is the role of a junction box in residential electrical wiring?
8. Discuss the importance of load balancing in residential electrical systems.
9. How do you determine the load capacity of a residential electrical panel?
10. Explain the purpose of an AFCI (Arc Fault Circuit Interrupter) in residential wiring.
11. What safety precautions should be taken when installing electrical outlets in wet areas like bathrooms and kitchens?
12. Describe the function of a main switch in a residential electrical panel.
13. Discuss the importance of selecting the suitable conductor material for residential wiring.

14. What are the primary differences between series and parallel electrical circuits in residential applications?
15. Explain how surge protectors work and why they are essential in residential buildings.

Long Answer Type Questions:

1. Discuss the process of designing an electrical layout for a new residential building. Include considerations such as load calculations, wire sizing, and placement of outlets and switches.
2. Explain the importance of electrical codes and regulations in residential building electrification. Discuss some essential safety standards and requirements that must be followed during installation.
3. Describe the role of a distribution board or electrical panel in a residential building. Discuss its components, functions, and the considerations for selecting an appropriate panel for a given application.
4. Discuss the various wiring methods used in residential building electrification. Compare and contrast the advantages and disadvantages of surface wiring, concealed conduit wiring, and overhead wiring.
5. Explain the significance of grounding and bonding in residential electrical systems. Describe the methods for establishing adequate grounding and bonding, and discuss their importance for electrical safety.
6. Discuss the importance of load balancing in residential electrical systems. Explain how load balancing can help prevent overloads, reduce energy consumption, and ensure the efficient operation of electrical circuits.
7. Describe selecting and installing GFCI (Ground Fault Circuit Interrupter) outlets in residential buildings. Discuss the purpose of GFCI protection, the types of GFCI devices available, and the locations where they should be installed.
8. Discuss the role of surge protection in residential building electrification. Explain how surge protectors work to safeguard electrical appliances and devices from voltage spikes, and describe the different types of surge protection devices available.
9. Explain the concept of power factor correction in residential electrical systems. Discuss the factors that affect power factors, the importance of power factor correction, and the methods used to improve power factors in residential buildings.

10. Discuss the impact of energy-efficient lighting technologies on residential building electrification. Compare and contrast traditional incandescent lighting with energy-efficient alternatives such as LED (Light-Emitting Diode) and CFL (Compact Fluorescent Lamp) lighting, and explain the benefits of transitioning to energy-efficient lighting solutions.

Numerical Problems:

1. Calculate the total power consumption of a residential building with the following loads: 6 LED bulbs (9 watts each), a refrigerator (150 watts), a microwave oven (1200 watts), and a television (250 watts). (1654 watts)
2. Determine the total current drawn by a residential circuit with the following appliances connected: a washing machine (1200 watts), a dishwasher (1500 watts), and a water heater (3000 watts), all operating simultaneously. (24.78 A)
3. Given a residential building with a total load of 10,000 watts, calculate the primary circuit breaker size required, assuming a safety factor of 25%. (54.35 A)
4. A residential building has a lighting circuit with 10 lamps (each 60 watts) and a power circuit with appliances totaling 5000 watts. Determine the minimum size of the copper conductor required for wiring, assuming a voltage drop limit of 3%.
5. Calculate the total power consumption of a residential kitchen equipped with the following appliances: refrigerator (200 watts), oven (1500 watts), microwave oven (800 watts), and dishwasher (1200 watts). (3700 watts)
6. Determine the maximum load (in watts) that can be safely connected to a residential outlet protected by a 15-ampere circuit breaker. (3450 watts)
7. A residential building has a lighting circuit with 15 LED bulbs (7 watts each) and a power circuit with appliances totaling 4000 watts. Calculate the minimum rating of the main switch required for the building. (4600 watts)
8. Given a residential circuit with a total load of 8000 watts, determine the minimum size of the protective switchgear required, considering future expansions and safety factors. (9600 watts)
9. Calculate the total power consumption of a residential living room with the following loads: television (150 watts), sound system (200 watts), and ceiling fan (75 watts), assuming all appliances operate simultaneously. (425 watts)

10. Determine the voltage drop across a residential wiring circuit that supplies power to a load of 6000 watts, given a wire length of 50 meters and a wire resistance of 0.1 ohms per meter.

Unit Summary

The unit summary of residential building electrification is written below

- ✓ Introduction to residential electrification
- ✓ Electrical codes and regulations
- ✓ Electrical power supply
- ✓ Electrical distribution system
- ✓ Wiring methods and layout
- ✓ Appliance wiring and circuits
- ✓ Electrical safety and maintenance

This unit provides a comprehensive overview of residential building electrification.

Practical

1. Design a wiring layout for a typical bedroom in a residential building, including placement of outlets, switches, and lighting fixtures. Consider load distribution, circuit requirements, and safety codes.
2. Calculate the total electrical load for a kitchen area, including a refrigerator, microwave, dishwasher, and lighting. Determine the required circuit capacities and breaker sizes for each appliance.
3. Design a lighting system for a living room that includes overhead lighting, wall sconces, and accent lighting. Specify the types of fixtures, lamp types, and control options (e.g., dimmer switches).
4. Identify and explain the critical safety devices and features required in a residential electrical system to meet code standards, such as AFCI breakers, GFCI outlets, and proper grounding.
5. Present a scenario where a circuit in a residential building is repeatedly tripping. Explain the steps you would take to troubleshoot and identify the cause of the problem.

These practical questions cover various topics within residential building electrification, testing students' knowledge of design principles, code compliance, safety measures, energy efficiency relevant to home electrical installations.

Know More

Sl No	Topic	QR Code
1	Guide for Using the National Building Code of India 2016	
2	Electrical Diagrams and Schematics	
3	Electrical Installation IS 10118-2 (1982)	
4	Switches for Domestic and Similar Purposes IS 3854:1997	
5	Electricity Distribution Network Planning Criteria	

References and Suggested Readings

1. H. Zerriffi, *Rural electrification: strategies for distributed generation*. Springer Science & Business Media, 2010.
2. D. Prasad, and M. Snow, *Designing with solar power: a source book for building integrated photovoltaics (BiPV)*, Routledge, 2014.
3. S.S. Kale, *Electrifying India: Regional political economies of development*. Stanford University Press, 2014.
4. M. Krarti, *Energy-Efficient Electrical Systems for Buildings*. CRC Press, 2017.
5. H.A. Gabbar, *Energy conservation in residential, commercial, and industrial facilities*. John Wiley & Sons, 2018.
6. P. Aalto, *Electrification: Accelerating the energy transition*. Academic Press, 2021.

UNIT 5 PROTECTION OF ELECTRICAL INSTALLATION

UNIT SPECIFICS

This unit elaborately discusses the following topics:

- ✓ Introduction to fuses in electric circuits for electrical wiring
- ✓ Classification of different types of fuses
- ✓ Exploring the circuit breakers for electrical wiring
- ✓ Various types of earthing equipment and their requirements
- ✓ Exploring various types of earthing as per IS 3043:1987

This unit gives an overview of the different fuses in electric circuits for electrical wiring. Elements, ratings, minimums, cut-offs, fuses, materials, fusing factors, etc., of fuses have also been covered. This section has also covered topics such as classifying fuses (e.g., rewirable vs. cartridge fuses) and choosing the suitable material for a fuse. The construction, principle rating, and uses of miniature circuit breakers (MCB) and earth leakage circuit breakers (ELCBs) are also reported here. This unit also addressed various significant aspects of electrical earthing, including the prerequisites for earthing systems and equipment, earthing techniques involving diverse earthing electrodes, earthing current, earthing terminal, earthing wire, earthing lead, calculation of fault current, leakage current, determination of earth resistance through the utilization of an earth tester, and strategies for mitigating earth resistance. Modern methods of earthing, such as driven pipe, pipe, and plate earthing, have all been covered in this chapter.

RATIONALE

The primary objective of electrical protection is to protect individuals, occupants, and property from electrical hazards such as electric shock, fires, and equipment damage. Proper protection measures reduce the likelihood of accidents and minimize the impact of electrical faults. Effective protection schemes incorporate devices and systems for fault detection, localization, and isolation. This enables quick identification of fault locations, facilitates troubleshooting and maintenance activities, and reduces system downtime during fault conditions. Electrical earthing is critical in safeguarding individuals, workers, and occupants against electric shock hazards. As a result of earthing, fault currents are able to safely flow to the ground, which helps to avoid electric shocks and lessens the likelihood of harm or death caused by coming into touch with live components or malfunctioning machinery.

PRE-REQUISITES

Basics of Electrical Engineering

UNIT OUTCOMES

The list of outcomes of this unit is as follows:

U5-O1: Describe various fuses in an electric circuit

U5-O2: Explain the classification of multiple fuses

U5-O3: Explain the diverse spectrum of circuit breakers

U5-O4: Describe about earthing equipment and its requirements

Unit-5 Outcomes	Expected Mapping with Programme Outcomes *(1- Weak Correlation; 2- Medium correlation; 3- Strong Correlation)*					
	CO-1	**CO-2**	**CO-3**	**CO-4**	**CO-5**	**CO-6**
U5-O1	3	3	1	2	3	3
U5-O2	3	3	1	2	3	3
U5-O3	3	3	1	2	3	3
U5-O4	3	3	1	2	3	3

Introduction:

Protection of electrical installations entails the implementation of numerous safeguards to ensure electrical systems' dependability and security. These precautions aim to mitigate potential electrical dangers, including fire, electric shock, and equipment damage, through adherence to pertinent regulations and standards. The following are essential elements in safeguarding electrical installations:

1) **Overcurrent Protection:** Overcurrent protection devices, such as circuit breakers and fuses, are implemented to inhibit excessive current flow in electrical circuits. They break the electrical circuit when the current surpasses the specified capability, safeguarding the wiring and equipment from harm and diminishing the likelihood of a fire.
2) **Ground Fault Protection:** Ground fault protection devices, such as ground fault circuit interrupters (GFCIs) and residual current devices (RCDs), identify discrepancies in the flow of electric current between the hot and neutral wires. They promptly interrupt the circuit upon the occurrence of a ground fault, thereby minimizing the potential for electric shock.
3) **Short Circuit Protection:** When a low-resistance channel forms between conductors, a short circuit occurs. In this situation, circuit breakers and fuses are designed to quickly stop the circuit. This avoids overcurrent flow and saves the electrical system from damage.
4) **Overvoltage Protection:** Surge protection devices (SPDs) are placed to protect electrical equipment from sudden increases in voltage caused by lightning strikes, utility switching, or other external events. Surge protective devices (SPDs) redirect surplus voltage to the ground, safeguarding delicate electronic components and averting equipment malfunction.
5) **Isolation and Disconnection:** Electrical installations must incorporate mechanisms for isolating and disconnecting power sources to enable maintenance, repair, and emergency shutdown processes. Switches, circuit breakers, and other isolation devices are employed to deactivate circuits securely.
6) **Equipment Protection:** Electrical equipment, such as motors, transformers, and control devices, should be chosen and placed based on their intended use and the surrounding environmental circumstances. Adequate safeguards, such as enclosures, insulation, and ventilation, are implemented to guarantee secure functioning and durability.

7) **Earthing and Bonding:** Establishing adequate earthing and bonding systems is crucial for ensuring electrical safety and preserving the system's integrity. Earthing, also known as grounding, establishes a pathway with low resistance to disperse fault currents and prevent electric shock safely. Bonding, conversely, guarantees that metallic items are connected to maintain equal voltage levels, reducing the chances of electric shock and fire hazards.
8) **Protection against Thermal Effects:** Electrical installations must be engineered to endure and disperse the heat produced by the flow of electric current, averting overheating and thermal harm. Ensuring sufficient airflow, correctly sizing conductors, and closely monitoring temperature are essential for reducing thermal impacts.
9) **Compliance with Regulations and Standards:** Electrical installations must adhere to relevant legislation, codes, and standards to guarantee safety and legal conformity. National and international standards establish electrical design, installation, and maintenance guidelines, including the National Electrical Code (NEC), International Electrotechnical Commission (IEC) standards, and municipal building regulations.
10) **Regular Inspection and Maintenance:** Regular examination, assessment, and upkeep of electrical systems are crucial for detecting possible dangers, confirming the correct operation of safety mechanisms, and guaranteeing ongoing safety and dependability. Standard maintenance involves inspecting connections, confirming insulating resistance, and testing protection devices.

By incorporating these safeguarding measures and following optimal procedures in electrical design and installation, the security and dependability of electrical systems can be improved, reducing potential hazards to individuals and assets while maintaining conformity with regulatory standards.

5.1 Fuse in Electric Circuit:

Fuse protection is ensured by an electrical, electronic, or mechanical apparatus that safeguards against excessive load and excessive current in circuits. In 1890, Thomas Alva Edison invented the electric fuse. Despite the variety of fuses available, their fundamental function remains consistent. This subject will comprehensively explore the various classifications of fuses, their structural composition, operational characteristics, and application in a diverse array of electrical and electronic systems. A fuse is a safeguarding apparatus employed in electrical circuits to avert harm to electrical equipment and wiring due to an excessive flow of electric current. It functions as a sacrificial component that disrupts the circuit when the current is beyond a predetermined threshold, thus safeguarding the circuit and linked devices against excessive current, short circuits, and electrical malfunctions. An electrical fuse is a safety mechanism to prevent excessive electrical circuit current flow.

The core component consists of a metallic strip or wire that undergoes melting when an excessive electrical current flows through it, interrupting or ceasing the flow of electricity. The sacrificial device in question is a fuse, which, depending on its type, will cause an open circuit when it operates. In such cases, the fuse must be either changed or rewired. A simple electrical circuit with a fuse is shown in Fig.5.1.

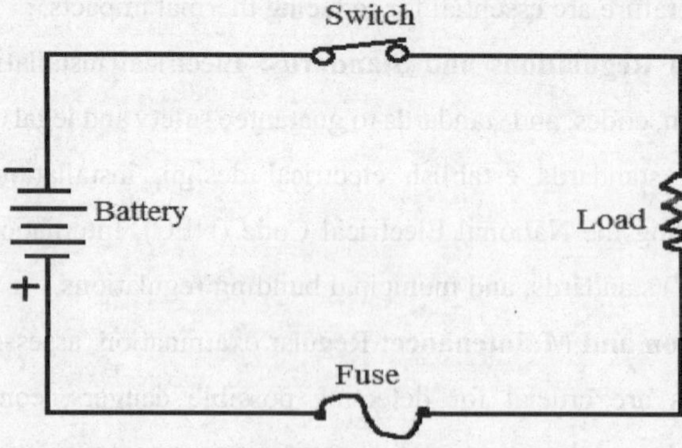

Figure 5.1 Simple electrical circuit with a fuse

- **Construction & Working of a Fuse:**

A fuse is made up of a non-combustible substance encasing a complete, low-resistance metallic wire. If a fuse or circuit breaker is missing, electrical equipment can sustain damage due to its inability to withstand the high current above its specified limitations. For circuits and devices that need protection from overcurrents and short circuits, this component is used in a sequential installation and connection process.

According to the "heating effect of current," a fuse melts its thin wire inside when a short circuit or mismatched load connection generates a high enough current to flow through it. Consequently, it cuts power to the linked device. The low resistance of the fuse wire ensures that it will not interfere with the normal functioning of the system that is linked to the power source.

- **Example: How do you select the proper rating size for a fuse?**

✓ It is important to select the appropriate fuse and corresponding rated size for electrical appliances, considering various circumstances and settings. However, the simple procedure below explains how to pick the suitable fuse size.

- ✓ The fuse rating is calculated as (Voltage / Power) times 1.25.
- ✓ For example, choose the right fuse size for a 10A two-pin socket.
- ✓ (1.25 × 1000W / 230V) = 5.4A
- ✓ In the example above, the main supply voltage is single phase 230V AC (120V AC in the US), with a power rating of 1kW controlled by a 2-pin socket.
- ✓ However, for safe and reliable circuit operation, utilize the maximum fuse rating of 6A rather than 5.4A.

- **Characteristics of a Fuse:** Fuse types can be classified based on the criteria listed below.
- ✓ The carrying capacity and current rating of the fuse
- ✓ Rating of the fuse's voltage
- ✓ Capacity of the fuse to break
- ✓ Value of the fuse's I^2t
- ✓ Fuse response characteristic
- ✓ Fuse-rated voltage
- ✓ Fuse packing size

- **Fuse Current Carrying Capacity:** A fuse's current carrying capability is the highest current it can safely transfer without causing the circuit to short out..
- **Breaking capacity:** A fuse's breaking capacity—the maximum current it can safely shut off—must exceed the current that could result from a short circuit.
- **Rated Voltage of Fuse:** Take into account the present capacity and the highest voltage rating that a fuse can safely endure. Fuses operating at varying voltage levels necessitate distinct levels of isolation.
- **I^2t Value of Fuse:** The term I^2t denotes a type of fuse commonly utilised in situations involving short circuits. Upon the resolution of an electrical problem, the fuse element facilitates the transfer of a specific amount of energy.
- **Response Characteristics of a Fuse:**
- ✓ The current flowing through a fuse's wire determines its blowing rate. The response time will rise as the current flows through the wire.
- ✓ The reaction characteristic indicates how rapidly an overcurrent occurrence happens. Ultrafast fuses, also known as Fast fuses, respond swiftly to an overcurrent situation and

are commonly used in semiconductor devices due to the rapid damage that overcurrent can cause.

- ✓ Slow burn fuses are another option. Switch fuses react slowly to overcurrent incidents, blowing after a few seconds. Because motors require significantly more voltage to start, these fuses are employed in motor control electronics systems.

Packaging Size:

- ✓ As previously stated, AC and DC fuses are packaged differently. Similarly, different applications necessitate careful placement of different packages in the circuit.
- ✓ Additional factors to take into account include labelling, temperature deceleration, voltage decrease, and velocity.

5.1.1 Fuse Element:

The fuse element is the component of a fuse that undergoes melting or breaking when subjected to an excessive amount of electricity, resulting in the fuse opening (blowing) and disrupting the circuit. It is an essential element that ensures overcurrent protection in electrical circuits. An integral component of an electrical fuse is a metallic wire or strip that undergoes melting when subjected to an excessive flow of electric current. Discontinuing or disrupting the flow of electric current contributes to safeguarding the gadget. Here is additional information regarding the fuse component:

- ✓ **Material:** Fuse components commonly contain materials with low melting points, such as copper, silver, aluminum, or alloys like tin-lead. These materials facilitate the rapid melting or vaporization of the fuse element when exposed to excessive current.
- ✓ **Construction:** The fuse element typically comprises a slender wire, strip, or ribbon from the chosen material. The design incorporates precise dimensions for the cross-sectional area and length, which is essential for ensuring optimal current-carrying capacity and thermal reactivity.
- ✓ **Encapsulation:** The fuse component is within a safeguarding enclosure, typically composed of glass, ceramic, or other insulating substances. The casing serves the purpose of providing structural support and insulation for the fuse element, as well as containing any molten material or debris that may be produced during the functioning of the fuse.

Types of Fuse Elements:

- ✓ **Wire Fuse Element:**

A slender wire composed of a conductive substance like copper or silver is frequently employed as a fuse component. This product is appropriate for applications with low to medium electrical currents and offers a dependable and cost-efficient option for protecting against excessive current.

✓ **Strip or Ribbon Fuse Element:**

Fusible links composed of flat conductive strips or ribbons are employed in some applications. These elements provide a greater capacity for carrying electric current and can be utilized in fuses built explicitly for heavy loads or industrial purposes.

✓ **Encapsulated Fuse Element:**

Occasionally, the fuse element could be enclosed in a protective covering or casing to improve its durability and safety. The encapsulation process safeguards the fuse element against oxidation, corrosion, or mechanical harm.

The fuse element must be appropriately selected and sized to guarantee the dependable operation of electrical circuits and adequate overcurrent protection. Selecting a fuse element with the requisite current-carrying capacity, response characteristics, and reliability is of utmost significance when considering the application and operating conditions. The fuse element for industries and substations is shown in Fig.5.2.

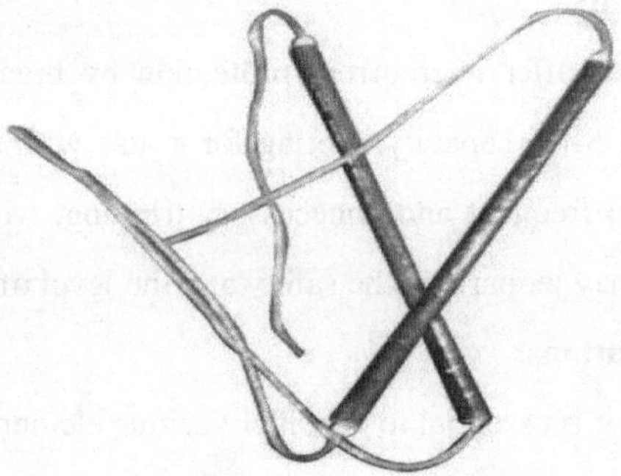

Figure 5.2 Fuse element for industries and substation

5.1.2 Fuse Current Ratings:

The fuse current rating called the ampere rating or current-carrying capacity, denotes the highest continuous current that a fuse can safely bear without causing the fuse element to open (blow) and disrupt the circuit. The crucial parameter is utilized to choose the suitable fuse for a particular application and to guarantee efficient protection against excessive current. Below are some crucial details regarding Fuse's current ratings:

✓ The maximum current that a fuse wire may withstand without undergoing melting.

✓ The term used to describe the maximum current that an electric fuse can handle is its current rating. Copper lead-acid and tin alloy are commonly used to produce an electric fuse.

✓ Usually, the measurement is expressed in milliamps (mA) or amps (A).Fuses with ratings of 1A, 2A, 3A, 5A, and 13A are frequently utilised.

✓ It is generally not advisable for the amp rating of a fuse to exceed the current carrying capacity of the circuit. For instance, if the conductor has a rating of 20A, the appropriate fuse to be used would be a 20A fuse, which is the greatest size recommended.

a) **Ampere (A) Rating:**

The current ratings of fuses are commonly denoted in amperes (A), the standard unit for measuring electrical current. The rating indicates the highest level of electric current that the fuse can sustain without any *interruption in its functioning*.

b) **Selection Criteria:**

The choice of a fuse with a suitable current rating is contingent upon the anticipated typical operating current of the circuit it safeguards. The fuse current rating should match or slightly exceed the anticipated maximum continuous current of the circuit.

c) **Overcurrent Protection:**

Fuses are engineered to offer overcurrent protection by breaking the circuit when the current exceeds the specified capacity. Opting for a fuse with a current rating lower than necessary might lead to frequent and unnecessary tripping, whereas selecting one with a more excellent rating may jeopardize the safety and the level of protection provided.

d) **Application Considerations:**

When choosing a fuse, it is essential to consider various elements such as the nature of the load, the initial currents, the surges in current, the surrounding temperature, and the circumstances of the environment. These variables have the potential to impact both the ability of the fuse to transport electrical current and its overall performance.

e) **Derating Factors:**

Reducing the fuse current rating in specific applications may require considering issues like temperature increase, voltage fluctuations, or continuous operation at high temperatures. Consult relevant standards and guidelines to obtain derating recommendations.

- **Example:** For instance, if a circuit has a maximum continuous current of 10 amperes, a fuse with a current rating of 10 amperes would be appropriate for safeguarding against excessive current. Opting for a fuse with a slightly elevated rating, such as 15 amperes, can offer an extra buffer and capacity to handle momentary surges or brief instances of excessive load.

f) **Importance of Proper Selection:**

The careful choice of the fuse current rating is essential to provide efficient protection against excessive current, reduce the likelihood of equipment harm and fire threats, and uphold the dependability and safety of the electrical system. Following relevant standards, rules, and manufacturer recommendations is crucial when choosing fuses for specific applications.

g) **Standards and Guidelines:**

Organizations such as the National Electrical Code (NEC), International Electrotechnical Commission (IEC), and Underwriters Laboratories (UL) provide standards and guidelines that offer advice and requirements for selecting, installing, and applying fuses. Referring to these resources can guarantee adherence to regulations and secure functioning.

5.1.3 Minimum Fusing Current:

The term "minimum fusing current" commonly denotes the lowest amount of electric current necessary to effectively liquefy the fuse component and trigger the fuse to disconnect, breaking the circuit. Considering this value is crucial when choosing a fuse for a particular application, as it guarantees its efficient operation in the presence of excessive current. To establish the minimum fuse size, it is generally recommended to utilize 125% of the circuit's total load current. Non-inductive loads should be protected with fast-acting fuses, while inductive loads require time-delay fuses. The minimum fuse rating refers to the lowest amperage capacity a fuse must have to use safely in a specific electrical circuit. The minimum fuse rating is often determined by the conductive capacity of the circuit's wires to carry electric current. Using a fuse with a lower rating than the wire's capacity can result in frequent blowing and unwanted interruptions to the circuit. However, using a fuse with a rating higher than the wire's capacity may not provide adequate protection against excessive current.

- **Calculation of minimum fuse rating:**

The maximum ampere rating of an electrical fuse is initially established by applying the following calculation to get the minimum current rating of the fuse:

Current, I = amps (rated power/rated voltage)

Increasing the maximum ampere rating by 125% will result in the minimum fuse rating.

The minimum fusing current can be determined by considering the following factors:

- ✓ **Dependence on Fuse Design:** The minimum fusing current is contingent upon several elements, such as the fuse element's design, construction, and materials. Various fuses, such as fast-acting, time-delay, and current-limiting fuses, may have distinct specifications for the minimum current required to cause them to fuse.
- ✓ **Application Considerations:** When choosing a fuse for a particular function, it is crucial to verify that the anticipated overcurrent situations are higher than the minimum fusing current of the fuse. Alternatively, if the fuse does not function consistently, it could result in possible safety risks or harm to the equipment.
- ✓ **Margin of Safety:** Design engineers commonly use fuses with fusing currents far lower than the anticipated overcurrent levels to ensure a safety buffer. This guarantees consistent fuse performance throughout a broad spectrum of operating situations, preventing unnecessary triggering or the circuit from being unprotected.
- ✓ **Derating Considerations:** Various factors, including temperature, voltage, and environmental conditions, might influence the minimum fusing current. To maintain optimal fuse performance in all circumstances, it is crucial to consider derating factors and adhere to the manufacturer's guidelines.
- ✓ **Importance of Proper Selection:** Thoroughly considering the cut-off current is crucial to guarantee the dependability and efficiency of the overcurrent protection offered by the fuse. Choosing a fuse with a cut-off current that is too low can jeopardize safety, but selecting one with a rating that is too high can lead to extra expenses or restrictions.
- ✓ **Manufacturer Specifications:** Fuse makers supply specifications and datasheets that encompass details regarding the cut-off current and other crucial aspects such as rated current, voltage rating, and time-current characteristics. Engineers and designers should use these specifications before choosing fuses for their applications.

5.1.4 Cut-off current of a fuse:

A fuse's "cut-off current" denotes the highest current level at which the fuse is intended to open (blow) and break the circuit consistently. The term refers to the maximum threshold at which the fuse must function to safeguard the circuit against an excessive flow of electric current. When the fuse undergoes melting, it signifies attaining the highest possible fault current value.

- ✓ **Definition:** The cut-off current, referred to as the "breaking capacity," is the utmost current that a fuse may safely interrupt without incurring harm or presenting a safety risk. It signifies the fuse's capacity to effectively manage fault currents without undergoing disastrous breakdown.
- ✓ **Safety Consideration:** The cut-off current is an essential element that guarantees the safety and dependability of electrical systems. It guarantees that the fuse can safely stop abnormal electric currents, such as short circuits or overloads, without creating sparks, excessive heat, or other hazardous situations.
- ✓ **Dependence on Fuse Design:** The cut-off current of a fuse is contingent upon multiple elements, including its design, construction, materials, and rating. Various fuses, such as fast-acting, time-delay, and current-limiting fuses, may possess distinct cut-off current specifications.
- ✓ **Application Considerations:** When choosing a fuse for a particular function, it is crucial to ensure that the anticipated fault currents do not exceed the cut-off current rating of the fuse. Otherwise, the fuse may fail to break the circuit safely, resulting in potential safety issues or equipment damage.
- ✓ **Coordination with Protection Devices:** Fuses are frequently chosen and synchronized with other safeguarding mechanisms, such as circuit breakers and protective relays, to guarantee thorough protection against excessive current. The cut-off current of the fuse must be suitable for the interrupting capability of the other protective devices in the circuit.

The following condition can determine the cut-off value:

- ✓ **Current rating of the fuse:** The current rating of a fuse refers to the highest amount of current it can sustain without causing significant degradation to its fuse element. Exceeding the current rating could cause the switch's contacts to fuse or bond.
- ✓ **Value of prospective current:** When a three-phase supply is available, the PSC is often between the line conductors. To determine the PSC, multiply the single phase reading by the square root of three, 1.73.
- ✓ **Asymmetry of the fault current:** A systemic short circuit can occur due to an external occurrence or an internal dielectric breakdown. It introduces a switching device into an

already defective system, such as an earthing connection that was unintentionally left intact during maintenance.

Engineers can ensure electrical systems' safety and integrity by selecting fuses that offer dependable overcurrent protection, considering the cut-off current and other significant factors.

5.1.5 Fusing Factor:

The phrase "fusing factor" commonly denotes a factor employed to ascertain a fuse's suitable rating or dimensions for a particular application. It considers multiple variables that can impact the performance and functioning of the fuse in specific circumstances. Although the definition of a fusing factor may vary across different applications, it often includes elements like derating, safety margins, and environmental considerations. The fusing factor is defined as the quotient obtained by dividing the minimal fusing current by the current rating of the fusing element.

- ✓ Fusing factor = minimum fusing current/current rating of fusing element.
- ✓ Its value is always more than 1 one.

5.1.6 Fuse Material:

The choice of fuse material is essential in defining their performance attributes, such as their ability to transport current, thermal properties, speed of response, and dependability. Various materials are used for the fuse elements, casings, and other components of different fuses. Fuse wire is typically made using lead and tin alloys. For a chemical to be suitable for forming a fuse, it must possess these two essential characteristics:

- ✓ **High resistance**: An electrical conductor that exhibits a significant electrical resistance hinders the passage of an electric current; the reciprocal of this measurement is referred to as electrical conductance. The ohm (Ω) is the standard unit of measurement for electrical resistance in the International System of Units (SI).
- ✓ **Low melting point**: Helium possesses the most minimal melting point among all chemical elements, whereas carbon exhibits the highest melting point. The Celsius (C) scale measures the temperature at which a substance melts.

In an ideal scenario, the fuse would rapidly melt when there is a slight increase in current and continue to carry its designated current indefinitely. The element must exhibit long-term stability without undergoing oxidation or alterations in its properties, and it must also be resistant to minor, harmless surges in current. It is feasible to manipulate the fuse elements to amplify the heating effect. Steel or nichrome wires can support fuse elements and prevent them from experiencing stress. However, adding a spring can expedite the separation of the individual components of the element. Fuse material (wire) in various fuses is shown in Fig.5.3.

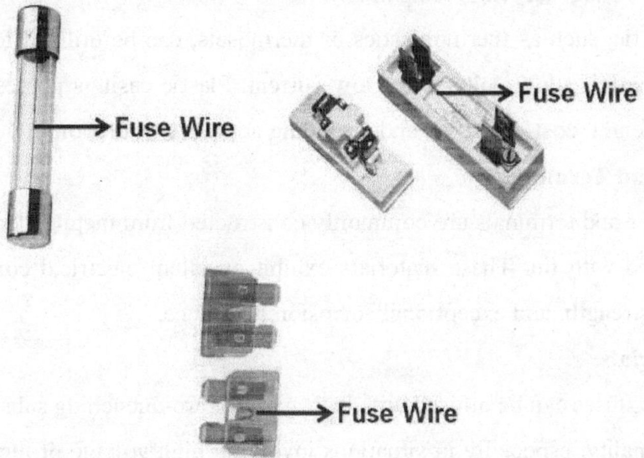

Figure 5.3 Fuse material (wire) in various types of fuses

- Here are some common materials used in fuses:

a) **Fuse Element Material:** The fuse element undergoes melting or vaporization upon encountering excessive current, resulting in the circuit being disconnected. Typical materials utilized for fuse components comprise:

✓ **Copper:** Copper is widely used for fuse elements because of its exceptional conductivity and low melting point. It enables swift liquefaction and dependable functioning.

✓ **Silver:** Silver is a frequently utilized material for fuse elements. It provides excellent electrical conductivity and exceptional resistance to oxidation and corrosion.

✓ **Aluminum:** Aluminium is occasionally employed to fuse elements when weight and cost considerations are significant. Nevertheless, aluminum exhibits more electrical resistance than copper or silver, leading to slower response times.

✓ **Alloys:** Multiple alloy compositions can be employed to enhance the performance of fuse elements by achieving a harmonious balance between variables such as conductivity, melting point, and mechanical properties.

b) **Fuse Casing Material:** A fuse enclosure offers physical safeguarding and electrical shielding for the inside elements. Typical materials utilized for fuse casings comprise:

✓ **Glass:** Glass casings are translucent, enabling visual examination of the fuse element. They offer exceptional electrical insulation and thermal stability.

✓ **Ceramic:** Ceramic casings possess exceptional mechanical strength, thermal resistance, and electrical insulation qualities. They are frequently employed in applications that involve high voltage and high temperature.

✓ **Plastic:** Plastic, such as thermoplastics or thermosets, can be utilized for fuse casings in situations involving low voltage and low current. Plastic casings possess the qualities of being lightweight, cost-effective, and providing adequate insulation.

c) **End Caps and Terminals:**

Fuse end caps and terminals are commonly constructed from metals like brass, copper, or copper coated with tin. These materials exhibit excellent electrical conductivity, robust mechanical strength, and exceptional corrosion resistance.

d) **Filler Materials:**

Occasionally, fuses can be infused with inert gases or arc-quenching substances to enhance their functionality, especially in situations involving high voltage or high power. Typical filler chemicals include nitrogen, sulfur hexafluoride (SF6), and different types of powders or granular solids.

The choice of fuse materials is contingent upon various aspects, including the specific demands of the application, the prevailing operating circumstances, the voltage levels, the current ratings, and the environmental considerations. Engineers and designers meticulously select fuse materials to guarantee the utmost efficiency, dependability, and security of electrical systems. In addition, fuse manufacturers offer guidelines and standards regarding the materials utilized in their products to aid in the appropriate selection and application.

5.2 Types of Fuses:

Electrical fuses are essential in electrical circuits to safeguard against excessive current situations. There are different varieties available, each possessing unique properties and uses. Fuses can be classified based on their utilization in various applications. The user must manually replace single-use fuses containing a metallic wire that burns out in the event of an overcurrent, overload, or mismatched load connection. On the contrary, switch fuses are cost-effective and widely employed in nearly all electrical and electronic systems. In contrast, the resettable fuse will automatically reset itself following the method in case of a system fault. The current limiting fuse exhibits a brief time of high resistance. In contrast, the non-current limiting fuse generates an arc when subjected to high current flow, interrupting and restricting the current in associated and interconnected circuits. Below are several prevalent varieties of electrical fuses:

- **Rewirable Fuses**:

Rewirable fuses (Fig.5.4), sometimes called kit-kat fuses or fuse links, are specific kinds of electrical fuses that may be rewired or substituted when they function and break the circuit due to excessive current conditions. Their composition comprises a foundation or fuse holder and a fuse carrier that contains a replacement fuse element.

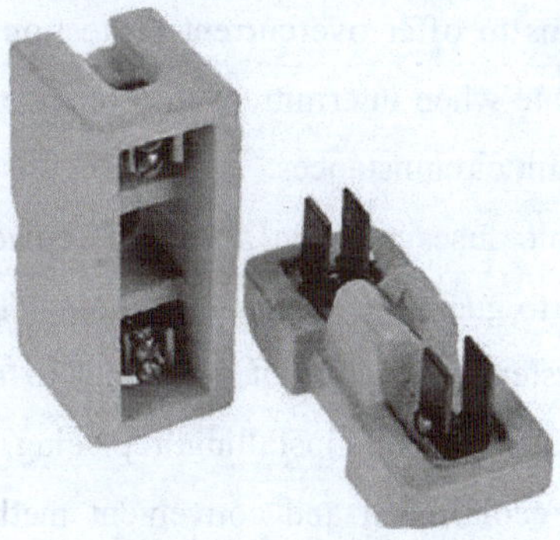

Figure 5.4 Rewirable fuse

- ✓ **Construction:** Rewirable fuses generally comprise a base or fuse holder made of ceramic or porcelain and a fuse carrier. The fuse carrier houses a fuse element, often composed of a conductive substance with a low melting point, such as copper or tin, linked between two terminals. The fuse carrier is fastened to the base by the use of screws or clips.
- ✓ **Fuse Element:** A rewirable fuse's fuse element is engineered to melt or break upon encountering excessive current, effectively terminating the circuit and safeguarding equipment or wiring from potential harm. Once the fuse element is activated, it can be readily substituted or reconnected with a fresh fuse element that has an identical rating.
- ✓ **Types:** Various types and configurations of rewirable fuses can be used for applications and current ratings. Typical varieties comprise BS 88 rewirable fuses, DIN-type fuse linkages, and other specifications specific to certain regions or countries. The selection of the rewirable fuse is contingent upon various aspects, including the individual application, voltage level, and current rating.
- ✓ **Installation:** Rewirable fuses are placed in fuse bases or holders, which offer a stable attachment and electrical linkage. The fuse carrier is put into the fuse base and fastened in

position using screws or clips. Accurate installation is crucial to guarantee dependable functioning and efficient overcurrent safeguarding.

- ✓ **Replacement:** When a rewirable fuse is activated and disrupts the circuit due to excessive current, the fuse element can be effortlessly substituted or rewired with a new fuse element with an identical rating. Using the appropriate replacement fuse element is essential for preserving the integrity and safety of the electrical system.
- ✓ **Applications:** Rewirable fuses are frequently employed in residential, commercial, and industrial electrical systems to offer overcurrent protection for circuits, equipment, and wiring. They are appropriate when intermittent fuse replacement may be required due to overcurrent incidents or fault circumstances.
- ✓ **Safety:** Although rewirable fuses provide a simple replacement, they require careful handling and maintenance to guarantee secure functioning. To prevent electrical risks and maintain the electrical system's integrity, it is crucial to adhere to the manufacturer's guidelines and industry standards while installing, replacing, or rewiring rewirable fuses.

Rewirable fuses offer an economical and convenient method for safeguarding against excessive electrical current in various applications.

Cartridge Fuses:

Cartridge fuses (Fig.5.5) comprise a fuse element contained within a cylindrical or tubular casing constructed from ceramic, glass, or fiber. These products are offered in various sizes and current ratings and come with choices for fast-acting, time-delay, and current-limiting features. Cartridge fuses are widely utilized in industrial, commercial, and residential settings.

Figure 5.5 Cartridge fuse

✓ **Construction:** Cartridge fuses often comprise a fuse element composed of a conductive substance, such as copper or silver, with a relatively low melting point. The fuse element is contained within a cylindrical or tubular enclosure composed of ceramic, glass, or fiber materials. The shell offers both mechanical reinforcement and electrical insulation.

✓ **Types:** Cartridge fuses come in different varieties, such as fast-acting (also called quick-blow or fast-blow), time-delay (also known as slow-blow or time-delay), and current-limiting. Rapid-responding fuses promptly react to excessive current situations and are appropriate for circuits containing delicate components. Time-delay fuses exhibit a delayed reaction to excessive current situations, enabling them to endure brief periods of excessive load. Current-limiting fuses regulate the fault current flow, enhancing the safeguarding of downstream equipment. Cartridge fuses can be classified into two primary types according to design and characteristics: HRC (High Rupturing Capacity) fuses and LRC (Low Rupturing Capacity) fuses.

a) **HRC (High Rupturing Capacity) Fuses:**

- HRC fuses are specifically engineered to safely endure and interrupt high levels of electrical current during fault conditions.
- They are commonly employed when anticipating high fault currents, such as in industrial power distribution systems, motor control centers, and utility substations.
- HRC fuses possess a sturdy design and may effectively interrupt fault currents without generating excessive arcing or causing harm to the fuse element.
- These fuses are specifically engineered to offer dependable protection against excessive current and guarantee the safety and dependability of the electrical system under situations of high fault.
- HRC fuses are commonly known for their high interrupting ratings, representing the most significant fault current that the fuse can safely interrupt without sustaining any harm.

b) **LRC (Low Rupturing Capacity) Fuses:**

- LRC fuses are engineered explicitly for usage in situations where the fault currents are of a lower or moderate magnitude.
- They are frequently employed in home and commercial electrical circuits, appliances, lighting circuits, and other low-voltage applications.

- LRC fuses provide less interrupting capability than HRC fuses, rendering them appropriate for circuits with smaller fault currents.
- These fuses offer efficient protection against excessive current in situations where the fault currents fall within their designated ratings.
- LRC fuses are commonly distinguished by their lower interrupting ratings compared to HRC fuses, indicating their ability to interrupt smaller fault currents safely.

Both HRC and LRC cartridge fuses offer the necessary protection against excessive electrical currents in systems. However, they are designed explicitly for distinct applications and fault current levels. HRC fuses are appropriate for situations with high fault current levels, whereas LRC fuses are appropriate for situations with low to moderate fault current levels.

✓ **Applications:** Cartridge fuses are employed in many electrical circuits, encompassing industrial machinery, motor control systems, power distribution panels, appliances, lighting circuits, and electronic devices. These devices safeguard against excessive loads, brief circuits, and electrical malfunctions that may harm equipment or present safety risks.

✓ **Voltage and Current Ratings:** Various voltage and current ratings offer cartridge fuses to accommodate diverse applications and operating situations. The voltage rating denotes the highest voltage the fuse may safely interrupt the circuit. In contrast, the current rating describes the maximum continuous current the fuse can handle without breaking the circuit.

✓ **Installation:** Cartridge fuses are commonly placed in fuse holders or blocks, ensuring a stable installation and electrical connection. The fuses are inserted into the fuse holder or block and fastened in position using screws, clips, or alternative methods. Accurate installation is crucial to guarantee dependable functioning and efficient overcurrent safeguarding.

✓ **Replacement:** When a cartridge fuse is triggered and interrupts the circuit due to excessive current, it must be substituted with a new fuse with the same kind and rating. Using the appropriate replacement fuse is essential for preserving the integrity and safety of the electrical system.

Cartridge fuses are highly adaptable and extensively employed to safeguard electrical circuits against excessive current.

Blade Fuses:

Blade fuses (Fig.5.6), often called automobile or spade fuses, are flat and rectangular. They have metal blades on both ends, which make them easier to insert into fuse blocks or holders. These devices are extensively utilized in automotive, marine, and low-voltage DC applications. They are called so due to their flat, rectangular structure featuring metal blades on both ends, mimicking the blade of a knife or spade. Blade fuses are available in standard sizes, including micro, regular, and maxi.

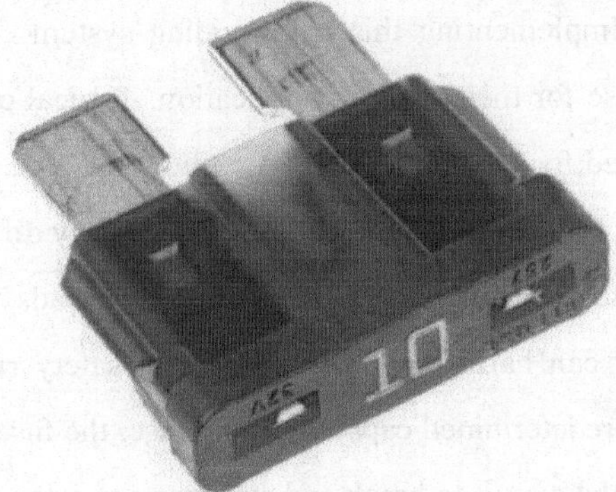

Figure 5.6 Blade fuse

- ✓ **Construction:** Blade fuses comprise a plastic casing with metal terminals on both ends and an inside fuse element. The fuse element commonly comprises a conductive substance, such as zinc, copper, or silver, with a comparatively low melting point. The plastic casing offers structural reinforcement and electrical isolation, while the metallic blades are contact points for electrical linkages.
- ✓ **Types:** Blade fuses are available in several sizes and styles, such as tiny, regular, and maxi. Mini-blade fuses are smaller and more condensed, rendering them appropriate for applications with restricted space. Standard blade fuses are the predominant variety and find extensive use in many automotive and marine applications. Maxi blade fuses are characterized by their bigger size and higher current ratings, which make them well-suited for heavy-duty applications.
- ✓ **Voltage and Current Ratings:** Various voltage and current ratings offer block fuses for various applications and operating situations. The voltage rating denotes the highest voltage the fuse may safely interrupt the circuit. In contrast, the current rating describes the maximum continuous current the fuse can handle without breaking the circuit.

- ✓ **Colour Coding:** Blade fuses commonly have housings with color-coded markings that reflect their current ratings. Implementing this color-coding system enables users to promptly discern the suitable fuse for their specific application. Typical color choices for different amperage ratings are red for 10A, blue for 15A, yellow for 20A, and so forth. However, it is essential to note that the specific color-coding scheme may differ among manufacturers.
- ✓ **Installation:** Blade fuses commonly have housings with color-coded markings that reflect their current ratings. Implementing this color-coding system enables users to promptly discern the suitable fuse for their specific application. Typical color choices for different amperage ratings are red for 10A, blue for 15A, yellow for 20A, and so forth. However, it is essential to note that the specific color-coding scheme may differ among manufacturers.
- ✓ **Protection:** Blade fuses offer to safeguard against overloads, short circuits, and other electrical failures that can harm equipment or create safety risks. When the electrical current surpasses the predetermined capacity of the fuse, the fuse element becomes heated and liquefies, causing the circuit to break and stopping the current flow.

Blade fuses are used to safeguard against excessive current in automotive, marine, and low-voltage DC systems. They are small, easy to install, and offer dependable performance.

Resettable Fuses (Polymeric Positive Temperature Coefficient - PPTC Fuses):

Resettable fuses, commonly called polymeric positive temperature coefficient (PPTC) or poly fuses, are a specific form of electrical safeguard that automatically restores its functionality after being triggered due to excessive current flow. Their approach involves utilizing a polymer material with a positive temperature coefficient (PPTC), demonstrating a distinct rise in resistance when subjected to high current and subsequent heating. Resettable fuses, in contrast to conventional fuses, revert to their initial low-resistance state after the fault situation is eliminated, eliminating the need for replacement. Resettable fuses are frequently employed in electronic circuits, battery protection systems, and USB ports.

- ✓ **Construction:** Resettable fuses, commonly called polymeric positive temperature coefficient (PPTC) or poly fuses, are a specific form of electrical safeguard that automatically restores its functionality after being triggered due to excessive current flow. Their approach involves utilizing a polymer material with a positive temperature coefficient

(PPTC), demonstrating a distinct rise in resistance when subjected to high current and subsequent heating. Resettable fuses, in contrast to conventional fuses, revert to their initial low-resistance state after the fault situation is eliminated, eliminating the need for replacement. Resettable fuses are frequently employed in electronic circuits, battery protection systems, and USB ports.

- ✓ **Operation:** When an excessive current situation arises, leading to the heating up of the PTC material and an increase in resistance, the resettable fuse successfully breaks the circuit and halts the current flow. This safeguards the circuit and its interconnected components against potential harm caused by an overabundance of electric current. The PTC material gradually loses heat after the fault condition is resolved and the current falls below the threshold level. It reduces resistance, enabling the resettable fuse to return to its initial low-resistance state. Including this automated resetting function obviates the necessity for human substitution of the fuse, hence diminishing periods of inactivity and expenses associated with upkeep.
- ✓ **Applications:** Resettable fuses are widely employed in many electronic and electrical devices and systems, such as power supplies, battery packs, USB ports, telecommunication equipment, automobile electronics, and consumer electronics. These devices offer dependable overcurrent protection and aid in preventing harm to delicate components and circuits.
- ✓ **Advantages:** Resettable fuses provide numerous benefits compared to conventional fuses. These advantages include automatic resetting, eliminating manual replacement requirements, rapid response time to overcurrent situations, small size and lightweight construction, and compatibility with automated manufacturing procedures.
- ✓ **Selection:** When choosing resettable fuses for a specific use, essential factors to take into account are the voltage rating, current rating, trip current (the level at which the fuse activates), hold current (the level below which the fuse stays in the reset state), maximum operating temperature, and physical dimensions. Choosing a resettable fuse with suitable specifications is crucial for ensuring efficient overcurrent protection and dependable performance in the desired application.

Resettable fuses offer a practical and efficient option for protecting electronic and electrical systems from excessive current. They provide automated resetting and dependable performance in small and cost-effective packaging.

Miniature Fuses:

Miniature fuses (Fig.5.7) are developed explicitly for situations with limited space or a need for low current capacity. They are offered in several physical configurations, including glass tube, ceramic tube, and surface-mount (SMD) variations. Miniature fuses are smaller than conventional cartridges and are frequently employed in electrical, automotive, and consumer equipment.

- ✓ **Size and Construction:** Miniature fuses are characterized by small dimensions and often possess cylindrical or rectangular shapes with metallic end caps. The fuse component, composed of a conductive substance with a relatively low melting point, such as silver or copper, is contained within the structure. The fuse is enclosed within a non-conductive substance, such as glass or ceramic, to offer structural reinforcement and electrical isolation.
- ✓ **Types:** Miniature fuses are available in different varieties, such as fast-acting (quick-blow), time-delay (slow-blow), and current-limiting. Rapid-acting fuses promptly react to excessive current situations and are appropriate for circuits containing delicate components. Time-delay fuses have a delayed reaction to overcurrent situations, enabling them to endure brief periods of excessive current. Current-limiting fuses regulate fault current flow, safeguarding downstream equipment by minimizing magnitude.
- ✓ **Applications:** Miniature fuses are frequently employed in electronic devices, appliances, automotive systems, telecommunications equipment, power supply, and control circuits. These devices safeguard against overloads, short circuits, and other electrical malfunctions that can harm equipment or create safety risks.
- ✓ **Voltage and Current Ratings:** Miniature fuses are available at different voltages and with other current ratings to suit various applications and operating conditions. The voltage rating specifies the maximum voltage the fuse can safely interrupt, while the current rating requires the maximum current that the fuse can carry continuously without opening.
- ✓ **Colour Coding:** Miniature fuses commonly incorporate color-coded end caps to reflect current ratings. Implementing this color-coding scheme facilitates the rapid identification of the suitable fuse for users' specific applications. Typical hue choices include blue for 2A, yellow for 5A, white for 10A, and so on, although the color scheme may differ according to the manufacturer.

✓ **Installation:** Miniature fuses are commonly placed in fuse holders or blocks, offering a reliable attachment and electrical linkage. The fuses are put into the fuse holder or block and fastened in position using clips or alternative methods. Ensuring the proper placement of tiny fuses is crucial for dependable operation and efficient overcurrent protection.

Miniature fuses provide a small size, dependable functionality, and a significant degree of safety in diverse electronic and electrical applications.

Figure 5.7 Miniature fuse

High-Voltage Fuses:

High-voltage fuses (Fig.5.8) are developed explicitly for electrical systems operating at voltage levels higher than 600 volts. They are offered in several arrangements: expulsion, liquid-filled, and current-limiting fuses. High-voltage fuses are employed in utility distribution networks, substations, and industrial power systems.

✓ **Voltage Rating:** High-voltage fuses are specifically engineered to manage the elevated voltages in electrical systems. Their voltage ratings often span from a few hundred to several thousand volts, contingent upon the specific application and operating circumstances.

✓ **Construction:** High-voltage fuses are designed with a strong and durable structure to endure the pressures of high-voltage usage. Typically, they comprise a conductive fuse element housed within a housing constructed of ceramic, glass, or epoxy. The housing serves the dual purpose of providing structural support and electrical insulation to the fuse element.

✓ **Types:** High-voltage fuses are available in various types and designs to accommodate diverse applications and operating conditions. Typical varieties comprise expulsion, liquid-filled, current-limiting, and semiconductor fuses. Each type possesses distinct qualities and benefits, including rapid response, the capacity to limit current, and compatibility with various voltage levels.

- ✓ **Expulsion Fuses:** Expulsion fuses are specifically engineered to swiftly extinguish the electric arc during fuse operation, usually by employing a gas or air blast to suppress the arc. They are frequently used in outdoor and high-current scenarios, such as utility distribution networks and substations.
- ✓ **Liquid-Filled Fuses:** Liquid-filled fuses utilize a liquid, such as oil or silicone, to offer cooling and arc extinguishing characteristics. They are commonly employed in high voltage and current applications, necessitating effective heat dissipation.
- ✓ **Current-Limiting Fuses:** Current-limiting fuses are specifically engineered to limit fault current flow, offering supplementary safeguarding to equipment located farther downstream. They are extensively employed in high-voltage systems to minimize harm and alleviate the consequences of short circuits and failures.
- ✓ **Semiconductor Fuses:** Semiconductor fuses are explicitly engineered to safeguard delicate semiconductor components, including diodes, transistors, and thyristors, against excessive current in high-voltage circuits. They provide rapid and effective safeguarding to avert harm to semiconductor components.
- ✓ **Installation and Maintenance:** Ensuring high-voltage fuses' reliable operation and efficacy requires proper installation and maintenance. This entails carefully selecting the suitable fuse type and rating for the individual application, properly installing fuses in compliance with the specifications provided by the manufacturer and relevant regulations and standards, and regularly inspecting and testing fuses to verify they are working correctly.

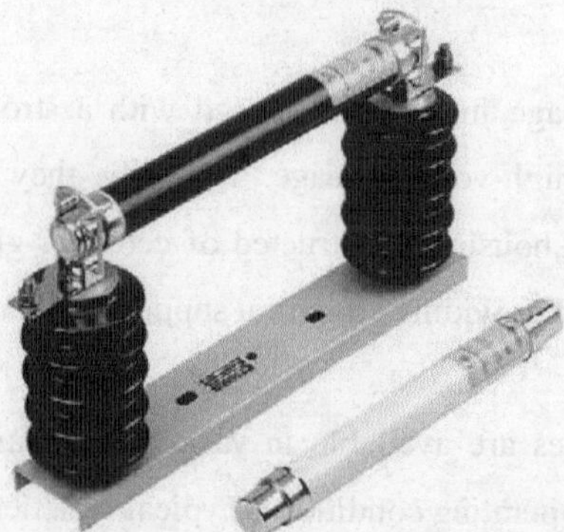

Figure 5.8 High-voltage fuse

High-voltage fuses are essential for safeguarding electrical systems and equipment against excessive current and assuring the dependability and safety of high-voltage installations.

- **Specialty Fuses:**

Specialty fuses (Fig.5.9) encompass a diverse array of fuse types that are specifically engineered for purposes or situations. Illustrative instances encompass semiconductor fuses (used for safeguarding semiconductor equipment), photovoltaic (PV) fuses (utilized in solar power systems), and arc-flash fuses (formulated to alleviate arc-flash perils).

Figure 5.9 Specialty fuse

- ✓ **Semiconductor Fuses:** Semiconductor fuses, alternatively referred to as high-speed fuses or fast-acting fuses, are specifically engineered to safeguard delicate semiconductor components, including diodes, transistors, and thyristors, against excessive current situations. They provide rapid and effective safeguarding to avert harm to semiconductor components in high-power applications, such as motor drives, power converters, and industrial control systems.
- ✓ **Photovoltaic (PV) Fuses:** PV fuses are engineered explicitly for implementation in photovoltaic (solar power) systems to safeguard solar panels, inverters, and other components against excessive electrical current. The fuses are designed to handle the distinctive features of solar power systems, such as high voltage and current levels, and are suitable for use with direct current (DC).
- ✓ **Arc-Flash Fuses:** Arc-flash fuses are specifically engineered to reduce the risks associated with arc-flash incidents in electrical systems. These fuses utilize arc-quenching technology to extinguish arcs rapidly and securely, minimizing the potential for harm to individuals

and equipment. Arc-flash fuses are frequently employed in industrial facilities, utility substations, and other high-risk locations with the potential for arc-flash hazards.

- ✓ **Medical Fuses:** Medical fuses are specifically engineered to adhere to rigorous safety and dependability standards for medical equipment and systems. These fuses are commonly employed in vital healthcare applications like patient monitors, diagnostic imaging systems, and life-support equipment. Medical fuses are produced according to rigorous quality standards and may include minimal temperature increase and high capacity for halting electrical currents.
- ✓ **Military-Grade Fuses:** Military-grade fuses are specifically engineered fuses that are made to withstand the rigorous demands of military and aerospace applications. These fuses are designed to endure challenging environmental conditions, such as severe temperatures, impact, oscillation, and moisture. Military-grade fuses are employed in military vehicles, aircraft, naval vessels, and defense systems where the utmost importance is placed on reliability and longevity.
- ✓ **Instrumentation Fuses:** Instrumentation fuses are very accurate, created explicitly for delicate electronic instruments and control circuits. These fuses offer accurate overcurrent protection with minimum voltage loss and negligible impact on circuit performance. Instrumentation fuses find frequent applications in test and measurement equipment, process control systems, and scientific instruments.
- ✓ **Marine Fuses:** Marine fuses are engineered explicitly for utilization in marine settings, where they need to endure the effects of saltwater, moisture, and corrosion. These fuses are designed to adhere to marine safety regulations and are appropriate for installation in boats, yachts, and other types of marine vessels. Marine fuses offer dependable safeguards against excessive current for electrical equipment installed on a ship or boat.
- ✓ **HVAC Fuses:** HVAC fuses are specifically engineered for use in HVAC systems to safeguard motors, compressors, and other components against excessive current situations. These fuses are designed to handle the elevated voltage and current levels typically encountered in HVAC equipment. HVAC fuses guarantee the secure and dependable functioning of heating and cooling systems in residential, commercial, and industrial structures.

Fuse Links:

Fuse links are discrete replaceable elements employed in fuse holders or fuse bases inside high-voltage electrical systems to offer overcurrent protection. Fuses comprise a fuse element enclosed within a protective shell and are provided in different current ratings and varieties. Fuse linkages are widely utilized in industrial and commercial settings that necessitate frequent replacement of fuses. Fuse links are specifically engineered to disrupt the passage of electric current when it is beyond a predetermined threshold. This safeguards equipment and electrical circuits from harm caused by excessive loads, malfunctions, and abrupt connections.

- ✓ **Construction:** Fuse links generally comprise a fuse element composed of a conductive substance with a low melting point and contained within a protective housing. The fuse component is precisely engineered to undergo melting or vaporization upon encountering excessive electric current, causing the circuit to open and discontinuing the current passage. The housing offers mechanical reinforcement and electrical isolation while safeguarding against environmental elements such as humidity, particulates, and oscillation.
- ✓ **Types:** Fuse linkages are available in various types and configurations to accommodate diverse applications and voltage levels. Typical categories comprise high-voltage fuse links, medium-voltage fuse links, and low-voltage fuse links. Every kind is engineered to function within a precise voltage range and current rating, possessing characteristics customized to meet the demands of the electrical system.
- ✓ **Voltage and Current Ratings:** Various voltage and current ratings offer fuse connections for different applications and operating situations. The voltage rating refers to the maximum voltage that the fuse connection can safely interrupt, whereas the current rating indicates the maximum current that the fuse link can sustain without opening. Choosing fuse links with suitable ratings is crucial for ensuring efficient overcurrent protection and compliance with the electrical system.
- ✓ **Installation:** Fuse links are placed within fuse bases or holders, offering reliable means of attachment and electrical linkage. The fuses are fitted into the fuse base or holder and fastened in position using clips, screws, or alternative methods. Ensuring the correct placement of fuse links is crucial for dependable operation and efficient overcurrent protection.

✓ **Replacement:** When a fuse link is triggered and interrupts the circuit due to excessive current, replacing it with a new fuse link with the same kind and rating is necessary. Using the appropriate replacement fuse link is essential for preserving the integrity and safety of the electrical system.

✓ **Applications:** Fuse linkages are frequently employed in high-voltage electrical systems, such as utility distribution networks, industrial power systems, substations, and electrical machinery. These devices offer crucial protection against excessive current to ensure the safety of equipment, circuits, and individuals by preventing potential dangers.

Plug Fuses:

Plug fuses, alternatively referred to as Edison-base fuses, are fuses that are screwed into place and suited explicitly for home electrical systems. These devices are immediately inserted into screw-type fuse sockets and are offered in both time-delay and fast-acting variations. Plug fuses safeguard branch, lighting, and appliance circuits inside residential settings.

✓ **Design:** Plug fuses possess a cylindrical structure with screw threads on one end, enabling them to be securely fastened into a fuse socket or receptacle. The other extremity of the fuse generally features a metallic cap or contact point to establish an electrical connection.

✓ **Rating:** Plug fuses are offered in a range of current ratings, usually spanning from a few amperes to approximately 30 amperes. The present rating denotes the highest permissible current the fuse may bear without triggering an opening.

✓ **Blow-Out Element:** The plug fuse contains a fuse element composed of a conductive substance, such as copper or zinc, with a low melting point. When the electric current surpasses the predetermined capacity of the fuse, the fuse element becomes heated and liquefies, causing the circuit to break and stopping the current flow.

✓ **Types:** There are two primary types of plug fuses: Type S and Type T. S-type fuses possess a time-delay attribute, allowing them to endure brief periods of excessive load without triggering an opening. T-type fuses have rapid response and promptly react to overcurrent situations. The selection of the appropriate fuse is contingent upon the particular application and the specific requirements of the circuit.

✓ **Installation:** Plug fuses are inserted into fuse sockets or receptacles in electrical panels or fuse boxes. The bulbs are securely fastened into the socket, establishing an electrical connection with the circuit. Correct installation is crucial to guarantee a stable connection and dependable overcurrent safeguarding.

✓ **Replacement:** When a plug fuse is triggered and interrupts the circuit due to excessive current, it is necessary to replace it with a new fuse with the same kind and rating. Using the appropriate replacement fuse is essential for preserving the integrity and safety of the electrical system.

✓ **Safety:** Plug fuses are crucial for safeguarding household electrical circuits by providing protection against excessive current, reducing the risk of electrical fires and equipment damage. Nevertheless, if the fuses are not correctly sized or changed with fuses that have higher current ratings, they can pose a safety risk. Adhering to correct safety protocols and seeking guidance from a certified electrician is imperative while handling plug fuses.

Nevertheless, circuit breakers are progressively supplanting them in contemporary electrical systems owing to their constraints and safety considerations.

5.3 Fuse Material Selection of Fuse

The choice of fuse material is essential for ensuring the optimal performance and dependability of a fuse inside an electrical circuit. The fuse material must exhibit specific properties to interrupt the circuit efficiently under an overcurrent situation while also enduring normal working conditions. When choosing fuse material, it is essential to take into account several things.

✓ **Melting Point:** The fuse material should have a melting point below the anticipated maximum current that will pass through the circuit during regular operation. When the electric current surpasses the designated capacity of the fuse, the fuse material undergoes melting, causing the circuit to open and the current to be interrupted.

✓ **Conductivity:** The fuse material must possess excellent electrical conductivity to facilitate the flow of electricity through the fuse without experiencing significant voltage drop or power dissipation. Fuse construction usually employs materials with high electrical conductivity, such as copper, silver, or aluminum.

✓ **Thermal Conductivity:** The fuse material must possess sufficient thermal conductivity to effectively disperse heat produced during regular operation and instances of excessive current flow. Effective thermal dissipation prevents the fuse from experiencing excessive heat and ensures consistent and dependable performance.

✓ **Arc Extinction:** When the fuse is activated and breaks the circuit, electricity discharge may occur between the melted fuse components, creating an arc. The fuse material must be able to rapidly and efficiently suppress the arc to safeguard the neighboring components and guarantee secure functioning.

✓ **Corrosion Resistance:** The fuse material must possess corrosion and oxidation resistance to preserve its electrical and mechanical characteristics throughout its lifespan. Fuse construction frequently incorporates corrosion-resistant elements, such as copper alloys or tin-plated copper, to guarantee enduring dependability.

✓ **Mechanical Strength:** The fuse material must possess adequate mechanical resilience to endure the mechanical pressures of installation, operating, and short-circuit circumstances. Avoid using materials susceptible to cracking or fracturing when subjected to stress.

✓ **Cost:** The cost of the fuse material should be affordable and cost-effective, considering the application requirements and the required performance characteristics. Although high-performance materials may possess exceptional electrical and mechanical characteristics, they may also come at a higher cost.

Commonly used materials for fuse elements include:

✓ **Copper:** Copper is commonly selected as a fuse element because of its exceptional electrical conductivity, minimal resistance, and favorable thermal characteristics. Copper fuses provide dependable functionality and are appropriate for diverse uses.

✓ **Silver:** Silver, a metal renowned for its exceptional conductivity, is frequently included in constructing fuse elements. Silver fuses have a low resistance and exceptional arc-extinction characteristics, rendering them appropriate for high-current applications and crucial circuits.

✓ **Aluminum:** Aluminium fuses possess a low weight and are economically advantageous, rendering them appropriate for applications involving low voltage and low current. Nevertheless, aluminum exhibits worse conductivity and more resistance than copper and silver, potentially constraining its application in high-performance fuses.

✓ **Alloys:** Additionally, fuse construction incorporates diverse copper alloys, such as brass or bronze, to augment mechanical robustness, resistance to corrosion, and other desirable characteristics. Tin-plated copper alloys are commonly employed to enhance the ability to solder and minimize the occurrence of oxidation.

The choice of fuse material is determined by criteria such as the application's specific needs, the conditions under which it will operate, the current rating, and cost considerations. It is crucial to select a fuse material that satisfies the application's particular requirements and offers dependable protection against excessive current while also guaranteeing safety and effectiveness.

IS 13703-1 (1993): LV Fuses	Indian Standard IS 2086:1993 : Electric Fuses

5.4 Miniature Circuit Breaker (MCB)

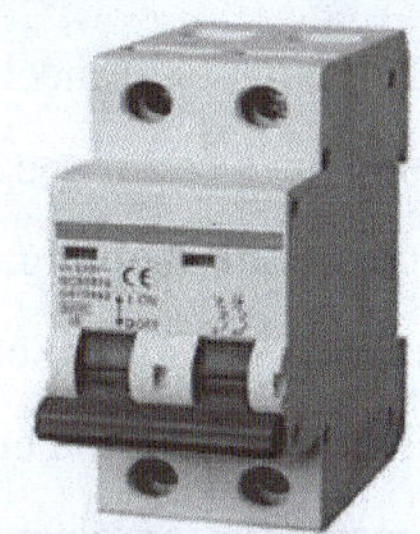

Figure 5.10 Miniature Circuit Breaker

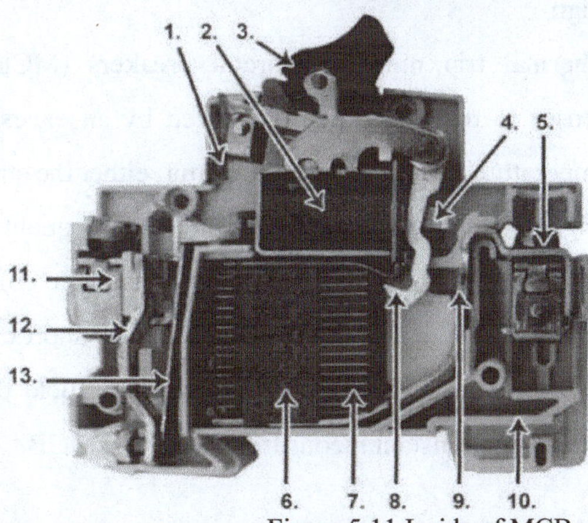

1. Latch
2. Solenoid
3. Switch
4. Plunger
5. Incoming Terminal
6. Arc Chutes Holder
7. Arc Chutes
8. Dynamic Contact
9. Fixed Contact
10. Din Rail Holder
11. Outgoing Terminal
12. Bi-metallic Strip Carrier
13. Bi-metallic Strip

Figure 5.11 Inside of MCB

A Miniature Circuit Breaker (MCB) is an electrical apparatus designed to safeguard electrical circuits against excessive loads and short circuits. Due to its myriad benefits, it is extensively employed in residential, commercial, and industrial settings as a substitute for conventional fuses. The MCB is shown in Fig.5.10, and the inside of the MCB is demonstrated in Fig.5.11.

Construction:

A Miniature Circuit Breaker (MCB) is constructed using many essential components specifically developed to offer dependable overcurrent protection and guarantee secure functioning in electrical circuits. Although there may be variations in designs across different manufacturers and models, the following are the typical components and their respective functions in a Miniature Circuit Breaker (MCB):

- ✓ **Enclosure or Housing:** A miniature circuit breaker (MCB) 's casing offers both mechanical protection and insulation for its internal components. It is commonly constructed from high-impact thermoplastic or flame-retardant materials to endure environmental conditions and ensure electrical safety.
- ✓ **Operating Mechanism:** An MCB's operational mechanism comprises a handle or lever that enables manual breaker control. Users can manually activate or deactivate the MCB using this technique for maintenance, troubleshooting, or circuit separation.
- ✓ **Trip Mechanism:** The trip mechanism detects instances of excessive current and activates the miniature circuit breaker (MCB) to interrupt the circuit. The composition of the MCB can vary, incorporating thermal, magnetic, or electronic elements, or a mix of these, depending on its type and design.
- ✓ **Thermal Trip Element:** Thermal trip miniature circuit breakers (MCBs) utilize a bimetallic strip or thermal sensor to detect the heat produced by an excessive flow of electrical current. Once the temperature surpasses a specific limit, either the strip undergoes deformation or the sensor is triggered, activating the MCB and its subsequent interruption.
- ✓ **Magnetic Trip Element:** Magnetic excursion MCBs employ a magnetic coil or solenoid to identify abrupt surges in electric currents, such as those induced by short circuits. Upon detecting a substantial electric current, the coil generates a magnetic field that promptly triggers the trip mechanism, resulting in instantaneous tripping of the MCB.

- ✓ **Current-Carrying Components:** The MCB has current-carrying elements, such as conductive bars, contacts, and terminals, which provide electrical connections within the circuit. Typically, these components are constructed using copper or other materials with high conductivity to reduce resistance and heat production.
- ✓ **Arc Chamber:** The arc chamber is precisely engineered to suppress the electrical discharge that occurs when the miniature circuit breaker (MCB) interrupts the current flow in the circuit. The MCB incorporates arc-suppressing elements, such as plates, baffles, or magnetic blowout coils, that effectively and securely extinguish the arc, safeguarding the MCB and its adjacent components from harm.
- ✓ **Contacts:** MCBs have mobile contacts that establish or interrupt the electrical circuit when the MCB is activated or deactivated. The purpose of these contacts is to provide dependable switching and minimize power loss and heat generation by maintaining low contact resistance.
- ✓ **Trip Indicator:** Certain miniature circuit breakers (MCBs) are furnished with a trip indication, such as a flag or window, which visually indicates when the breaker has tripped due to excessive current. This indication offers a visual confirmation of the breaker's status and assists users in identifying and resolving electrical faults.

Principle rating and uses:

- ✓ **Rated Voltage (Ue):** The rated voltage of a miniature circuit breaker (MCB) signifies the highest voltage level at which the breaker is engineered to function securely. MCBs commonly have rated voltages of 230V for single-phase systems and 400V for three-phase systems.
- ✓ **Rated Current (In):** The rated current of a miniature circuit breaker (MCB) indicates the highest amount of continuous electric current that the breaker can handle without activating its tripping mechanism. Miniature Circuit Breakers (MCBs) are offered in a wide range of current ratings, often spanning from a few amperes to several hundred amperes, depending on the specific use case. Typical ratings are 6A, 10A, 16A, 20A, 32A, and so on.
- ✓ **Rated Breaking Capacity (Icn or Ics):** The rated breaking capacity of a miniature circuit breaker (MCB) signifies the highest magnitude of fault current that the breaker may safely interrupt without incurring any damage. Choosing an MCB with a breaking capacity that is higher than the predicted short-circuit current (PSC) of the electrical system is crucial. This represents the highest magnitude of fault current that the miniature circuit breaker

(MCB) may safely interrupt without harm. The unit of measurement commonly used to express it is kA (kiloamperes). Typical breaking capacity ratings include 6kA, 10kA, 16kA, etc.

✓ **Tripping Characteristics:** MCBs are offered several tripping characteristics, including B, C, D, K, and Z curves, which dictate the speed at which the breaker reacts to excessive current situations. The choice of the suitable tripping feature is contingent upon the specific demands of the application and the nature of the load being safeguarded. MCBs can possess several tripping characteristics, including Type B (suitable for general applications), Type C (designed for motors and inductive loads), and Type D (specifically for high inrush current loads).

✓ **Pole Configuration:** Miniature Circuit Breakers (MCBs) can function as single-pole, double-pole, or multipole devices. The MCB's capacity to disconnect several wires simultaneously determines the arrangement of poles.

Working Principle Miniature Circuit Breaker:

✓ Each instance of a sustained flow of excessive current via MCB results in the bimetallic strip's heating and subsequent bending and deflection. The deflection of this bi-metallic strip triggers the release of a mechanical latch.

✓ When the mechanical latch, connected to the operating mechanism, opens the contacts of the MCB, it interrupts the flow of electricity in the circuit. To restore the current flow, the MCB must be activated manually.

✓ The purpose of this device is to prevent short circuits and failures caused by excessive current or overload.

✓ An MCB is fundamentally elementary, user-friendly, and seldom requires repair. In essence, the act of replacing it is more straightforward. The trip unit is the main component responsible for ensuring proper operation. There are two primary sorts of trip mechanisms.

✓ An electromagnet safeguards against electric short-circuit current, whereas a bi-metal protects against overload current. The working principle of MCB is shown in Fig.5.12.

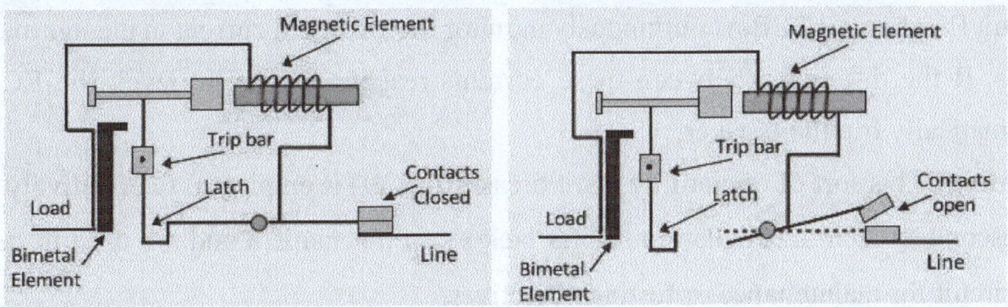

Figure 5.12 Working principle of MCB

Different Types of MCBs

Miniature Circuit Breakers (MCBs) can be classified into many categories depending on factors such as current rating, voltage rating, and trip characteristics. Several prevalent varieties of miniature circuit breakers (MCBs) include:

- ✓ **Thermal:** This miniature circuit breaker (MCB) is designed to trip when there is an increase in temperature due to the current passing through the circuit. Thermal miniature circuit breakers (MCBs) have a bimetallic strip that flexes and triggers the breaker when the temperature exceeds a specific limit.
- ✓ **Magnetic:** This miniature circuit breaker (MCB) is designed to trip when the current passing through the circuit generates a magnetic pull. Magnetic miniature circuit breakers (MCBs) are equipped with a solenoid that exerts a force on the trip mechanism, causing the breaker to trip when the magnetic force is beyond a specific threshold.
- ✓ **Hybrid:** This MCB integrates the characteristics of both thermal and magnetic MCBs. Hybrid miniature circuit breakers (MCBs) incorporate both a bimetallic strip and a solenoid, enabling them to trip in response to either the increase in temperature or the magnetic force produced by the electrical current.
- ✓ **Electronic:** An electronic MCB utilizes electronic components to monitor the current and activate the breaker. Electronic miniature circuit breakers (MCBs) exhibit more sensitivity and offer enhanced speed and precision in tripping, in contrast to conventional thermal and magnetic MCBs.
- ✓ **Differential:** This miniature circuit breaker (MCB) is specifically designed for use in direct current (DC) circuits. Its primary function is to protect against both earth faults and short circuits. Differential miniature circuit breakers (MCBs) are designed to monitor the electrical current in both the live and neutral wires. These breakers are programmed to trip and interrupt the circuit when the current discrepancy surpasses a specific threshold.
- ✓ **Residual Current Circuit Breaker (RCCB):** This miniature circuit breaker (MCB) is designed to protect against electric shock and fire from earth faults. Residual Current

Circuit Breakers (RCCBs) continuously monitor the electrical current in the live and neutral wires. If the difference between these currents reaches a specific level, the RCCB will automatically trip the breaker.

✓ **Isolation:** This sort of miniature circuit breaker (MCB) is employed to effectively separate or disconnect a circuit. Isolation MCBs lack a trip mechanism and are used to deactivate the circuit for maintenance or testing objectives.

Different types of MCBs used in Electrical Protection Systems

Within Miniature Circuit Breakers (MCBs), the designations Type A, Type B, Type C, Type D, Type E, and Type F denote distinct degrees of safeguarding offered by the apparatus.

1) **Type A:** MCBs of Type A are specifically engineered to safeguard against excessive current. They are appropriate for use in circuits where the highest anticipated current is known and is reasonably consistent, such as circuits for lighting.
2) **Type B:** Type B miniature circuit breakers (MCBs) are specifically engineered to safeguard against excessive current and abrupt electrical faults. They are appropriate for use in circuits with fluctuating loads, such as those seen in motor circuits.
3) **Type C:** Circuit breakers of Type C are specifically engineered to offer safeguards against excessive current and ground fault currents. They are appropriate for utilization in circuits with a significant likelihood of earth fault currents, such as those driven by direct current (DC) or those incorporating delicate electronic equipment.
4) **Type D:** Type D miniature circuit breakers (MCBs) are specifically engineered to safeguard against excessive current and ground fault currents, featuring a tripping threshold greater than that of Type C MCBs. They are appropriate for use in circuits with a significant likelihood of earth fault currents but where the fault current is anticipated to exceed the protection capabilities of Type C MCBs.
5) **Type G:** G MCBs are specifically engineered to offer safeguarding against excessive current and ground fault currents in residual current devices (RCDs) employed in electrical systems.
6) **Type H:** H MCBs are specifically engineered to safeguard against excessive current and earth fault currents in electrical systems energized by direct current (DC).
7) **Type K:** K-type miniature circuit breakers (MCBs) are designed to safeguard against excessive current and short circuits in electrical systems characterized by elevated fault levels.

- **Overcurrent Protection:**

An MCB's primary purpose is to offer safeguards against excessive current situations. The MCB functions by automatically tripping and disconnecting the circuit when the current surpasses the rated capacity, safeguarding the electrical wiring, appliances, and other connected equipment from potential harm.

✓ **Manual Operation:** MCBs generally feature a manual switch that enables users to activate or deactivate the circuit manually. This function offers convenience for maintenance, troubleshooting, and circuit isolation, as and when required.

✓ **Trip Mechanism:** MCBs have a trip mechanism that detects and reacts to excessive current situations. The trip mechanism of an MCB can be either thermal, magnetic, or a hybrid of both, depending on its specific kind. Thermal trip mechanisms detect overloads by sensing the heat produced from high current flow, while magnetic trip mechanisms detect short circuits by detecting the quick increase in current.

✓ **Compact Design:** MCBs provide a compact and space-efficient design, rendering them well-suited for installations with limited area. These devices can be readily put in electrical distribution panels, consumer units, and switchboards.

✓ **Modular Installation:** MCBs are engineered explicitly for modular installation, enabling effortless insertion and removal from electrical panels or enclosures without rewiring. The modular nature of this system allows for the rapid and effective installation, upkeep, and expansion of electrical circuits.

✓ **High Breaking Capacity:** MCBs have a substantial breaking capacity, enabling them to effectively halt the current in the event of a malfunction. This guarantees dependable overcurrent protection and reduces the likelihood of electrical fires, equipment harm, and physical harm.

✓ **Reliability:** MCBs provide dependable performance and sustained durability, ensuring uninterrupted safeguarding against overloads and short circuits over their operational lifespan. The products are produced according to strict quality criteria and undergo a thorough examination to guarantee adherence to safety legislation and standards.

✓ **Uses:**

- **Overload Protection:** Miniature Circuit Breakers (MCBs) safeguard electrical circuits against overload situations when the current surpasses the circuit's rated capacity due to high load or faulty circumstances. The MCB is designed to trip and interrupt the circuit to safeguard against potential harm to the wiring, appliances, and other interconnected equipment.

- **Short Circuit Protection:** Miniature Circuit Breakers (MCBs) offer to safeguard against short circuits, which entail an abrupt and excessive surge in electric current resulting from a malfunction in the electrical system. The MCB promptly detects the excessive current and immediately interrupts the circuit to isolate the malfunctioning circuit and avoid additional harm.
- **Circuit Isolation:** MCBs enable the manual disconnection of electrical circuits for maintenance, service, or troubleshooting. By deactivating the MCB (Miniature Circuit Breaker), the circuit can be securely disengaged without shutting down the entire electrical system.
- **Distribution Protection:** MCBs are frequently employed in electrical distribution panels, consumer units, and switchboards to offer overcurrent protection for individual circuits or clusters of circuits. They contribute to maintaining the electrical distribution system's safety and dependability by averting overloads and short circuits.
- **Residential, Commercial, and Industrial Applications:** MCBs are utilized in many contexts, encompassing residential structures, commercial establishments, and industrial factories. These devices safeguard lighting circuits, power outlets, motor circuits, HVAC systems, control panels, and other electrical installations.

5.5 Earth Leakage Circuit Breaker (ELCB)

An ELCB (Fig.5.13), alternatively referred to as a Ground Fault Circuit Interrupter (GFCI) in certain areas, is an electrical safety apparatus specifically engineered to safeguard against electric shock and electrical fires resulting from earth faults or earth faults.

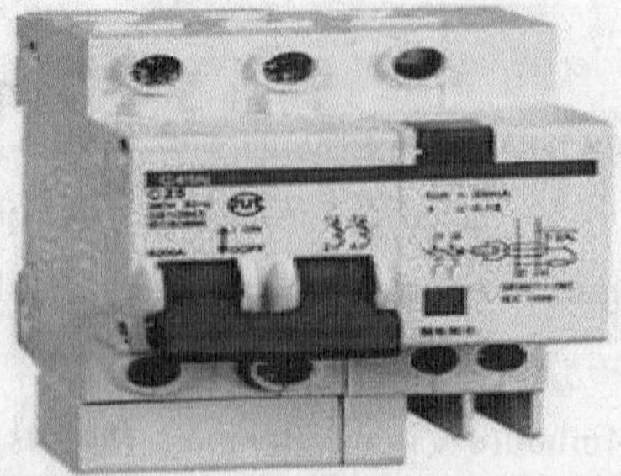

Figure 5.13 Earth leakage Circuit breaker

Construction:

An ELCB is constructed with many essential components and mechanisms specifically designed to identify earth faults and promptly interrupt electrical circuits in case of such failures. Below is a detailed explanation of the standard structure of an ELCB:

✓ **Enclosure:** ELCBs, like other circuit protection devices, are contained in an enclosure constructed from robust and non-conductive materials such as thermoplastics or moulded insulating materials. The enclosure safeguards the interior components by offering mechanical protection and insulation to avoid any contact with live parts.

✓ **Sensing Circuit:** The sensing circuit of an Earth Leakage Circuit Breaker (ELCB) is tasked with identifying earth defects by constantly monitoring the current imbalance between the live and neutral conductors. This circuit generally comprises a highly responsive or differential current transformer that quantifies the residual current passing through the conductors.

✓ **Core and Windings:** The sensing circuit comprises a magnetic core with windings coiled around it. When the currents flowing through the live and neutral wires are equal, the magnetic flux generated in the core by the windings is likewise equal. Yet, when there is an earth fault, an asymmetry in the electric currents generates a residual current, producing a magnetic field in the core.

✓ **Trip Mechanism:** The trip mechanism of an ELCB reacts to the measured residual current and triggers the tripping process to interrupt the circuit. The trip mechanism of an ELCB can be either instantaneous or time-delayed, depending on its kind. Instantaneous tripping promptly activates upon fault detection, but time-delayed tripping introduces a delay before activation to meet specific loads or temporary situations.

✓ **Contacts:** ELCBs are equipped with contacts that establish or interrupt the electrical connection when the device is triggered. The contacts commonly comprise high-conductivity materials, such as silver or copper alloys, to minimize resistance and provide dependable switching. When the Earth Leakage Circuit Breaker (ELCB) is triggered, the electrical contacts separate to break the circuit and isolate the electrical fault.

✓ **Reset Mechanism:** Certain ELCBs have a reset mechanism that enables users to reset the device following a trip manually. The reset mechanism may have a push-button or lever that restores the contacts to their closed state after the problem is resolved and the circuit is ready to be re-energized.

✓ **Test Button:** Several contemporary ELCBs have a test button that enables users to conduct regular assessments of the device's functionality. A simulated earth fault condition is created by pressing the test button, which triggers the ELCB to trip and confirms its proper functionality. Regular testing is essential to guarantee the dependability and efficiency of the ELCB in identifying problems and safeguarding against electric shock risks.

Operation Principle:

ELCBs function based on the premise of detecting residual current. To continuously monitor the current flowing through the live and neutral conductors, they employ either a sensitive or differential current transformer. Any disparity in the currents indicates an earth fault, causing the ELCB to trip and interrupt the circuit. The Earth Leakage Circuit Breaker (ELCB) operates by detecting earth faults or ground faults in electrical circuits and then promptly cutting off the power supply to prevent the occurrence of electric shock dangers.

a) **Residual Current Sensing:**

✓ ELCBs function by detecting residual current through a sensor mechanism. They consistently observe the current balance between an electrical circuit's live and neutral conductors.

✓ In typical circumstances, the electric current passing through the live conductor (phase) should be equivalent to the current returning through the neutral conductor.

✓ Nevertheless, when an earth fault occurs, a fraction of the current can pass through the earth connection, resulting in an asymmetry in the currents.

✓ The ELCB identifies the residual current, which is the disparity between the currents in the live and neutral conductors, signifying the existence of an earth fault.

b) **Tripping Mechanism:**

✓ When an imbalance in the currents is detected, the tripping mechanism of the ELCB is engaged.

✓ The tripping mechanism of the ELCB can be determined by either thermal, magnetic, or electronic principles, which vary depending on the design and kind of the device.

✓ Thermal ELCBs utilize a bimetallic strip or thermal sensor to detect and respond to the heat produced by the residual current, resulting in the tripping of the ELCB.

✓ Magnetic ELCBs utilize a magnetic coil or solenoid to detect the magnetic field produced by the residual current, resulting in immediate tripping.

✓ Certain contemporary ELCBs integrate thermal and magnetic components to enhance sensitivity and dependability.

c) **Circuit Interruption:**

✓ The ELCB functions by tripping and opening its contacts, thus interrupting the electrical circuit, and disconnecting the power supply to the protected circuit.

✓ The ELCB promptly interrupts the electrical current upon detecting a ground fault, thereby preventing the occurrence of electric shock dangers, and reducing the likelihood of electrical fires.

d) **Manual Reset and Test:**

✓ Following a trip, certain ELCBs are equipped with a manual reset function that enables users to reinstate power to the circuit once the fault has been resolved and the underlying cause of the failure has been dealt with.

✓ Furthermore, numerous ELCBs are equipped with a test button that allows users to simulate an earth fault situation to conduct regular tests on the device's functionality. This enhances the dependability and efficiency of the ELCB in identifying malfunctions and safeguarding against the risks of electric shock.

- **Detection of Earth Faults:** ELCBs detect and monitor the disparity in current between the live and neutral wires in an electrical circuit. During an earth fault, such as when a human touches a live conductor or when a fault occurs in an appliance, a fraction of the electric current may pass through the ground. The Earth Leakage Circuit Breaker (ELCB) identifies this disparity and disrupts the electrical circuit to mitigate the risk of electric shock dangers.
- **Tripping Characteristics:** ELCBs have several tripping characteristics, including instantaneous and time-delayed tripping. ELCBs with instantaneous tripping capabilities promptly react to earth faults, causing the circuit to trip nearly immediately upon detecting an imbalance in current. ELCBs with time-delayed tripping provide a built-in delay before tripping, allowing them to handle specific loads or transitory circumstances.
- **Types of ELCBs:**

• **Voltage-Operated ELCB (VO-ELCB):** This Earth Leakage Circuit Breaker (ELCB) identifies earth faults by detecting voltage imbalances between the live and neutral wires. Today, its usage is less frequent due to its limitations in identifying specific problems and susceptibility to interference from external factors.

- **Current-Operated ELCB (CO-ELCB):** CO-ELCBs are the prevailing variant of ELCBs currently in widespread use. These devices' function using the residual current theory and have a higher level of sensitivity to faults in the earth. CO-ELCBs are offered in both instantaneous and time-delayed tripping variants.

- **Installation:** ELCBs are commonly placed in the primary distribution board or electrical panel, offering protection for numerous circuits or the complete setup. They are linked in series with the live wire and can be fitted for specific circuits or as part of a residual current device (RCD) for complete protection.

- **Applications:** ELCBs are widely employed in residential, commercial, and industrial electrical systems to offer safeguards against the risks of electric shock. These devices are placed in areas with a potential for direct or indirect exposure to live electrical components, such as restrooms, kitchens, outdoor spaces, and equipment with metal casings.

- **Code Requirements:** ELCBs are mandated by numerous countries' electrical safety rules and building codes to ensure electrical safety and adhere to standards. ELCBs are essential components of electrical safety systems and serve a critical function in preventing electric shock incidents and safeguarding individuals and assets.

- **Uses:**

a) **Protection against Electric Shock:** The primary purpose of ELCBs is to safeguard against the risks of electric shock. They identify minute leakage currents that can arise when an electrical device or wiring touches moisture or a conductive surface, such as a person or the ground. ELCBs effectively prevent electric shocks and potential injury by quickly interrupting the circuit when they detect leakage currents.

b) **Ground Fault Protection:** ELCBs protect ground faults by identifying ground faults or earth faults in electrical circuits. These failures arise when the electrical current from the live conductor inadvertently passes to the ground due to insulation failure, device malfunction, or accidental contact with conductive surfaces. ELCBs, or Earth Leakage Circuit Breakers, are crucial in preventing electrical fires and equipment damage by promptly terminating the circuit when ground faults are detected.

c) **Residual Current Protection:** ELCBs provide residual current protection by constantly monitoring the current balance between an electrical circuit's live and neutral wires. The ELCB is activated and interrupts the circuit whenever there is an uneven flow of currents, indicating the existence of residual or leakage current. This measure mitigates the risk of insulation failure, wiring defects, and other electrical irregularities that may result in electrical mishaps.

d) **Compliance with Electrical Codes and Standards:** ELCBs are mandated by electrical safety rules and building codes in numerous countries to ensure compliance with safety standards and regulations. ELCBs are typically put in locations with an elevated likelihood of electric shock dangers, such as restrooms, kitchens, outdoor outlets, and swimming pools. Adhering to these rules guarantees the safety of individuals and reduces the likelihood of electrical mishaps.

e) **Enhancement of Electrical System Safety:** ELCBs, or Earth Leakage Circuit Breakers, improve the safety and dependability of electrical systems in homes, businesses, and factories by effectively safeguarding against electric shock dangers, ground faults, and residual currents. They are vital in reducing the likelihood of electrical mishaps, averting harm to property, and safeguarding life.

f) **Protection for Portable and Temporary Installations:** ELCBs are also utilized to offer ground fault protection for portable and temporary electrical installations, such as building sites, outdoor events, and temporary power distribution systems. Portable ELCBs, also called portable GFCIs, are small, plug-in devices that can be conveniently placed between electrical devices and power sources to offer protection against ground faults immediately.

IS/IEC 60898-1 (2002): Circuit-breakers

IS/IEC 60898-1:2015 -Circuit Breakers

5.6 System and Equipment Earthing and its Requirements:

Earthing, or grounding, is essential for electrical and electronic systems to guarantee safety and optimal operation. It entails establishing a pathway with minimal resistance for the flow of electric electricity into the ground in case of a malfunction. System and equipment grounding are vital components of electrical installations to guarantee safety, safeguard equipment, and enable optimal operation of electrical systems. There are two primary categories of earthing: system earthing and equipment earthing.

- **System Earthing:**

a) **Purpose:** System earthing, or grounding, refers to connecting electrical systems to the earth or ground. The main objective of a grounding system is to establish a secure route for fault currents to be directed toward the ground, effectively eliminating the occurrence of electric shock dangers and reducing the likelihood of electrical fires. The main objective of system earthing is to establish a fixed point of reference for the voltage within the system. It aids in maintaining a stable system voltage and guarantees that all conductive components of the system are at a predetermined voltage level relative to the ground.

b) **Types of System Earthing:**

✓ **Neutral Earthing:** In low-voltage electrical systems, the neutral conductor is commonly linked to the earth at a specific location, referred to as neutral earthing or system grounding. *This guarantees that* the system maintains a state of equilibrium or proximity to the earth's potential and establishes a fixed point of reference for voltage measurements. Earth grounding is commonly employed in power distribution systems to establish a fixed point of reference for system voltages at the neutral point of transformers. Typical techniques comprise solid grounding, resistance grounding, and reactance grounding.

✓ **Solid Grounding:** The neutral is directly linked to the ground with minimal resistance.

✓ **Resistance Grounding:** The neutral is linked to the ground via a resistor to restrict fault currents.

✓ **Reactance Grounding:** A reactor is installed in the circuit between the neutral and ground to restrict the flow of fault currents.

✓ **Equipment Earthing:** Electrical equipment, including transformers, generators, and switchgear, can be connected to the earth for safety and operational purposes.

c) **Requirements:**

✓ **Safety:** System grounding is crucial for ensuring the safety of personnel, as it effectively diverts fault currents away from conductive surfaces and equipment enclosures, hence minimizing the risk of electric shock risks.

✓ **Fault Protection:** Effective system earthing ensures a clear path for fault current, enabling rapid operation of protective devices like fuses and circuit breakers to isolate faulty circuits.

✓ **Equipment Protection:** System grounding safeguards electrical equipment and insulation from excessive voltages and sudden surges caused by lightning strikes, switching processes, or other external influences.

Equipment Earthing:

a) **Purpose:** Equipment earthing establishes a connection between metallic enclosures, frames, and conductive components of electrical equipment and the earth. The main objective of a low-impedance channel is to facilitate the flow of fault currents, enabling protective mechanisms to operate swiftly and reducing the likelihood of electric shock and equipment harm.

b) **Requirements:**

✓ **Low Impedance:** Equipment earthing connections should possess a low impedance to expedite the swift dissipation of fault currents. This generally entails utilizing conductors that are sufficiently sized and made of appropriate materials while also assuring the use of competent installation techniques.

✓ **Continuity:** To provide a consistent flow of fault currents, the equipment earthing conductors must be uninterrupted and devoid of fractures or corrosion.

✓ **Bonding:** To maintain a uniform grounding system and avoid any discrepancies, it is necessary to connect and secure metallic enclosures, frames, and conductive components of equipment together.

c) **Methods of Equipment Earthing:**

✓ **Direct Earthing:** The equipment's metallic components are directly linked to earth electrodes through earthing wires.

✓ **Indirect Earthing:** The equipment's metallic components are linked to the ground using earthing conductors, which are securely attached to the primary earthing system or earth electrode network.

d) **Testing and Maintenance:** Regular testing and maintenance of equipment grounding systems are crucial to ensure their efficacy. This may encompass assessments of earth resistance, verification of continuity, visual examinations, and implementation of appropriate corrective measures.

- **General Earthing Requirements:**

✓ **Safety:** The main objective of earthing is to optimize safety by reducing the likelihood of electric shock and guaranteeing the effective operation of preventive mechanisms.

✓ **Compliance:** All earthing systems must adhere to local and international standards, norms, and regulations.

✓ **Maintenance:** Periodic examination, evaluation, and upkeep of earthing systems are crucial to guarantee their ongoing efficiency.

✓ **Documentation:** Thorough documentation of the earthing system, which encompasses design specifications, test outcomes, and maintenance records, is crucial to reference and ensure adherence to regulations.

It is crucial to acknowledge that the precise criteria may differ depending on the electrical system's type, local legislation, and the characteristics of the equipment in question. Engaging in discussions with electrical experts and adhering to relevant standards is crucial when designing and executing efficient earthing systems.

- **Earth:**

In the context of electrical installations, the term "earth" pertains to grounding. Grounding is an essential safety precaution that safeguards individuals, machinery, and buildings against electrical malfunctions. In this context, "earth" is frequently used synonymously with "ground". Grounding fulfills multiple functions in electrical systems, including ensuring safety, safeguarding equipment, and providing a reference point.

- **Earth Electrode:**

An earth electrode is a conductive structure or device that directly links an electrical or electronic system and the Earth. Earth electrodes serve the objective of establishing a pathway with minimal resistance for fault currents to disperse into the ground, guaranteeing the electrical system's safety and optimal operation. Earth electrodes play a vital role in the comprehensive grounding system of electrical systems. Common earth electrodes include ground rods, ground plates, ground electrodes, pipe electrodes, and chemical electrodes.

- **Earth Current:**

Earth current is the movement of electric current that passes through the grounding system and into the Earth. This phenomenon occurs in electrical faults when an inadvertent link is formed between a live conductor and a conductive element that is either exposed or connected to the Earth.

Earth Terminal:

An earth terminal is a specific location inside an electrical system where a direct connection is established to the grounding or earthing system. Typically, an earth electrode is attached to a metal termination or wire.

Earthing Wire:

An earthing wire, sometimes called a grounding wire or conductor, is a conductive line that establishes a connection between electrical equipment, enclosures, or exposed conductive portions and the grounding system. It facilitates the safe movement of fault currents into the ground.

Earthing Lead:

An earthing lead is a conductor designed to connect an electrical system or equipment and an earth electrode or grounding point. It is an integral component of the whole grounding system.

Fault Current:

Fault current refers to the electric current in an electrical system during a defect, such as a short circuit or ground fault. The fault current naturally follows the path of least resistance to return to its source.

Leakage Current:

Leakage current refers to the unintentional movement of electric current from a conductor to the surrounding insulation or the earth. It can manifest in many electrical equipment and systems and is commonly assessed to guarantee safety and optimal operation.

Measurement of Earth Resistance Using Earth Tester:

An earth tester or ground tester is used to measure the Earth's resistance. The tester administers a predetermined voltage to the ground and gauges the ensuing current. The resistance of the earth electrode system can be determined using Ohm's Law, which states that resistance is equal to voltage divided by current.

Methods of Reducing Earth Resistance:

- ✓ **Increasing Electrode Length:** Increasing the length of electrodes typically decreases the resistance of the earth.
- ✓ **Increasing Electrode Surface Area:** Increasing the surface area by employing several ground rods or plates can reduce resistance.

✓ **Adding Ground Enhancement Materials:** Chemical compounds or backfills can enhance soil conductivity and decrease resistance.

✓ **Parallel Grounding:** Connecting numerous electrodes in parallel helps reduce the total resistance.

✓ **Deep Grounding:** Increasing the depth of the electrodes can decrease electrical resistance.

Ensuring proper earthing and grounding is crucial to ensure the safety and dependability of electrical equipment. Adhering to local electrical codes and standards is vital when developing and implementing grounding systems. Consistent testing and maintenance are crucial to sustain the system's long-term effectiveness. It is advisable to get advice from competent electrical experts for intricate installations.

5.7 Methods of Earthing as per IS 3043: 1987:

IS 3043: 1987 is an Indian Standard that offers instructions for selecting and installing earthing systems in diverse applications. The standard delineates various techniques of grounding, each appropriate for particular circumstances. The following are the techniques for earthing according to IS 3043: 1987, along with a concise summary of their processes:

IS 3043 (1987): Code of Practice for Earthing

5.7.1 Driven Pipe Earthing:

Driven pipe earthing is a grounding technique that entails the installation of a metal pipe in the ground to establish a pathway with low resistance for the dissipation of fault currents. This technology is frequently employed in electrical systems to guarantee equipment and personnel safety.

Driven Pipe Earthing Procedure:

a) **Selection of Materials:**

✓ Select an appropriate material for the pipe. Typical materials used are galvanized iron (GI) or copper.

✓ Copper pipes are favoured for their superior conductivity.

b) **Determination of Pipe Length:**

✓ The length of the pipe is a crucial determinant of the efficacy of the earthing system.

✓ The answer varies based on the soil resistivity at the specific location. As the pipe is driven more profound, the soil resistance decreases.

c) **Driving the Pipe:**

✓ Utilise specialized machinery, such as a pipe-driving machine or a manual driver, to put the pipe efficiently and effectively into the ground.

✓ The pipe is inserted vertically into the soil until it reaches the appropriate depth. The depth is commonly established by considering the soil resistivity and the desired earth resistance.

d) **Backfilling the Pit:**

✓ Once the pipe is aligned correctly, an excavation is made around it.

✓ Fill the pit with a combination of salt and charcoal. This blend improves the conductivity of the soil that surrounds the pipe.

✓ The backfill material must be uniformly distributed and densely packed around the pipe.

e) **Connection to Equipment:**

✓ The upper portion of the pipe, typically protruding above the surface, is linked to the equipment or structure that necessitates grounding.

✓ Ensure that suitable connectors are used to establish a connection with minimal resistance between the equipment and the pipe.

f) **Testing and Verification:**

✓ After the installation is complete, earth resistance testing will be performed using an earth tester or other appropriate measurement devices.

✓ The test findings must adhere to the defined requirements and standards for the specific installation.

- **Plate Earthing:**

✓ A copper or galvanized iron plate is inserted vertically into the ground.

✓ The plate should be interred in a pit of precise dimensions, containing a blend of charcoal and sodium chloride to augment conductivity.

✓ The plate must be buried at a minimum depth of 3 meters below ground level.

- **Pipe Earthing:**
  - ✓ A vertical galvanized iron or copper pipe is inserted into the earth.
  - ✓ The pipe must possess a minimum diameter, while the soil resistivity chooses the length of the pipe.
  - ✓ The pipe is filled with a mixture of charcoal during backfilling.
- **Plate and Pipe Combination Earthing:**
  - ✓ This method involves a combination of both plate and pipe earthing systems.
  - ✓ It is employed when the soil conditions or resistivity are not conducive to a singular technique.

5.8 Modern methods of earthing :

Contemporary techniques for grounding are constantly developing to satisfy the growing requirements of electrical systems, tackle environmental issues, and improve safety. Below are a few contemporary techniques for grounding:

- **Chemical Ground Enhancements:**
  - ✓ Specialised chemical compounds or backfill materials are employed to augment the conductivity of the soil surrounding the earth electrode, hence diminishing earth resistance.
  - ✓ Advantages: Chemical ground improvements can significantly enhance the efficiency of the earthing system, particularly in regions with high-resistivity soil.
- **Grounding with Conductive Concrete:**
  - ✓ Conductive concrete, which incorporates conductive components such as metal fibers or particles, is a substitute for conventional backfill material surrounding grounding electrodes.
  - ✓ Advantages: Conductive concrete enhances the uniformity and conductivity of the medium surrounding the electrode, hence decreasing earth resistance.
- **Grounding using Grounding Mats:**
  - ✓ Grounding mats are composed of conductive materials, such as copper strips or meshes, directly in contact with the earth.
  - ✓ Advantages: Grounding mats offer an expansive surface area for grounding and are especially beneficial in areas with restricted space.

- **Microbial Electrochemical Grounding:**
  - ✓ This novel method utilizes microbial fuel cells to augment microbial activity in the soil, enhancing its conductivity.
  - ✓ Advantages: Microbial electrochemical grounding offers a viable and environmentally friendly approach to improve soil conductivity.
- **Deep Grounding Electrodes:**
  - ✓ Electrodes, such as rods or pipes, are inserted further into the earth to access soil layers with reduced resistivity.
  - ✓ Advantages: Deep grounding electrodes can attain a lower level of earth resistance than shallow electrodes.
- **Grounding using Engineered Materials:**
  - ✓ Engineered materials with improved conductivity, such as conductive polymers or graphene, are employed as grounding electrodes.
  - ✓ Advantages: These materials exhibit enhanced electrical conductivity and may be appropriate for particular applications.
- **Grounding with Exothermic Welding:**
  - ✓ Exothermic welding is a process that utilizes a chemical reaction to generate significant heat, resulting in the formation of a high-quality and low-resistance connection between conductors.
  - ✓ Advantages: Exothermic welding is commonly employed to establish dependable and long-lasting connections in grounding systems.
- **Smart Grounding Systems:**
  - ✓ Smart grounding systems utilize monitoring and diagnostic technology to evaluate the condition and efficiency of the grounding system consistently.
  - ✓ Advantages: Real-time monitoring has the advantage of proactive maintenance, which can improve the overall reliability of the earthing system.

It is crucial to acknowledge that choosing a particular earthing method is contingent upon several aspects, such as soil resistivity, installation type, and local legislation. Engaging in discussions with electrical engineers and strictly following applicable regulations and rules is essential when using contemporary earthing techniques. Furthermore, consistent testing and maintenance are crucial to guarantee the continuous efficiency of the grounding system.

Solved Numerical Problem

Example 5.1:

Determine the minimum rating for a circuit breaker required to protect a 5 kW, 230V single-phase motor with a power factor of 0.8 and an efficiency of 85%.

Solution:

Given: P = 5 kW, V = 230V, Power Factor (pf) = 0.8, Efficiency (η) = 85%

ate the current using $I = \frac{P}{\sqrt{3} \times V \times Pf \times \eta}$

Substitute the given values, $I = \frac{5000}{\sqrt{3} \times 230 \times 0.8 \times 0.85} \approx 16.83A$

Therefore, a circuit breaker rated at least 16.83 A is required.

Example 5.2:

A 3-phase motor draws a maximum current of 80A. Determine the minimum rating of a circuit breaker required for its protection.

Solution:

Given: Maximum current (I) = 80A

The circuit breaker rating should equal or exceed the maximum current. Therefore, an 80A circuit breaker is required.

Example 5.3:

A lighting circuit has a total load of 12kW, with a power factor of 0.9. Calculate the minimum rating of the circuit breaker for protection.

Solution:

Given: P = 12 kW, Power Factor (pf) = 0.9

Using the formula, $I = \frac{P}{\sqrt{3} \times V \times Pf}$, where V is the voltage

Assuming a voltage of 230V, $I = \frac{12000}{\sqrt{3} \times 230 \times 0.9} \approx 33.55A$

Therefore, a circuit breaker rated at least 33.55 A is required.

Example 5.4:

A 415V, 3-phase induction motor has a full load current of 30A. Determine the appropriate size of a circuit breaker to protect the motor.

Solution: Given: Voltage (V) = 415V, Full-load current (I) = 30A The circuit breaker should be rated equal to or slightly higher than the full load current. Therefore, a 30A or higher-rated circuit breaker is required.

Example 5.5:

A 240V, single-phase motor has a full load current of 20A. Determine the appropriate size of a circuit breaker to protect the motor.

Solution:

Given: Voltage (V) = 240V, Full-load current (I) = 20A The circuit breaker should be rated equal to or slightly higher than the full load current.

Therefore, a 20A or higher-rated circuit breaker is required.

Example 5.6:

A 480V, the 3-phase motor has a full load current of 40A. Determine the minimum rating of the circuit breaker required to protect the motor.

Solution:

Given: Voltage (V) = 480V, Full-load current (I) = 40A The circuit breaker should be rated equal to or slightly higher than the full load current.

Therefore, a 40A or higher-rated circuit breaker is required.

Example 5.7:

A 220V, single-phase motor has a full load current of 25A. Determine the appropriate size of a circuit breaker to protect the motor.

Solution:

Given: Voltage (V) = 220V, Full-load current (I) = 25A The circuit breaker should be rated equal to or slightly higher than the full load current.

Therefore, a 25A or higher-rated circuit breaker is required.

Example 5.8:

A 600V, the 3-phase motor has a full load current of 60A. Determine the minimum rating of the circuit breaker required to protect the motor.

Solution:

Given: Voltage (V) = 600V, Full-load current (I) = 60A The circuit breaker should be rated equal to or slightly higher than the full load current.

Therefore, a 60A or higher-rated circuit breaker is required.

Example 5.9:

A 240V, single-phase motor has a full load current of 15A. Determine the appropriate size of a circuit breaker to protect the motor.

Solution:

Given: Voltage (V) = 240V, Full-load current (I) = 15A The circuit breaker should be rated equal to or slightly higher than the full load current. Therefore, a 15A or higher-rated circuit breaker is required.

Example 5.10:

An electrical circuit has a total load of 10 kW and operates at 240V. Determine the minimum rating of the circuit breaker required for its protection.

Solution:

Given: P = 10 kW, V = 240V

Calculate the current using $I = \frac{P}{\sqrt{3} \times V}$

Substitute the given values, $I = \frac{10000}{\sqrt{3} \times 240} \approx 24.24A$

Therefore, a circuit breaker rated at least 24.24 A is required.

Exercise

Multiple Choice Questions:

1. What is the primary purpose of protection devices in electrical installations?

   a) To regulate voltage levels b) To prevent overload and short circuits c) To control the frequency of electrical power d) To increase the efficiency of electrical equipment
2. Which of the following is not commonly used protection device in electrical installations?

   a) Fuse b) Circuit breaker c) Voltage regulator d) Relays
3. What is the function of a fuse in an electrical circuit?

   a) To regulate voltage fluctuations b) To disconnect the circuit in case of overcurrent c) To amplify the electrical signal d) To prevent voltage surges
4. How does a circuit breaker differ from a fuse?

   a) A circuit breaker is reusable, while a fuse needs replacement after tripping.

   b) A fuse is manually operated, while a circuit breaker is automatically operated.

   c) A circuit breaker is only used in residential installations, while a fuse is used in industrial installations.

   d) A fuse has a higher breaking capacity than a circuit breaker.
5. What is the purpose of a ground fault circuit interrupter (GFCI)?

   a) To protect against overvoltage conditions b) To protect against ground faults and electric shocks c) To regulate the frequency of electrical power d) To prevent short circuits

6. what does "overload" refer to in an electrical installation?

   a) A situation where the voltage exceeds the maximum permissible limit

   b) A situation where the current exceeds the rated capacity of the circuit

   c) A situation where the frequency of the electrical power fluctuates

   d) A situation where there is a ground fault in the circuit

7. Which of the following conditions can lead to a short circuit?

   a) Excessive current flowing through the circuit b) Voltage fluctuations within the permissible range c) Grounding of the electrical equipment d) Use of high-quality insulation materials

8. What role do relays play in the protection of electrical installations?

   a) Relays regulate the voltage in the circuit. b) Relays protect against overload and short circuits by tripping the circuit. c) Relays provide backup power in case of a blackout. d) Relays regulate the frequency of electrical power.

9. What is the purpose of coordination between protection devices in an electrical system?

   a) To ensure that protection devices operate independently of each other

   b) To ensure that protection devices operate in a coordinated sequence to isolate faults

   c) To reduce the efficiency of the protection system

   d) To increase the likelihood of electrical accidents

10. Which of the following factors should be considered when selecting a protection device for an electrical circuit?

    a) The color of the protection device b) The cost of the protection device c) The rated current and voltage of the circuit d) The location of the circuit breaker in the building

11. What is the primary purpose of an arc-fault circuit interrupter (AFCI)?

    a) To regulate voltage fluctuations b) To protect against ground faults and electric shocks c) To prevent arc faults that can lead to fires d) To control the frequency of electrical power

12. Which of the following is a characteristic of a residual current device (RCD)?

    a) It protects against overvoltage conditions b) It detects insulation faults in the circuit c) It prevents short circuits in the electrical system d) It trips the circuit if it detects a leakage current to the ground

13. what does "discrimination" refer to in electrical installations?

a) The process of selecting protection devices based on their color b) The coordination of protection devices to isolate faults effectively c) The adjustment of voltage levels in the electrical system d) The use of relays to regulate the frequency of electrical power

14. What is the advantage of using a circuit breaker over a fuse?

    a) Circuit breakers are less expensive than fuses, b) Circuit breakers offer better protection against short circuits, c) Fuses can be reset multiple times, while circuit breakers need replacement after tripping, and d) Fuses are more durable and long-lasting compared to circuit breakers.

15. What is the function of a surge protector in an electrical system?

    a) To regulate the voltage in the circuit b) To prevent short circuits c) To protect against sudden increases in voltage caused by lightning or power surges d) To regulate the frequency of electrical power.

16. Which of the following conditions can lead to an overload in an electrical circuit?

    a) A sudden increase in voltage b) Excessive current flowing through the circuit c) A ground fault in the circuit d) The presence of high-quality insulation materials

17. What is the purpose of zone-selective interlocking (ZSI) in electrical protection systems?

    a) To prevent the occurrence of overcurrents b) To coordinate the operation of protection devices in different zones of the electrical system c) To regulate the frequency of electrical power d) To increase the likelihood of electrical accidents

18. Which of the following protection devices is typically used to protect sensitive electronic equipment?

    a) Fuse b) Surge protector c) Circuit breaker d) Residual current device (RCD)

19. What does the time-current characteristic of a protection device indicate?

    a) The voltage rating of the protection device b) The coordination between different protection devices c) The relationship between the operating time and the magnitude of current for the device d) The cost of the protection device.

20. Which of the following types of faults is typically detected by a differential protection relay?

    a) Overload b) Short circuit c) Ground fault d) Voltage fluctuation

Answer:

1. b) To prevent overload and short circuits
2. c) Voltage regulator

3. b) To disconnect the circuit in case of overcurrent
4. a) A circuit breaker is reusable, while a fuse needs replacement after tripping.
5. b) To protect against ground faults and electric shocks
6. b) A situation where the current exceeds the rated capacity of the circuit
7. a) Excessive current flowing through the circuit
8. b) Relays protect against overload and short circuits by tripping the circuit.
9. b) To ensure that protection devices operate in a coordinated sequence to isolate faults
10. c) The rated current and voltage of the circuit
11. c) To prevent arc faults that can lead to fires
12. d) It trips the circuit if it detects a leakage current to the ground
13. b) The coordination of protection devices to isolate faults effectively
14. b) Circuit breakers offer better protection against short circuits
15. c) To protect against sudden increases in voltage caused by lightning or power surges
16. b) Excessive current flowing through the circuit
17. b) To coordinate the operation of protection devices in different zones of the electrical system
18. b) Surge protector
19. c) The relationship between the operating time and the magnitude of current for the device
20. c) Ground fault

Short Answer Type Questions:

1) What is the purpose of overload protection in electrical circuits?
2) Define short circuits and explain why they are hazardous to electrical systems.
3) What is the function of a fuse in an electrical circuit?
4) Describe the operation of a circuit breaker during an overcurrent event.
5) Explain the difference between a thermal-magnetic circuit breaker and an electronic circuit breaker.
6) What is the role of a residual current device (RCD) in electrical safety?
7) How does a ground fault circuit interrupter (GFCI) protect against electric shocks?
8) Define arc fault and explain the significance of arc fault detection in electrical systems.
9) What are surge protectors, and how do they safeguard electrical devices?
10) Discuss the coordination between protection devices in an electrical system.
11) Explain the concept of discrimination in electrical protection systems.

12) What is the significance of time-current characteristics in protection devices?

13) Describe the operation of a differential protection relay.

14) How does zone-selective interlocking enhance the performance of protection systems?

15) Discuss the importance of periodic electrical installation maintenance and testing of protection devices.

Long Answer Type Questions:

1) Discuss the importance of protection devices in electrical installations. Explain the consequences of inadequate protection and the risks associated with overcurrent events.

2) Explain the operation of a fuse in detail. Describe the different types of fuses and their applications in electrical systems. Discuss the advantages and limitations of fuses as protection devices.

3) Compare and contrast thermal-magnetic circuit breakers and electronic circuit breakers. Discuss their construction, operation principles, and applications in electrical installations.

4) Describe the function of residual current devices (RCDs) in electrical safety. Explain how RCDs detect leakage currents and discuss their role in protecting against electric shock hazards.

5) Discuss the principles of operation of ground fault circuit interrupters (GFCIs). Explain how GFCIs detect ground faults and describe their applications in residential and commercial electrical systems.

6) Define arc faults and explain their associated risks in electrical installations. Describe the operation of arc fault circuit interrupters (AFCIs) and discuss their role in preventing electrical fires.

7) Explain the concept of coordination between protection devices in electrical systems. Discuss the factors influencing coordination and the methods used to achieve effective coordination.

8) Describe the purpose and operation of surge protectors in electrical installations. Explain how surge protectors mitigate the effects of voltage surges and transient overvoltages.

9) Discuss the principles of operation of differential protection relays. Explain how differential relays detect internal faults in electrical equipment and discuss their applications in transformer and motor protection.

10) Explain the concept of selective coordination in protection systems. Discuss the importance of selective coordination in minimizing downtime and improving the reliability of electrical installations. Provide examples of selective coordination schemes in practical applications.

Numerical Problems:

1. A 240V single-phase motor has a full load current of 30A. Calculate the minimum rating of the circuit breaker required to protect the motor. (30A)
2. Determine the maximum allowable load for a 20A circuit protected by a fuse. The voltage rating of the circuit is 120V. (2400 watt)
3. A 480V, three-phase motor draws a maximum current of 60A. Determine the appropriate size of the circuit breaker required to protect the motor. (60A)
4. Calculate the total load in a residential panel board protected by a 100A main circuit breaker. The loads include lighting (10A), HVAC (30A), kitchen appliances (40A), and general outlets (20A). (100A)
5. A lighting circuit operates at 120V and has a total load of 2 kW. Determine the minimum rating of the circuit breaker required for its protection. (16.67A)
6. Calculate the total load demand for a 3-phase electrical installation with three motors rated at 10 kW, 8 kW, and 12 kW, respectively. The system voltage is 415V. (54KW)
7. Determine the maximum current to flow through a 14-gauge copper wire protected by a 15A circuit breaker. (15A)
8. A 3-phase transformer has a primary voltage of 480V and a secondary voltage of 240V. If the primary current is 50A, calculate the secondary current. (100A)
9. Determine the rating of the overcurrent protection device required for a 240V, 3-phase induction motor with a full load current of 80A. (80A)
10. A residential panel board has a total connected load of 15 kW and operates at 240V. Calculate the minimum rating of the main circuit breaker required for its protection. (62.5A)

Unit Summary

The unit summary for the protection of electrical installation is written below

- ✓ Introduction to Electrical Installation Protection
- ✓ Types of Protective Devices
- ✓ Overcurrent Protection
- ✓ Earth Fault Protection

✓ Transient Overvoltage Protection

✓ Protection Coordination and Selectivity

✓ Testing and Maintenance of Protective Devices

✓ Safety Standards and Regulations

This unit provides a comprehensive understanding of the principles, devices, and systems that protect electrical installations, emphasizing safety, reliability, and compliance with industry standards and regulations.

Practical

1. Conduct a practical test to demonstrate the tripping action of an RCD by simulating an earth fault using a test button or fault simulator.
2. Demonstrate how to select the appropriate rating for a circuit breaker or fuse based on the connected load and wiring specifications.
3. Create a protection coordination diagram for a distribution system that includes various circuits and protective devices such as circuit breakers and fuses.
4. Demonstrate the testing of surge protection devices (SPDs) using surge generators or transient voltage sources to simulate lightning or switching surges.
5. Conduct insulation resistance testing on a circuit to verify the insulation integrity and safety of the installation.
6. Perform trip testing on a circuit breaker to ensure its proper operation and tripping characteristics under overload or short circuit conditions.
7. Identify the key responsibilities of electrical professionals in ensuring compliance with safety standards during the design, installation, and maintenance of electrical systems.
8. Create and simulate fault scenarios such as overloads, short circuits, or earth faults in a circuit. Ask students to analyze the fault and determine the appropriate protective device response.

Know more

Sl. No	Topic	QR Code
1	National Electrical Code of India-2023	
2	Electrical Safety and National Electrical Code 2011	
3	Electrical Switchgear and Other Electrical Products/Accessories	
4	Electrical Protection and Control	

References and Suggested Readings

1. J.M. Gers, and E.J. Holmes, *Protection of Electricity Distribution Networks*, Vol. 47, IET, 2004.
2. C.L. Wadhwa, *Electrical Power Systems*, New Age International, 2006.
3. G. Stokes, *Handbook of Electrical Installation Practice*, John Wiley & Sons, 2008.
4. B. Scaddan, *Electrical Installation Work*, Routledge, 2011.
5. B. Atkinson, R. Lovegrove, and G. Gundry, *Electrical Installation Designs*, John Wiley & Sons, 2013.

6. C. Christopoulos, and A. Wright, *Electrical Power System Protection*, Springer Science & Business Media, 2013.
7. S. Corsi, *Voltage Control and Protection in Electrical Power Systems: from System Components to Wide-Area Control*, Springer, 2015.
8. J.A. Momoh, *Electric Power Distribution, Automation, Protection, and Control*, CRC press, 2017.
9. Y.G. Paithankar, and S.R. Bhide, *Fundamentals of Power System Protection*, PHI Learning Pvt. Ltd, 2022.
10. P.M. Anderson, C.F. Henville, R. Rifaat, B. Johnson, and S. Meliopoulos, *Power System Protection*, John Wiley & Sons, 2022.
11. F. Asadi, *Electric Circuits Laboratory Manual*, Springer Nature, 2023.

UNIT 6
ILLUMINATION IN RESIDENTIAL INSTALLATION

UNIT SPECIFICS

This unit elaborately discusses the following topics:

- ✓ Introduction to the concept of illumination
- ✓ Discussion on laws of illumination
- ✓ Exploring the factors of illumination
- ✓ Exploring various types of light sources

Illumination is an essential part of lighting design, and engineering is crucial in creating functional, comfortable, and aesthetically pleasing environments. Luminous efficiency, space-height ratio, flux, luminous intensity, lumen, illumination, illuminance, lux, utilisation factor, depreciation factor, and luminous efficiency are all defined in this unit. It also discusses the significance of laws of illumination, such as the inverse square law and cosine law, and their different aspects. This unit also discussed different types of lighting sources and their various arrangements.

RATIONALE

This unit focuses on illumination in residential installations, which stems from its critical role in providing occupants with comfortable, functional, and safe living environments. Illumination design in residential installations must adhere to relevant standards and building codes to ensure safety, efficiency, and compliance with regulatory requirements.

Considerations such as lighting levels, uniformity, colour rendering, glare control, and accessibility for maintenance are addressed in standards like the National Electrical Code (NEC) and local building regulations. While ensuring adequate illumination is essential, optimizing energy efficiency is equally crucial. Using energy-efficient lighting fixtures, such as LED (Light-Emitting Diode) bulbs, CFLs (Compact Fluorescent Lamps), and intelligent lighting controls, residential installations can minimize energy consumption and reduce electricity bills without compromising lighting quality.

PRE-REQUISITES

Basics of Electrical Engineering

UNIT OUTCOMES

The list of outcomes of this unit is as follows:

U6-O1: Describe various definitions of illumination

U6-O2: Explain the laws of illumination

U6-O3: Explore different factors of illumination

U6-O4: Describe various types of light sources and lighting arrangements

Unit-6 Outcomes	Expected Mapping with Programme Outcomes (*1- Weak Correlation; 2- Medium correlation; 3- Strong Correlation*)					
	CO-1	CO-2	CO-3	CO-4	CO-5	CO-6
U6-O1	3	3	3	3	1	2
U6-O2	2	2	3	2	1	2
U6-O3	3	3	2	3	1	2
U6-O4	3	3	3	3	1	2

6.1 Introduction:

Illumination in residential installations pertains to artificial lighting to guarantee sufficient visibility and provide a pleasant living environment within a home. Optimal lighting design in home settings is crucial for multiple reasons, encompassing safety, aesthetics, and functionality. The following are essential factors and elements to consider about lighting in home installations:

6.1.1 Types of Lighting:

1) **Ambient Lighting:** Ambient lighting illuminates a room, making moving and completing tasks easier. Ceiling-mounted fixtures, chandeliers, and recessed lighting are familiar sources of illumination.
2) **Task Lighting:** Task lighting is directed towards particular places where specific activities like working, reading, or cooking. Illustrative instances encompass desk lamps and pendant lights positioned over kitchen islands.
3) **Accent Lighting:** Accent lighting highlights specific features or items, creating visual attention. This may comprise track lighting, wall-mounted fixtures, or spotlights.
4) **Natural Lighting:** Optimizing the presence of sunlight by strategically positioning windows, skylights, and glass doors can augment the overall brightness of a residential area and add to a delightful ambiance.

6.1.2 Lighting Design Considerations:

- ✓ **Room Functionality:** When designing lighting, consider each room's specific purpose. Various spaces, such as the living room, kitchen, bedroom, and bathroom, may possess unique lighting needs.
- ✓ **Lighting Levels:** Ensure optimal illumination levels for various tasks. For instance, specific places such as kitchen countertops or reading nooks may require more intense lighting than the overall living spaces.
- ✓ **Colour Temperature:** Select light sources that have suitable colour temperatures. Living rooms frequently benefit from warmer tones, which have lower Kelvin, while task-oriented areas may be better suited for cooler tones, which have higher Kelvin.
- ✓ **Fixture Styles:** Choose fittings that harmonize with the overall design and decor of the home. Consider the design, finish, and size of fixtures to consider the room's aesthetics.

- ✓ **Energy Efficiency:** Implement energy-effective lighting options, such as LED bulbs, to reduce energy usage and utility expenses.
- ✓ **Dimmers and Controls:** Implement dimmer switches to regulate illumination levels according to the time of day or specific tasks. Intelligent lighting controls can provide more versatility and ease of use.

6.1.3 Room-Specific Considerations:

- ✓ **Kitchen:** Ensure sufficient task lighting in food preparation facilities. Installing under-cabinet lighting, pendant lights above islands, and recessed lights can improve visibility.
- ✓ **Living Room:** Utilize a blend of ambient illumination and task illumination to facilitate reading or highlight artwork. Floor lamps, table lights, and wall sconces enhance the ambiance by creating a warm and inviting setting.
- ✓ **Bedroom:** Integrate a combination of ambient and task lighting. Bedside table lamps, wall-mounted reading lights, or pendant lights can establish a soothing atmosphere.
- ✓ **Bathroom:** Provide bright, shadow-free lighting surrounding the vanity mirror for grooming needs. Consider installing waterproof fixtures and sconces to increase safety.
- ✓ **Home Office:** Provide adequate task illumination for work surfaces. Desk lighting and adjustable settings can help decrease eye strain and increase productivity.

6.1.4 Lighting Controls and Automation:

- ✓ **Dimmers:** Dimmer switches enable occupants to control lighting settings, creating a varied and customizable environment.
- ✓ **Smart Lighting:** Intelligent lighting solutions provide the remote manipulation and automation of lighting using smartphones or voice commands. This enhances convenience and promotes energy efficiency.
- ✓ **Timers and Sensors:** Timers and motion sensors can be utilized to automate lighting, activating and deactivating lights according to occupancy or predetermined schedules.

6.1.5 Safety Considerations:

a) **Emergency Lighting:** Implement emergency lighting installation in strategic locations to guarantee optimal visibility in case of a power failure. This may encompass signage indicating exits and emergency lights powered by batteries.

b) **Exterior Lighting:** Enhance nighttime safety and security by illuminating outside pathways, entryways, and vital places around the property.

c) **Stair Lighting:** It is advisable to place lights on staircases to enhance safety. Options for lighting can encompass step lights or bulbs set on walls.

6.1.6 Regulatory Compliance:

✓ **Building Codes:** Comply with local construction norms and regulations for lighting systems to ensure adherence to safety requirements.

✓ **Energy Codes:** Explore energy-efficient lighting choices that comply with energy codes and support sustainability goals.

An intelligently crafted lighting scheme in residential settings enhances the living area's comfort, attractiveness, and functionality. It is crucial to consider each area's requirements, integrate energy-saving solutions, and apply lighting controls to create a personalized and flexible environment. Furthermore, adherence to safety standards and building rules is vital to guarantee a secure and protected household lighting installation.

6.2 Definition:

6.2.1 Concepts of Luminous Flux (Φ):

The study of lighting and photometry relies on the idea of luminaous flux. A light source's total incident light intensity is the amount of visible light it emits in all directions. In other words, it measures how much light the eye can perceive. The precise measurement of the brightness emitted by a light source is conveyed by luminaous flux, which is measured in lumens (lm).

✓ **Definition: Luminous Flux (Φ)** denotes the quantity of visible light discharged by a light source within a specific period. The unit of measurement for brightness is lumens (lm).

✓ **Explanation:** Luminous flux measures the illumination intensity produced by a light source. More excellent luminous flux results in increased seeming illumination to the human eye. It considers the complete range of visible light, focusing on the overall amount of light produced.

6.2.2 Luminous Intensity (I):

Luminous intensity is a photometric parameter that quantifies the quantity of visible light emitted or transmitted in a specific direction by a light source. Luminous intensity is a crucial factor in lighting design and is quantified in candelas (cd). Luminous intensity quantifies the level of brightness emitted by a light source as observed by the human eye in a particular direction.

- ✓ **Definition:** Luminous intensity (I) quantifies the magnitude of luminous flux emitted in a specific direction. The unit of measurement is candelas (cd).
- ✓ **Explanation:** Luminous intensity is a degree of the quantity of light concentrated in a specific direction. A greater luminous intensity signifies the concentration of light in a specific region, enhancing the perception of brightness in that particular direction.

6.2.3 Candela (cd):

- ✓ **Definition:** One common way to measure the brightness of a light source is in candelas. One candela is the luminosity of a single-color light source having a 540 terahertz frequency and 1/683 watts per steradian radiant intensity.
- ✓ **Symbol:** cd

Relationship to Luminous Flux:

Luminous intensity is related to luminous flux(Φ) by the equation: $I = \Phi/\Omega$

Where:

I is luminous intensity in candelas(cd).

Φ is luminous intensity in lumens (lm).

Ω is the solid angle in steradians (sr).

6.2.4 Lumen:

The lumen (symbol: lm) is the unit of amount used to quantify the total luminous flux emanated by a light source per unit of time. Luminance is a photometric unit that incorporates the human eye's sensitivity to various light wavelengths, prioritizing light's perceived brightness rather than its overall energy production. Lumens quantify the luminosity of light as experienced by the human visual system.

- ✓ **Definition:** The lumen is the widely recognised unit of measurement for the amount of light that is transmitted. The concept of luminosity flux refers to the total quantity of visible light emitted by a light source in all directions.
- ✓ A lumen refers to the luminous intensity emitted by a light source, which uniformly distributes one candela of light over a solid angle of one steradian.
- ✓ A lumen is equivalent to one candela-steradian (cd·sr).
- ✓ **Explanation:** The lumen is a fundamental unit employed for quantifying the efficacy of a light source in emitting visible light. Luminance is a measure of the perceived brightness as experienced by the human eye, irrespective of the direction of the emitted light.

Relation to Candela and Steradian:

The formula gives the relationship between lumens, candelas, and steradians:

Luminous Flux(lm) – Luminous Intensity(cd)×solid Angle(sr)

6.2.5 Solid Angle:

A solid angle quantifies how much a three-dimensional angle covers or subtends space. In geometry and physics, this term denotes the magnitude of a region resembling a cone or pyramid in three-dimensional space. Solid angle is a crucial notion in optics and illumination, as it enables accurate measurement of the amount of light emitted or received by a surface. The unit of measurement for solid angle is the steradian (sr).Here are critical concepts related to solid angle:

✓ **Definition:**

- **Solid Angle (Ω):** The solid angle is precisely defined as the magnitude or dimension of the three-dimensional angle that arises when the sides of a cone or pyramid intersect at a single point. The unit of measurement is steradians (sr).
- **Definition:** The steradian is the fundamental unit of measurement for solid angles. A steradian is the unit of solid angle formed when a sphere is intersected by two cones sharing a vertex so that the intersection area on the sphere is equal to the square of the sphere's radius.
- **Symbol:** sr

Mathematical Representation:

The formula for calculating solid angle (Ω) in terms of the subtended area (A) and the square of the radius (r^2) is given by: $\Omega = A/r^2$

6.2.6 Illumination or Illuminance

Illuminance quantifies the light intensity that reaches a surface per unit area. Luminance is a vital factor in lighting design and photometry as it offers precise information about the brightness of a given environment. Illuminance is commonly quantified in lux (lx), corresponding to one lumen per square meter. The formula for illuminance (E) is:

E = Luminous flux (lm) / Area(m^2)

Here's a breakdown of the critical components:

1. **Luminous Flux (lm):** Luminous flux is a measure that quantifies the total amount of visible light emitted by a source of illumination. Lumens (lm) is used to represent the measurement.
2. **Area (m²):** The region where the luminous flux is spread out. The unit of measurement is square meters (m²).
3. **Lux (lx):** Lux is a measurement of illuminance, defined as one lumen per square meter.

- **Recommended Illuminance Levels (Approximate Values):**

✓ **General Office Space:** 300 - 500 lux

✓ **Classrooms:** 300 - 500 lux

✓ **Supermarkets:** 750 - 1,000 lux

✓ **Reading or Writing:** 500 lux

✓ **Precision or Detailed Work (e.g., Laboratories):** 1,000 - 2,000 lux or more

✓ **Street Lighting:** Varies based on the type of street and safety requirements

✓ **Outdoor Daylight:** Approximately 10,000 lux (on a sunny day)

6.2.7 Lux:

Illuminance, defined as the quantity of visible light (luminous flux) falling on a surface as a function of area, is commonly measured in lux (symbol: lx). When the quantity of light falling on a specific surface is of the utmost importance, illumination becomes an integral aspect of lighting design. Light intensity, measured in lux, is one lumen per square metre. The formula for illuminance (*E*) is:

$$E = \Phi/A$$

Where:

✓ E is illuminance in lux(lx),

✓ Φ is luminous in lumens(lm),

✓ A is the area of the surface in square meters(m²)

- **Definition:** Illuminance, quantified in lux (lx), denotes the quantity of light flux on a particular area. It measures the intensity of light that is incident on a surface.
- **Explanation:** Lux is essential for comprehending an observer's perceived brightness of a surface or region. It considers the overall luminous flux and the surface area across which the light is spread. Lux is commonly employed to delineate the lighting needs for various purposes.

- **Units Relationship:**

  One lux is equivalent to one lumen per square meter:

  1 lux = 1 lumen/1 square meter

6.2.8 Space-Height Ratio:

✓ **Definition:** The space-height ratio is a metric that evaluates the correlation between a space's vertical dimension and its illumination dispersion.

✓ **Explanation:** The space-to-height ratio is significant in architectural and lighting design. It impacts the uniform distribution of light throughout various regions. When dealing with areas that have tall ceilings, it can be not easy to achieve even lighting. Designers can use the space-height ratio to optimize the distribution of light.

✓ **Key Considerations:**

- **Application in Lighting Design:** These concepts are fundamental in lighting design. Designers utilize luminous flux and intensity to ascertain suitable brightness for various places and activities.
- **Real-world Example:** When choosing a light bulb for a reading lamp, it is essential to examine the luminous flux, which refers to the total amount of light emitted, and the illuminance, which measures the brightness of the light on the reading surface. These factors are crucial in ensuring the illumination is sufficient for the task.

Comprehending these essential lighting ideas is crucial for creating lighting systems that fulfill specific residential, commercial, or industrial requirements.

6.2.9 Utilization factor:

The Utilization Factor, commonly referred to as UF, is a metric employed in lighting design to evaluate the effectiveness of a lighting installation in delivering and utilizing the luminous flux (light output) generated by the light sources. This considers the reflecting characteristics of the surfaces present in a given area, along with the effectiveness of the lighting fixtures in directing light to the desired locations. The utilization factor is denoted as a percentage and is determined by applying the following formula:

UF = (Utilized Luminous Flux/Total Luminous Flux)×100%

✓ **Definition:** The utilization Factor is a quantitative measure that indicates the proportion of the overall luminous flux generated by light sources that are efficiently employed to illuminate the designated work surface or area of operation. The utilization factor is a metric that quantifies the effectiveness of a lighting system in utilizing light within a particular area. It considers variables such as reflectance, room geometry, and luminaire effectiveness.

✓ **Explanation:**

- Reflectance refers to the ability of surfaces in a given area to absorb or reflect light, influencing the amount of light in that space. Increased reflectivity leads to an elevated utilization factor.
- **Room Shape**: A room's spatial configuration affects light distribution. Rooms with non-uniform shapes can lead to unequal light distribution, impacting the efficiency of light utilization.
- **Luminaire Efficiency**: The efficiency of the light fixtures utilized in the system directly affects the utilization factor. Well-designed luminaires enhance efficiency.

✓ **Key Considerations:**

- **Optimizing Utilization:** Designers strive to maximize the utilization factor by considering variables such as room reflectance, luminaire selection, and the overall design of the lighting system.
- **Maintenance Strategies:** Considering the depreciation factor helps formulate maintenance methods that ensure lighting systems consistently satisfy illuminance needs throughout their lifespan.
- **Energy-Efficient Design:** Energy-efficient design maximizes luminous efficiency in lighting systems to minimize power consumption while ensuring sufficient illumination.

I is comprehending and enhancing these crucial factors aid in developing efficient, eco-friendly, and enduring lighting designs for diverse applications.

6.2.10 Depreciation Factor:

In lighting design and calculations, the Depreciation Factor is a coefficient employed to compensate for the decline in illuminance levels over time caused by factors including lamp depreciation, dirt buildup on luminaires, and various climatic conditions. As lighting installations age, the brightness of lights often diminishes, reducing the intensity of illumination. The Depreciation Factor is utilized to modify the initial design illuminance values to account for the anticipated decline in light output as time progresses.

The formula for calculating illuminance (E) with the Depreciation Factor (DF) is

$I_{corrected} = E_{initial} \times DF$

Here are key points related to the Depreciation Factor:

✓ **Definition:** The Depreciation Factor is a scalar multiplied by the starting illuminance level to compensate for the gradual reduction in light production and the influence of ambient conditions. The depreciation factor quantifies the decrease in luminous flux over time caused by factors such as lamp aging, dirt collection, and environmental conditions.

✓ **Explanation:**

- **Lamp Lumen Depreciation**: As time passes, lamps undergo a reduction in the amount of light they emit, which is caused by factors such as the wearing down of the filament in incandescent lamps or the deterioration of the phosphor in fluorescent lamps.
- **Environmental Conditions:** Factors such as temperature, humidity, and vibrations can impact the durability and efficiency of bulbs, hence impacting their depreciation rate.
- **Maintenance:** Routine maintenance, such as cleansing fixtures and replacing deteriorating bulbs, reduces the effects of depreciation.

6.2.11 Luminous Efficiency:

A light source's ability to produce visible light is called its luminous efficiency or luminous efficacy. The efficiency with which a light source transforms electrical energy into light that can be seen is called its luminance efficiency. Luminous efficacy is a way to assess how efficient a light source is in terms of energy consumption. It is calculated as the ratio of lumens to watts (lm/W). A greater luminous effectiveness signifies that a light source generates a more significant amount of visible light relative to the electrical power consumed.

The formula for calculating luminous efficiency:

Luminous Efficacy (η) = Luminous Flux (Φ)/Power Input (P)

✓ **Definition:** Luminous efficiency quantifies the effectiveness with which a light source transforms electrical energy into visible light. The luminous efficacy is defined as the quotient of radiant flux (measured in lumens) divided by power input (measured in watts).

✓ **Explanation:**

- **Energy Conversion:** Luminous efficiency measures the efficiency of a light source in turning electrical energy into visible light. Greater luminous efficiency indicates a higher amount of light produced for each unit of electrical power consumed.
- **Energy Conservation**: Using efficient light sources aids energy conservation by generating a more significant amount of light while consuming less power. LED technology is renowned for its exceptional luminous efficiency.
- **Sustainability**: In sustainable lighting design, luminous efficiency is key since it helps with efforts to reduce energy consumption and lessen environmental damage.

IS 16106 (2012): Method of Electrical and Photometric Measurements

6.3 Laws of Illumination

The illumination rules delineate the behavior and interaction of light with surfaces, offering fundamental concepts applicable in the domains of lighting design, photography, computer graphics, and numerous other disciplines. These rules facilitate the prediction and comprehension of how light illuminates objects and surfaces. Below are a few fundamental principles governing the phenomenon of illumination:

6.3.1 Inverse Square Law:

Explanation:

- ✓ In the context of light and illumination, the Inverse Square Law states that as the square of the distance from the light source to the place of measurement increases, the light's brightness decreases proportionally. There is an inverse relationship between the square of the distance from a point source of light and its illuminance, or brightness.
- ✓ The inverse square law is mathematically formulated as $I \alpha 1/r^2$, where I is inversely proportional to the square of the distance r.
- ✓ Mathematically, the law is expressed as $I = P/4\pi r^2$

Where:

I stands for the light's illuminance or intensity,

P for the light source's radiant flux or luminous power (total power output)

r for the distance from the light source,

π for pi, a mathematical constant(approximately 3.14159)

- **Significance:**

✓ As light propagates at increasing distances from its origin, it disperses across a greater surface area, reducing illuminance. Understanding how light levels decrease with distance is essential for understanding this law.

✓ Comprehending and utilizing the Inverse Square Law is crucial for attaining the intended degrees of illumination in diverse fields, such as architectural lighting, photography, and astronomy. It offers a numerical foundation for forecasting the variation in illuminance when the distance from the light source changes.

6.3.2 Cosine Law:

The Cosine Law, also known as Lambert's Cosine Law, is a fundamental principle in physics and optics that specifies the relationship between a surface's illuminance (brightness) and the angle of incidence of light. The law holds significant importance within the realm of lighting design and photometry. The term "Lambert's Cosine Law" is derived from Johann Heinrich Lambert, who initially formulated this principle during the 18th century.

- **Explanation:**

✓ The Cosine Law is used to compute the illuminance on a surface by considering the angle of incidence between the direction of light and the surface's normal (perpendicular).

✓ Mathematically, it is expressed as follows

The cosine law is mathematically expressed as follows:

$E = E_0 . \cos(\theta)$

Where:

E is the illuminance on the surface,

E_0 represents the surface illuminance when the light is normal, or perpendicular, to the surface, and θ represents the incidence angle between the average direction of the incident light and the surface.

- **Significance:**

✓ The Cosine Law is essential in architectural and lighting design, particularly when considering the interaction of light with surfaces at different angles. It facilitates the attainment of consistent illumination dispersion.

✓ Understanding and comprehending the Cosine Law is essential for effective forecasting and devising lighting conditions in diverse applications. It aids in maximizing the evenness of light dispersion on surfaces by considering the direction and angle at which light sources hit them.

6.3.3 Illumination Received directly underneath:

- **Explanation:**

✓ Illumination Received is determining the amount of light that falls on a surface, considering both the light source's brightness and the surface's size.

✓ Mathematically, it is expressed as

$E = \Phi/A$

Where:

E is illuminance in lux(lx),

Φ is luminous in lumens (lm) emitted by a light source,

A is the area in square meters (m^2)

- **Significance:**

✓ To calculate the illuminance, it is necessary to have information on the luminous flux of the light source and the surface area that the light is being directed towards.

✓ This legislation is crucial for establishing the precise degree of illuminance necessary for particular tasks or in various situations. It guarantees that the lighting design adheres to the requisite criteria for visibility and comfort.

- **Key Considerations:**

✓ **Practical Application:** Comprehending these rules is essential for lighting designers to ensure that the design satisfies the illuminance prerequisites for a particular area.

✓ **Architectural Considerations:** Architectural lighting design involves adhering to the illumination principles to determine the best placement and alignment of light fixtures to generate specific lighting effects.

✓ **Energy Efficiency:** Complying with these regulations enables designers to maximize the efficiency of lighting systems, guaranteeing sufficient brightness while minimizing energy usage.

To summarize, the laws of illumination establish the fundamental principles for constructing lighting systems that efficiently illuminate environments, considering parameters such as distance, angles, and luminous flux.

6.4 Factors Affecting Illumination:

Many different factors influence the amount of light in a room. These factors include the lighting systems' design and the characteristics of the surroundings. Below is a comprehensive analysis of the primary factors:

- **Light Source Characteristics:**

✓ **Luminous Flux:** The aggregate quantity of visible light discharged by a light source.

✓ **Color Temperature:** The degree of warmth or coolness of light, quantified in Kelvin (K). It has an impact on the perceived hue of items.

✓ **Color Rendering Index (CRI):** Measures how well the light source mimics natural light in terms of colour reproduction.

- **Fixture and Luminaire Design:**

✓ **Type of Fixture:** Various fixtures (recessed, pendant, and track lighting) offer diverse lighting distributions and effects.

✓ **Luminaire Efficiency:** The luminaire's ability to distribute light efficiently and minimize losses.

✓ **Light Distribution Patterns:** How light is dispersed, concentrated, spread out, or directed.

- **Room and Surface Characteristics:**

✓ **Room Size and Shape:** Affect how light is distributed and how evenly it is spread throughout the space.

✓ **Reflectance of Surfaces:** The hue and reflectivity of walls, ceilings, and floors impact the amount of light absorbed or reflected.

- **Human Factors:**

✓ **Task Requirements:** The illuminance levels needed in space are determined by the nature of the tasks. Reading, cooking, and precise jobs require distinct requirements.

✓ **Age and Vision:** As individuals age, their eyes may necessitate increased illumination levels. Visual tasks differ among persons.

- **Natural Lighting:**

✓ **Daylight Availability:** The quantity of natural light that enters a room impacts the requirements for artificial lighting.

✓ **Window Size and Placement:** Ensuring that windows are appropriately sized and positioned contributes to achieving a well-balanced illumination.

- **Lighting Control Systems:**

✓ **Dimming Systems:** allow for the adjustment of light levels according to specific requirements, hence promoting energy efficiency

✓ **Smart Lighting Controls:** Integration with automation systems to enable dynamic control and scheduling.

- **Environmental Factors:**

✓ **Ambient Light Levels:** The brightness of artificial lighting is influenced by the existing light levels in the environment.

✓ **Light Pollution:** Unwanted or excessive artificial light that can negatively affect visibility and disturb natural light cycles.

- **Energy Efficiency and Sustainability:**

✓ **Type of Light Source:** Energy-efficient light sources like LEDs minimize electricity usage.

✓ **Lighting Design:** Expertly crafted systems maximize the amount of light produced while minimizing energy use and environmental adverse effects.

- **Safety and Regulations:**

✓ **Emergency Lighting Requirements:** Specifies the necessary illumination levels to be maintained during power outages or other emergencies.

✓ **Regulatory Standards:** Compliance with prescribed regulations and standards for specific applications or areas.

- **Maintenance Considerations:**

✓ **Reliability:** The durability and consistency of light sources and fixtures over an extended period.

✓ **Cleaning and Upkeep:** Consistent maintenance to guarantee peak performance and cleanliness.

- **Adaptability and Flexibility:**

✓ **Adjustability:** the capacity to modify illumination settings to suit various activities or moods.

✓ **Future Modifications:** Implementation of a design that facilitates effortless modifications or additions to the lighting system.

Comprehending these aspects is essential for lighting designers to construct surroundings that fulfill functional necessities, improve visual comfort, and correspond with energy efficiency objectives. The interaction of these factors enhances the overall achievement of a lighting design project.

6.5 Types of Lighting Arrangements:

Lighting arrangements involve deliberately positioning and arranging light fixtures in a given area to accomplish particular visual, functional, and aesthetic objectives. Various lighting configurations are utilized to provide different effects and fulfill multiple needs. Presented below are several prevalent lighting configurations, each designed to achieve a distinct objective:

1. **Ambient Lighting:**

✓ **Purpose:** Offers comprehensive lighting to establish a pleasant and brightly illuminated ambiance across an area.

✓ **Fixtures:** Ceiling-mounted fixtures, recessed lights, chandeliers, and wall sconces.

✓ **Distribution:** Uniformly disperses light to eradicate shadows and establish a pervasive feeling of luminosity.

2. **Task Lighting:**

✓ **Purpose:** Focuses light on specific work areas or tasks, providing ample illumination for home activities.

✓ **Fixtures:** Desk lamps, pendant lights, under-cabinet lights, and adjustable track lighting.

✓ **Placement:** Provides focused illumination to the task area while reducing glare and shadows.

3. **Accent Lighting:**

✓ **Purpose:** Accent lighting serves to illuminate and draw attention to particular items, architectural elements, or artworks, enhancing their visual appeal and emphasizing their significance as focal points.

✓

✓ **Fixtures:** Spotlights, track lighting with adjustable heads, wall-mounted fixtures.

✓ **Techniques:** Utilizes highlighting specific aspects while leaving others in obscurity to create contrasting effects.

4. **Decorative Lighting:**

✓ **Purpose:** Enhances the visual attractiveness of an area, functioning as a decorative feature while also supplying light.

✓ **Fixtures:** Chandeliers, decorative wall sconces, and unique light sculptures, pendant lights.

✓ **Design Impact:** Enhances the overall aesthetic and concept of the interior design.

5. **Cove Lighting:**

✓ **Purpose:** Cove lighting is an indirect technique that effectively hides the light source, resulting in a gentle and dispersed illumination along ceilings or architectural elements.

✓ **Fixtures:** Concealed recessed lighting or LED strips *discreetly integrated into recesses or* ledges.

✓ **Effects:** Introduces a gentle, atmospheric illumination that heightens the aesthetic appeal of architectural features.

6. **Under-Cabinet Lighting:**

✓ **Purpose:** Provides illumination for counters and work surfaces in kitchens and other spaces focused on specific tasks.

✓ **Fixtures:** Install LED strips, puck lights, or linear fixtures underneath cabinets.

✓ **Functionality:** Enhances visibility during culinary or work activities on various surfaces.

7. **Wall Washing:**

✓ **Purpose:** To achieve a consistent and balanced illumination of vertical surfaces, such as walls, to create an aesthetically pleasing and uniformly lit background.

✓ **Fixtures:** Lights that are either recessed or surface-mounted, having wide beam angles.

✓ **Effects:** Reduces the presence of shadows and improves the feeling of depth.

8. **Up-Lighting and Down-Lighting:**

✓ **Purpose:** The objective is to direct light upward or downward to achieve various visual effects.

- ✓ **Fixtures:** Uplights, such as wall sconces or floor lamps, and downlights, including pendant or recessed lights.
- ✓ **Applications:** Up-lighting accentuates architectural features while down-lighting offers task or ambient illumination.

9. **Pendant Cluster Lighting:**

- ✓ **Purpose:** Utilizing groups of pendant lights placed in clusters to generate a visually captivating centerpiece.
- ✓ **Fixtures:** Several pendant lights hanging at different heights.
- ✓ **Aesthetic Impact:** Enhances the room's visual appeal by introducing a dynamic and decorative element.

10. **Daylight Integration:**

- ✓ **Purpose:** Effortlessly integrates natural daylight with artificial lighting to ensure an unvarying illumination level during the day.
- ✓ **Control Systems:** Automated systems regulate artificial lighting by adapting to the quantity of normal light present.
- ✓ **Energy Efficiency:** Decreases dependence on artificial lights during daylight hours.

11. **Pathway Lighting:**

- ✓ **Purpose:** Facilitates navigation across a specific area, such as corridors, garden trails, or footpaths.
- ✓ **Fixtures:** Lights installed in the ground, bollards, or small fixtures placed at a low level along the path.
- ✓ **Safety:** Enhances visibility and adds a decorative touch to outdoor places.

12. **Spot Lighting:** Spotlighting is a technique that utilizes concentrated beams of light to accentuate particular items or locations. Spotlights are versatile fixtures that can be precisely aimed.
13. **Linear Lighting:** Employs linear luminaires, such as LED strips or linear pendant lights, to generate uninterrupted lines of illumination. It is frequently employed for ornamental or structural purposes.
14. **Under Cabinet Lighting:** Mounted below kitchen cabinets to offer focused illumination for tasks performed on the countertop. It improves visibility in workspaces.

15. **Step Lighting:** Luminaires mounted on the vertical part of stairs to enhance safety and provide an aesthetic touch. Step lighting can be accomplished by utilizing recessed lights or LED strip lights.
16. **Flood Lighting:** Utilises wide-angle fixtures to achieve even and consistent illumination across a vast expanse. They are typically employed for illuminating outside spaces, such as sports grounds or architectural features.

The selection of the suitable lighting configuration is contingent upon the particular needs, tasks, and intended atmosphere in each area. An intelligently crafted amalgamation of various configurations enhances the lighting environment's aesthetics, functionality, and comfort.

6.6 Luminous Flux of Different Light Sources

Luminous flux measures the amount of visible light that a light source emits within a specific time frame. Luminous flux is quantified in lumens (lm) and measures the intensity of the light emitted. It is essential to comprehend the radiant flux emitted by different light sources to create lighting systems that satisfy specific illumination needs. Below is a clarification of the luminous flux emitted by various light sources:

1) **Incandescent Bulbs:**

✓ **Luminous Flux:** Incandescent bulbs (Fig.2.1(a)) generally emit a luminous flux that varies between 300 and 1,700 lumens, depending on the bulb's wattage and design. A conventional 60-watt incandescent bulb generally emits approximately 800 lumens, and a 100-watt bulb can emit roughly 1600 lumens.

✓ **Characteristics:** Incandescent bulbs produce a warm and vibrant light but have worse energy efficiency than recent technology.

2) **Halogen Lamps:**

✓ **Luminous Flux:** Halogen lamps (Fig.2.1(b)) offer a greater luminous flux than conventional incandescent bulbs, ranging from around 500 to 4,000 lumens. Halogen lamps exhibit higher efficiency compared to traditional incandescent bulbs. A halogen lamp with a power rating of 50 watts can generate around 900-1000 lumens of light.

✓ **Characteristics:** Halogen lamps emit a luminous and sharp light while demonstrating enhanced energy efficiency compared to incandescent bulbs.

3) **Fluorescent Tubes:**

✓ **Luminous Flux:** Fluorescent tubes (Fig.2.1(c)) exhibit a wide range of luminous flux, often ranging from 800 to 6,000 lumens or higher, contingent upon the dimensions and classification of the tube. A 4-foot T8 fluorescent tube emits roughly 2,600 lumens of luminous flux, whereas a T5 fluorescent tube can generate around 2,000 lumens.

✓ **Characteristics:** Fluorescent lighting is renowned for its high energy efficiency and is frequently employed in commercial and industrial environments.

4) **Compact Fluorescent Lamps (CFLs):**

✓ **Luminous Flux:** CFLs (Fig.2.1(d)) emit a luminous flux that varies between approximately 800 and 3,500 lumens, depending on the wattage and design. A standard 15-watt compact fluorescent lamp (CFL) can generate around 800 lumens, equivalent to the brightness emitted by a 60-watt incandescent bulb.

✓ **Characteristics:** CFLs exhibit greater energy efficiency than incandescent bulbs and provide a range of color temperatures.

5) **LED Lamps:** LEDs (Fig.2.1(e)) exhibit a broad spectrum of luminous flux, ranging from a few lumens for compact ornamental bulbs to more than 20,000 lumens for high-output LED fixtures. LED bulbs exhibit exceptional efficiency. A 9-watt LED bulb has a luminous flux of around 800 lumens, which is comparable to the illuminating power of a 60-watt incandescent bulb. Nevertheless, the luminous flux may fluctuate based on the particular LED brand and design.

✓ **Characteristics:** LED lighting exhibits exceptional energy efficiency and remarkable durability and is available in various color temperatures. It has extensive applications in home, commercial, and outdoor lighting.

6) **High-Intensity Discharge (HID) Lamps:** HID lamps are a distinct type of electric lamp that produce illumination through the generation of an electric arc between tungsten electrodes housed within a translucent or transparent arc tube containing gas and metal salts. HID lamps are renowned for their exceptional efficiency and powerful illumination, rendering them appropriate for various applications. HID lights, such as metal halide and high-pressure sodium lamps, are commonly employed for outdoor and industrial illumination. Their luminous flux can vary from a few thousand to tens of thousands of lumens, contingent upon the wattage and type. HID lights, such as metal halide and high-pressure sodium lamps, can provide a luminous flux varying from 5,000 to 100,000 lumens

or even higher. HID bulbs encompass various varieties, each distinguished by its unique attributes. The following are several prevalent varieties of HID lamps:

a) **Metal Halide Lamps:** Metal halide lamps (Fig.2.1(f)) are a specific form of High-Intensity Discharge (HID) lamp that incorporates metal halide compounds within the arc tube. LEDs emit a luminous white light with excellent color rendering capabilities, rendering them well-suited for various applications, including stadium, street, and indoor business lighting.

b) **High-Pressure Sodium (HPS) Lamps:** HPS lamps (Fig.2.1(g)) consist of an arc tube filled with sodium vapor. They produce a distinct yellow-orange illumination. High-pressure sodium (HPS) lights are commonly employed for street lighting, security illumination, and outdoor area lighting owing to their exceptional efficiency and extended lifespan.

c) **Low-Pressure Sodium (LPS) Lamps:** LPS lamps (Fig.2.1(h)) produce light limited to a restricted range of wavelengths, resulting in a single-color yellow light. While they possess exceptional energy efficiency, their color rendering capabilities are subpar. LPS lamps are employed in specialized contexts where color rendering is unimportant, such as astronomical observatories.

d) **Mercury Vapor Lamps:** Mercury vapor lamps (Fig.2.1(i)) were among the initial forms of High-Intensity Discharge (HID) lights. They produce a luminescent glow that appears bluish-green in color. Nevertheless, metal halide and high-pressure sodium lamps outperform them in terms of efficiency. Mercury vapor lamps are now less prevalent due to their diminished efficacy and environmental apprehensions associated with mercury.

✓ **Characteristics:**

- HID lights provide notable attributes such as high lumen output and efficiency, rendering them well-suited for applications necessitating powerful illumination across expansive spaces. These are frequently utilized in outdoor illumination, industrial establishments, sports arenas, and vast commercial areas. Nevertheless, HID bulbs often necessitate a warming period to achieve maximum brightness, and they exhibit a lengthier re-strike time in the event of a power interruption. HID bulbs are commonly employed for outdoor and industrial illumination owing to their substantial luminous output and efficiency.

- LED technology has become increasingly popular as an energy-efficient alternative to HID lighting. LED lighting provides equivalent or superior efficiency, extended lifespans, and the ability to instantly acquire full brightness without any warm-up period. Consequently, numerous applications are shifting towards LED lighting to enhance energy efficiency and performance.

7) **Candle:** A solitary candle (Fig.2.1(j)) emits a dim light, generating approximately 12 lumens. Candlelight is commonly employed for ornamental or atmospheric intentions.
8) **Sunlight:** The intensity of sunlight (Fig.2.1(k)) can vary depending on factors such as the time of day, geographical location, and atmospheric conditions. During a cloudless day at noon, sunlight can emit approximately 100,000 lux, a unit of measurement for light intensity, specifically illuminance, rather than the total amount of light emitted, known as luminous flux.

Fig.2.1(a): Incandescent bulb; Fig.2.1(b): Halogen lamp; Fig.2.1(c): Fluorescent Tube; Fig.2.1(d):Compact fluorescent lamps; Fig.2.1(e): LED lamp; Fig.2.1(f): Metal halide lamp; Fig.2.1(g): High-pressure sodium lamps; Fig.2.1(h): Low-pressure sodium lamps; Fig.2.1(i): Mercury vapor lamps; Fig.2.1(j): Candle light; Fig.2.1(k): Sunlight

Figure 1.1 Various screwdrivers

Comprehending the luminous flux of various light sources is crucial for choosing suitable lighting solutions according to desired brightness levels, energy efficiency, and unique application needs. The proliferation of LED lighting has dramatically broadened the range of choices for creating efficient and effective illumination, leading to notable advancements in lighting technology.

Acknowledging that the given luminous flux figures are approximate and subject to variation depending on specific product models, technological breakthroughs, and efficiency improvements is crucial. Moreover, how we perceive brightness is affected by other elements, including color temperature, color rendering, and the direction of light emitted by the source. When choosing lighting for a particular purpose, it is crucial to consider the luminous flux and other qualitative characteristics of the light source.

Indian standards on LED

6.7 Lux Level Requirements (as per SP 72: 2010):

The phrase "lux level" denotes the illuminance level quantified in lux, representing the quantity of light incident on a surface per unit area. Measuring lux levels is essential for establishing suitable lighting for different situations, guaranteeing that areas are sufficiently lighted for their intended use. The SP 72: 2010 is a standard published by the Bureau of Indian Standards (BIS) that offers lighting and illuminance criteria standards. Let us examine the specified lux level criteria as stated in SP 72: 2010:

1. **Residential Areas:**

✓ **Living Rooms:** It is advisable to maintain a lux level of approximately 100-300 lux in general living spaces where multiple activities may occur.

✓ **Bedrooms:** Bedrooms often require lower levels of illumination, ranging from 50-200 lux, to establish a pleasant and tranquil ambiance.

2. **Commercial and Office Spaces:**

✓ **General Offices:** Workstations and general office facilities typically require an illuminance level of approximately 300-500 lux to facilitate functions such as reading and writing.

✓ **Conference Rooms:** Higher lux levels, often ranging from 300-750 lux, to promote effective communication.

✓ **Reception Areas:** A lux level ranging from 200 to 500 is typically appropriate for spaces designated for welcoming and waiting.

3. **Educational Institutions:**

✓ **Classrooms:** The lux levels in classrooms typically range from 300 to 750 lux, providing ideal conditions for learning.

✓ **Libraries:** Higher lux levels, specifically about 500-1000 lux, are advised in library reading spaces.

4. **Healthcare Facilities:**

✓ **Patient Rooms:** The illumination levels in patient rooms are typically maintained at lower lux levels, ranging from 100 to 300, to provide a soothing atmosphere.

✓ **Operating Rooms:** Surgical and procedural areas require lux levels exceeding 1,000 lux.

5. **Industrial Spaces:**

✓ **Manufacturing Areas:** Workspaces in manufacturing environments may require illumination levels ranging from 200-750 lux, contingent upon the specific tasks.

✓ **Warehouses:** Warehouses generally require lower levels of illumination, ranging from 50 to 150 lux, as they often have less task-specific demands.

6. **Retail Spaces:**

✓ **Sales Floors:** Lux levels in retail spaces, such as shopping areas, are often maintained at 300-750 lux to optimize product visibility and establish an appealing ambiance.

7. **Outdoor Spaces:**

✓ **Street Lighting:** The illumination levels for street lighting might vary but typically range from 10-20 lux for residential areas and 20-50 lux for urban streets.

✓ **Sports Stadiums:** The lux levels required for sports lighting in these venues can be relatively high, typically ranging from 500 to 2000 lux. The specific lux level needed depends on the sport being played and the level of competition.

8. **Specialized Environments:**

✓ **Art Galleries and Museums:** The illumination levels in these environments typically vary from 50 to 300 lux, ensuring the preservation of artwork while facilitating optimal viewing conditions.

✓ **Theaters and Auditoriums:** Lux levels in performance spaces are typically customized to meet specific needs, with a 100-1000 lux range.

9. **Key Considerations:**

✓ Lux levels may differ depending on the specific activities in a particular area.

✓ When assessing the lighting requirements, it is essential to consider the availability of natural daylight.

✓ Advancements in energy-efficient lighting technology, such as LED, affect the necessary lux level and energy usage.

Adhering to the required lux levels specified in SP 72: 2010 guarantees that areas are adequately lit for their intended purpose, enhancing comfort, safety, and productivity. It is imperative to consider these recommendations while designing and implementing lighting systems for different applications. Acknowledging that these are overarching principles and precise criteria may differ is crucial.

It is advisable to consult the most recent edition of the applicable lighting standards for your area, as these standards are frequently altered and updated. To obtain the most accurate information about SP 72:2010, a local or specialized standard, it is advisable to contact the organization responsible for maintaining and enforcing that standard.

SP 72 (2010): National Lighting Code 2010

Solved Numerical Problem

Example 6.1:

Calculate the luminous flux emitted by a light source with a radiant intensity of 800 lumens and emits light uniformly in all directions.

Solution:

Luminous Intensity (I) = 800 lumens

As the light is emitted uniformly in all directions, the luminous flux (Φ) can be calculated using the formula:

$\Phi = 4\pi I$

$\Phi = 4\pi\times$ (800 lumens)

$\Phi = 3200\pi$ lumens

$\Phi \approx 10053.096$ lumens

Example 6.2:

A light source emits a luminous intensity of 1200 lumens.

What will the luminous intensity be at a distance of 4 meters from the light source if it is originally 2 meters away?

Solution:

The inverse square law of lighting states that the luminous intensity (I) exhibits an inverse relationship with the square of the distance (d).

$I_1 / I_2 = (d_2^2 / d_1^2)$

Given:

Initial distance (d_1) = 2 meters Luminous intensity at initial distance (I_1) = 1200 lumens New distance (d_2) = 4 meters Luminous intensity at new distance (I_2) = ?

$I_1 / I_2 = (4^2 / 2^2)$

$I_1 / I_2 = 16 / 4$

$I_1 / I_2 = 4$

$I_2 = I_1 / 4$

$I_2 = 1200$ lumens / 4

$I_2 = 300$ lumens

Therefore, the luminous intensity at a distance of 4 meters from the light source will be 300 lumens.

Example 6.3:

Calculate the illuminance produced by a 60-watt incandescent lamp placed 3 meters above a desk with an area of 2 square meters.

Solution:

The luminous flux emitted by a 60-watt incandescent lamp is approximately 800 lumens. Illuminance (E) is the luminous flux incident per unit area. $E = \Phi / A$

Given:

Luminous flux (Φ) = 800 lumens Area (A) = 2 square meters

E = 800 lumens / 2 square meters E = 400 lumens/square meter E = 400 lux

Therefore, the illuminance produced by the lamp is 400 lux.

Example 6.4:

A spotlight emits an illuminance of 600 lux on a surface 5 meters away. What is the luminous intensity of the spotlight?

Solution:

Luminous intensity (I) is the flux emitted per unit solid angle.

Illuminance (E) = Luminous intensity (I) / Distance squared (d^2)

Given:

Illuminance (E) = 600 lux Distance (d) = 5 meters

Rearranging the formula: $I = E \times d^2$

I = 600 lux × (5 meters)^2

I = 600 lux × 25

I = 15000 lumens

Therefore, the luminous intensity of the spotlight is 15000 lumens.

Example 6.5:

A fluorescent lamp consumes 40 watts of power and emits a luminous flux of 3000 lumens. Calculate its lighting efficiency.

Solution:

Lighting efficiency is the ratio of the luminous flux emitted by the lamp to the power consumed.

Efficiency = Luminous flux (Φ) / Power consumed

Given:

Luminous flux (Φ) = 3000 lumens

Power consumed = 40 watts

Efficiency = 3000 lumens / 40 watts

Efficiency = 75 lumens per watt (lm/W)

Therefore, the lighting efficiency of the fluorescent lamp is 75 lumens per watt.

Exercise

Multiple Choice Questions:

1. What is the SI unit of luminous flux?

a) Lumens (lm) b) Lux (lx) c) Candela (cd) d) Watt (W)

2. Which of the following is a room-specific consideration for illumination in residential installations?

   a) Ohm's Law b) Kirchhoff's Law c) Inverse Square Law d) Task lighting placement

3. What does luminous intensity measure?

   a) The brightness of a light source b) The color temperature of a light source c) The energy consumption of a light source d) The efficiency of a light source

4. Which law governs how light intensity diminishes as the distance from the source increases? a) Ohm's Law b) Lenz's Law c) Beer-Lambert Law d) Inverse Square Law

5. Which of the following is used to measure illuminance?

   a) Lumens (lm) b) Candela (cd) c) Lux (lx) d) Watt (W)

6. What does "task lighting" refer to in residential installations?

   a) Lighting used for decorative purposes b) Lighting used to illuminate specific tasks or activities c) Lighting controlled by automation systems d) Lighting used for outdoor spaces

7. What is the formula for calculating illuminance?

   a) $E = \Phi / A$ b) $E = I / d^2$ c) $E = \Phi * d^2$ d) $E = I * A$

8. Which component of lighting controls regulates the intensity of light emitted by a source?

   a) Dimmer b) Timer c) Sensor d) Switch

9. What is the unit of measurement for luminous intensity?

   a) Lumens (lm) b) Candela (cd) c) Lux (lx) d) Watt (W)

10. Which law states that the total luminous flux in a closed system remains constant?

    a) Beer-Lambert Law b) Ohm's Law c) Lenz's Law d) Conservation of Luminous Flux

11. What is the concept of "color temperature" in lighting?

    a) The temperature at which the light source operates b) The intensity of the light emitted by the source c) The hue of the light emitted by the source d) The brightness of the light emitted by the source

12. Which lighting control system adjusts light levels based on detected ambient light levels?

    a) Timer-based control b) Dimming control c) Daylight harvesting d) Motion sensor control

13. Which parameter represents the light emitted by a source per unit solid angle?

    a) Luminous flux (Φ) b) Luminous intensity (I) c) Illuminance (E) d) Lighting efficiency

14. Which type of lighting is suitable for highlighting artwork or architectural features?

    a) Task lighting b) Accent lighting c) Ambient lighting d) Decorative lighting

15. Which law states that the illuminance on a surface varies inversely with the square of the distance from the source?

    a) Ohm's Law b) Beer-Lambert Law c) Lenz's Law d) Inverse Square Law

Answer:

1. a) Lumens (lm)
2. d) Task lighting placement
3. a) The brightness of a light source
4. d) Inverse Square Law
5. c) Lux (lx)
6. b) Lighting used to illuminate specific tasks or activities
7. a) $E = \Phi / A$
8. a) Dimmer
9. b) Candela (cd)
10. d) Conservation of Luminous Flux
11. a) The temperature at which the light source operates
12. c) Daylight harvesting
13. b) Luminous intensity (I)
14. b) Accent lighting
15. d) Inverse Square Law

Short Answer Type Questions:

1) What are some critical room-specific considerations when designing lighting for residential installations?
2) Explain the concept of luminous flux (Φ) and its significance in residential lighting design.
3) How do lighting controls and automation systems contribute to energy efficiency and user convenience in residential lighting?
4) Define luminous intensity (I) and discuss its importance in determining the brightness of light sources.

5) Briefly explain the Inverse Square Law of illumination and its implications for residential lighting design.
6) What are some critical room-specific considerations when designing lighting for residential installations?
7) Explain the concept of luminous flux (Φ) and its significance in residential lighting design.
8) How do lighting controls and automation systems contribute to energy efficiency and user convenience in residential lighting?
9) Define luminous intensity (I) and discuss its importance in determining the brightness of light sources.
10) Briefly explain the Inverse Square Law of illumination and its implications for residential lighting design.

Long Answer Type Questions:

1) Discuss the importance of room-specific considerations in residential lighting design. Explain how room size, function, and layout influence lighting decisions.
2) Explain the concept of lighting controls and automation in residential installations. Discuss the benefits of using advanced lighting control systems and provide examples of how they can enhance energy efficiency and user comfort.
3) Describe the relationship between luminous flux (Φ) and luminous intensity (I). How do these parameters determine lighting solutions' overall brightness and effectiveness in residential settings?
4) Discuss the significance of the Inverse Square Law of illumination in residential lighting design. Provide examples of how this law affects the distribution and intensity of light in different areas of a residential space.
5) Explore the concept of lighting zoning and its application in residential installations. How can zoning strategies be used to create versatile and adaptable lighting environments that meet the diverse needs of homeowners?

Numerical Problems:

1) Calculate the illuminance produced by a 40-watt LED bulb placed 2 meters above a desk with an area of 1.5 square meters. The luminous efficacy of the LED bulb is 100 lumens per watt. (400 lux)
2) Problem: A spotlight emits a luminous intensity of 1500 lumens at a distance of 3 meters. Calculate the illuminance on a surface located 5 meters away from the spotlight. (180 lux)
3) Determine the luminous flux emitted by a fluorescent tube with a luminous efficacy of 80 lumens per watt if the tube consumes 30 watts of power. (2400 lumens)
4) A room measures 4 meters by 5 meters. If the room requires an average illuminance of 300 lux, calculate the total luminous flux needed to light the room adequately. (6000 lumens)
5) A chandelier has a total luminous flux of 5000 lumens. If the chandelier emits light uniformly in all directions, calculate the illuminance produced on the floor directly below it, which has an area of 3 square meters. (555 lux)
6) Problem: Determine the luminous intensity of a desk lamp that emits 600 lumens and spreads light uniformly in all directions. (150 candela)
7) A recessed LED light fixture has a luminous intensity of 1200 lumens. Calculate the illuminance produced on a desk located 2.5 meters below the fixture. (192 lux)
8) Calculate the luminous flux required to achieve an illuminance of 500 lux on a reading chair with an area of 0.8 square meters. The chair is located 1.5 meters from a table lamp with a luminous intensity of 800 lumens. (2500 lumens)
9) A room with an area of 25 square meters requires an average illuminance of 400 lux. If the room is illuminated by three ceiling-mounted LED panels, each emitting 4000 lumens, determine the spacing between the panels to achieve uniform lighting. (2.5 meters)
10) An outdoor pathway measuring 10 meters in length requires an illuminance of 50 lux. If the path is illuminated by six bollard lights emitting 800 lumens, calculate the spacing between the lights for uniform illumination. (Approximately 2 meters)

Unit Summary

The unit summary for illumination in residential installation is written below:

- ✓ Introduction to illumination
- ✓ Types of lighting fixtures

- ✓ Lighting design considerations
- ✓ Lighting control systems
- ✓ Lighting calculation and layout
- ✓ Lighting maintenance and safety
- ✓ Lighting design projects

This unit provides a comprehensive understanding of illumination in residential installations, covering lighting fixtures, design principles, control systems, energy efficiency, maintenance, and safety considerations.

Practical

1. Demonstrate how to calculate the required lumen output for a specific room based on its size, function, and recommended illuminance levels.
2. Create a lighting layout plan for a living room that includes ambient lighting, task lighting, and accent lighting. Specify fixture types, positions, and light levels.
3. Demonstrate the installation and programming of a dimmer switch for controlling the brightness of LED lights in a residential setting.
4. Calculate the illuminance levels (lux or foot-candles) required for task lighting in a kitchen workspace based on industry standards.
5. Demonstrate the proper procedure for replacing a burnt-out bulb in a recessed lighting fixture, ensuring safety precautions are followed.
6. Conduct a visual inspection of lighting fixtures, switches, and electrical connections to identify potential hazards or maintenance issues.

Know more

Sl. No	Topic	QR Code
1	Code of practice for Interior illumination IS 3646 (Part 1): 1992	
2	Technical Specification For Distribution Boards & Lighting Panels	

3	Energy Efficient Street Lighting Guidelines BEE	
4	Lectures on Illumination	

References and suggested readings.

1. R. Whitehead, *Residential lighting: a practical guide to beautiful and sustainable design*, John Wiley & Sons, 2008.
2. M. Corrodi, and K. Spechtenhauser, *Illuminating: natural light in residential architecture*, Vol. 3, Walter de Gruyter, 2014.
3. M. Steffen, *Residential lighting design*. Crowood., 2014
4. S. Isenstadt, M.M. Petty, and D. Neumann, *Cities of light: Two centuries of urban illumination*, Routledge, 2014.
5. E-recourse URL
    a. https://youtu.be/BOPCnvvp9jI
    b. https://youtu.be/4hLNaI4CZ_M
    c. https://youtu.be/IqZ8e6YrVrs
    d. https://youtu.be/gCfP-uKe1RI
    e. https://youtu.be/pwNmUEBz--k
    f. https://youtu.be/ZXEqVUVkX8E
    g. https://youtu.be/NupEhy0VQtM

APPENDICES

APPENDIX-A

Glossary of Terms

- Electrification: The process of converting systems, infrastructure, and equipment to use electricity instead of other energy sources such as fossil fuels.
- HVAC: Heating, Ventilation, and Air Conditioning.
- Net Zero Energy Building: A building that produces as much energy as it consumes on an annual basis.

Case Study 1: Green Office Building Electrification

❖ Background:

GreenScape Developers undertook a project to electrify a newly constructed office building designed with sustainability in mind. The goal was to implement energy-efficient electrical systems while ensuring optimal functionality and occupant comfort.

❖ Challenges:

1. Integrating renewable energy sources such as solar panels into the building's electrical grid.
2. Designing lighting and HVAC systems for maximum energy efficiency without compromising performance.
3. Ensuring compliance with green building standards and codes while managing costs effectively.

❖ Actions Taken:

1. Renewable Energy Integration:
   - Installed solar panels on the building's rooftop to harness solar energy for electricity generation.

- Designed a grid-connected solar PV system with inverters and battery storage to optimize energy utilization and reduce reliance on the main power grid.

2. Energy-Efficient Lighting and HVAC:
   - Implemented LED lighting fixtures throughout the building to reduce energy consumption and maintenance costs.
   - Integrated occupancy sensors and daylight harvesting controls to adjust lighting levels based on occupancy and natural light availability.
   - Designed a zoned HVAC system with variable-speed equipment and programmable thermostats for precise temperature control and energy savings.
3. Smart Building Technologies:
   - Deployed a building automation system (BAS) to monitor and control electrical systems, lighting, HVAC, and energy usage in real-time.
   - Utilized smart meters and energy management software to track electricity consumption, identify trends, and implement energy-saving strategies.

❖ Results:

1. Energy Savings: The integration of renewable energy and energy-efficient systems resulted in significant electricity savings for the office building.
2. Environmental Impact: The project contributed to reducing the building's carbon footprint and promoting sustainability practices.
3. Occupant Comfort: The smart technologies implemented enhanced occupant comfort by optimizing lighting, temperature, and air quality.
4. Compliance and Certification: The building met green building standards and obtained certifications for energy efficiency and environmental performance.

Case Study 2: Hospital Electrical System Upgrade

❖ Background:

A regional hospital required a comprehensive electrical system upgrade to meet growing patient care demands, enhance safety, and comply with healthcare facility regulations and standards.

❖ Challenges:

1. Ensuring uninterrupted power supply (UPS) for critical medical equipment and emergency systems.
2. Designing electrical layouts for different areas of the hospital, including patient rooms, operating theaters, laboratories, and administrative spaces.
3. Adhering to strict electrical safety protocols and infection control measures within healthcare facilities.

❖ Actions Taken:

1. Critical Power Systems:
   - Installed redundant backup generators and UPS systems to provide backup power during outages and maintain essential operations, including life-support equipment and critical care areas.
   - Implemented automatic transfer switches (ATS) to seamlessly transition between utility power and backup power sources.
2. Electrical Distribution Design:
   - Designed separate electrical circuits for sensitive medical equipment, ensuring proper grounding, isolation, and surge protection.
   - Implemented isolation transformers in patient care areas to minimize electrical noise and ensure patient safety.

3. Lighting and Safety:
    - Upgraded lighting systems to energy-efficient LEDs and incorporated lighting controls for adjustable brightness and occupancy sensing.
    - Installed emergency lighting and exit signs to comply with building codes and facilitate safe evacuation during emergencies.

❖ Results:

1. Enhanced Reliability: The upgraded electrical systems improved the hospital's reliability and resilience against power interruptions, critical for uninterrupted patient care.
2. Compliance and Safety: The project ensured compliance with healthcare facility regulations, electrical codes, and safety standards, enhancing patient and staff safety.
3. Efficiency and Cost Savings: Energy-efficient lighting and power management strategies resulted in reduced electricity consumption and operating costs for the hospital.

Case Study 3: Retrofitting a Residential Building with Electric Heat Pumps

Overview: This case study examines the process and outcomes of retrofitting a multi-unit residential building with electric heat pumps for space heating and cooling.

Challenges: High initial cost, compatibility with existing infrastructure.

Solutions: Financing options, technical support from manufacturers.

Results: Reduced energy consumption, improved comfort for residents.

Case Study 4: Implementing Electric Vehicle Charging Infrastructure in a Commercial Office Building

Overview: This case study explores the installation of electric vehicle charging stations in a commercial office building.

Challenges: Upfront investment, load management.

Solutions: Subsidies, demand response programs.

Results: Increased tenant satisfaction, reduced carbon emissions.

APPENDIX-B

Technical Specifications

(1) Electric Heat Pump System Specifications

- Manufacturer: ABC Heating & Cooling
- Model: XYZ-5000
- Heating Capacity: 20,000 BTU/hr
- Cooling Capacity: 18,000 BTU/hr
- COP (Coefficient of Performance): 3.5
- Voltage: 240V
- Refrigerant: R-410A

(2) Electric Vehicle Charging Station Specifications

- Manufacturer: ChargePoint
- Model: CP500
- Power Output: 7.2 kW
- Voltage: 208-240V
- Amperage: 32A
- Connector Type: SAE J1772

APPENDIX-C

Additional Resources

(1) **Websites:**

- U.S. Department of Energy - Building Technologies Office: Provides resources and programs for building electrification.
- International Living Future Institute: Offers guidance on achieving net-zero energy buildings and sustainable development practices.

(2) **Books:**

- "Electrify Everything: How to Decarbonize the Economy" by Saul Griffith
- "The New Carbon Architecture: Building to Cool the Climate" by Bruce King

ANNEXURES

Technical Drawings and Schematics

- Electrical Layout Diagram: A detailed drawing showing the layout of electrical circuits, outlets, switches, and appliances in a building.
- Heat Pump Installation Schematic: A schematic diagram illustrating the installation process of a heat pump system, including the positioning of components such as the compressor, evaporator, and condenser.

Energy Consumption Data

- Monthly Electricity Consumption Data: A table or graph presenting monthly electricity consumption data for a sample of residential or commercial buildings before and after electrification.
- Comparison of Energy Use: A comparison chart showing energy consumption patterns before and after the implementation of electrification measures, such as the installation of electric heat pumps or induction cooktops.

REFERENCES FOR FURTHER LEARNING

1. K.B. Raina, and S.K.Bhattacharya, Electrical Design Estimating and Costing, New Age International Ltd., New Delhi, ISBN 978-81-224-0363-3
2. Allagappan, N. S.Ekambarram, Electrical Estimating and Costing, New Delhi, ISBN-13: 9780074624784
3. Singh, Surjit, Electrical Estimating and Costing, Dhanpat Rai and Co. New Delhi, ISBN: 1234567150995
4. J B Gupta: A Course in Electrical Installation Estimating and Costing, S K Kataria and Sons,New Delhi, ISBN:978-93-5014-279-0
5. Bureau of Indian Standard, IS: 732-1989, Code of practice for electrical wiring installation
6. Bureau of Indian Standard, SP 30 National Electrical Code 2010
7. Bureau of Indian Standard, SP 72 National Lighting Codes 2010
8. E-REFERENCES:-
   - http://nptel.ac.in/courses/108108076/1, assessed on 18th January 2016
   - http://www.electrical4u.com, assessed on 18th January 2016
   - https://www.youtube.com/playlist?list=PLbMcCymQk6zYQR-REPv4pbRGyjEEhfy4f
   - https://www.youtube.com/playlist?list=PLbMcCymQk6zYmbzlHluQaWwJ2VT343DPe
   - https://www.youtube.com/playlist?list=PLbMcCymQk6zYQR-REPv4pbRGyjEEhfy4f
   - https://www.youtube.com/watch?v=A9KSGAnjo2U, assessed on 18th January 2016
   - http://www.electricaltechnology.org/2015/09, assessed on 30 Jan 2016
   - www.slideshare.net/bawaparam/made-by-paramassesed on 30 Jan2016
   - www.electricaltechnology.org/2013/09/electrical-wiring.html assessed on 16 March 2016.
   - https://youtu.be/BOPCnvvp9jI

- https://youtu.be/4hLNaI4CZ_M
- https://youtu.be/IqZ8e6YrVrs
- https://youtu.be/gCfP-uKe1RI
- https://youtu.be/pwNmUEBz--k
- https://youtu.be/ZXEqVUVkX8E
- https://youtu.be/NupEhy0VQtM

CO AND PO ATTAINMENT TABLE

Course outcomes (COs) for this course can be mapped with the programme outcomes (POs) after the completion of the course and a correlation can be made for the attainment of POs to analyze the gap. After proper analysis of the gap in the attainment of POs necessary measures can be taken to overcome the gaps.

Table for CO and PO attainment

Course Outcomes	**Expected Mapping with Programme Outcomes** *(1- Weak Correlation; 2- Medium correlation; 3- Strong Correlation)*						
	PO-1	**PO-2**	**PO-3**	**PO-4**	**PO-5**	**PO-6**	**PO-7**
CO-1	3	3	3	3	1	2	3
CO-2	3	3	3	3	1	2	3
CO-3	3	3	3	3	1	2	3
CO-4	3	3	2	3	1	2	3
CO-5	3	3	2	3	1	1	3
CO-6	3	3	2	3	1	1	3

The data filled in the above table can be used for gap analysis.

INDEX